W0050992

Advanced Courses in Mathematics
CRM Barcelona

Institut d'Estudis Catalans
Centre de Recerca Matemàtica

Managing Editor:
Manuel Castellet

Ken A. Brown
Ken R. Goodearl

Lectures on Algebraic Quantum Groups

Springer Basel AG

Authors' addresses:

Ken A. Brown
Department of Mathematics
University of Glasgow
Glasgow, G12 8QW
UK
e-mail: kab@maths.gla.ac.uk

Ken R. Goodearl
Department of Mathematics
University of California
Santa Barbara, CA 93106
USA
e-mail: goodearl@math.ucsb.edu

2000 Mathematical Subject Classification 16W35, 17B37, 20G42, 81R50

A CIP catalogue record for this book is available from the
Library of Congress, Washington D.C., USA

Deutsche Bibliothek Cataloging-in-Publication Data

Brown, Ken A.:
Lectures on algebraic quantum groups / Ken A. Brown ; Ken R. Goodearl. -
Basel ; Boston ; Berlin : Birkhäuser, 2002
   (Advanced courses in mathematics - CRM Barcelona)
   ISBN 978-3-7643-6714-5   ISBN 978-3-0348-8205-7 (eBook)
   DOI 10.1007/978-3-0348-8205-7

This work is subject to copyright. All rights are reserved, whether the whole or part of the material is concerned, specifically the rights of translation, reprinting, re-use of illustrations, recitation, broadcasting, repro-
duction on microfilms or in other ways, and storage in data banks. For any kind of use permission of the
copyright owner must be obtained.

© 2002 Springer Basel AG
Originally published by Birkhäuser Verlag in 2002
Cover design: Micha Lotrovsky, 4106 Therwil, Switzerland
Printed on acid-free paper produced from chlorine-free pulp. TCF ∞

9 8 7 6 5 4 3 2 1   www.birkhäuser-science.com

CONTENTS

PART III. QUANTIZED ALGEBRAS AT ROOTS OF UNITY

PREFACE

In September 2000, at the Centre de Recerca Matemàtica in Barcelona, we presented a 30-hour *Advanced Course on Algebraic Quantum Groups*. After the course, we expanded and smoothed out the material presented in the lectures and integrated it with the background material that we had prepared for the participants; this volume is the result. As our title implies, our aim in the course and in this text is to treat selected algebraic aspects of the subject of quantum groups. Several of the words in the previous sentence call for some elaboration. First, we mean to convey several points by the term 'algebraic' – that we are concerned with algebraic objects, the quantized analogues of 'classical' algebraic objects (in contrast, for example, to quantized versions of continuous function algebras on compact groups); that we are interested in algebraic aspects of the structure of these objects and their representations (in contrast, for example, to applications to other areas of mathematics); and that our tools will be drawn primarily from noncommutative algebra, representation theory, and algebraic geometry.

Second, the term 'quantum groups' itself. This label is attached to a large and rapidly diversifying field of mathematics and mathematical physics, originally launched by developments around 1980 in theoretical physics and statistical mechanics. It is a field driven much more by examples than by axioms, and so resists attempts at concise description (but see Chapter I.1 and the references therein). The algebras we shall study here are of two main types. On the one hand there are 'quantized coordinate rings', a term used in the literature to refer to various noncommutative algebras which are, informally expressed, deformations of the classical coordinate rings of algebraic groups or related algebraic varieties; the adjective 'quantized' usually indicates that some solution to the quantum Yang-Baxter equation is involved in the construction and/or the representation theory of the algebra. Unfortunately, to date no axiomatic definition of this family of algebras has been given, nor a complete formulation of properties an algebra should satisfy in order to qualify as a quantum analogue of a given classical coordinate ring. This lack of defining axioms and characteristics should, in our opinion, be highlighted as a major open problem in the area. The known algebras which, by general consensus, are considered to be quantized coordinate rings do share a substantial number of common features, many of which will be developed in the text. The second broad class consists of 'quantized enveloping algebras' – as their name suggests, these are certain deformations of the universal enveloping algebras of semisimple Lie algebras (or of affine Kac-Moody Lie algebras). This class of algebras is somewhat more tightly defined than the first, in that generators and relations are given by a standard process applied to the Serre relations for classical enveloping algebras. As in the classical setting, there is a duality between these quantized enveloping algebras and the quantized coordinate rings of the corresponding semisimple algebraic groups.

The last point in our stated aim needing further discussion is the phrase 'selected aspects' (beyond the obvious comment that the choices of material reflect our own preferences and interests). In a lecture course lasting 30 hours we could have no hope of achieving comprehensive coverage, even of the limited terrain mapped out above. We have made no attempt to rectify that defect here. We have, however, tried to begin from a point accessible to graduate students, and to provide sufficient signposts to the literature (and a sufficiently extensive bibliography) to guide the reader wishing to study any of the topics we cover in more detail than is possible here. As regards specific prerequisites, the reader is expected to have a basic background in noncommutative algebra, including some familiarity with, for example, noetherian rings, Lie algebras and their enveloping algebras, and homological algebra. This background can be found, for instance, in the texts [**83, 98, 158, 190, 191**]. Some exposure to the basics of Hopf algebras and (affine) algebraic geometry would also be useful. The former is summarized in Appendix I.9 (see [**163**] for many of the details), while the latter can be found, for example, in the early parts of the texts [**89, 99, 173**]. Summaries of certain background topics to which we refer are given in the appendices to Part I.

Here are a few words concerning the layout of the book. All quantized algebras involve one or more parameters, which may be scalars from a base field or extension field, or indeterminates. Typically, when these parameters are roots of unity, the quantized algebra is a finite module over its centre, and the theory of polynomial identity rings provides crucial tools for its study. The 'generic' case, when some or all of the parameters are non-roots of unity, exhibits very different and much more rigid structures. Our arrangement of the material reflects this dichotomy, with the generic case being treated in Part II and the root of unity case in Part III. Part I consists of material which, we believe, is fundamental to both aspects; as already mentioned, we've arranged it as a series of eight chapters, followed by eight appendices, in the expectation that most readers will progress from the chapters in Part I to either Part II or Part III, making use of the appendices to fill in background as and when needed. As this layout suggests, Parts II and III are to a considerable extent independent of each other. Nevertheless, there are many parallels and resonances between the two theories, some clearly visible but others still awaiting clarification, so that we believe there are strong reasons for studying these two aspects of the subject in parallel.

The alert reader will soon notice that this text is written in a mixture of several styles – portions are developed from basic definitions with full proofs (as in an introductory textbook), portions are discussed with sketches of proofs (as in lecture notes), while still other portions are merely quoted with no proofs (as in an overview article). This mixture of styles was chosen in order to keep reasonable bounds on the length of the book and on the time involved in preparing it; we hope that the transitions from one style to another will not prove too disconcerting. Another aim of our approach has been to provide something of interest both to newcomers and to experts. For instance, readers who are encountering quantum groups here for the first time may find in the early chapters of Parts I, II, III an accessible entry into our part of the subject. With non-expert readers in mind, we have provided sets of exercises in which selected items of the development are to be worked out. Finally, we hope that readers will be able to use what they find here as a basis for further explorations of quantum groups. The bibliography given at the end of the book provides a selection of papers and books on this area, but it is far from a complete list (which would have required many thousands of items). We have only sought to include references needed for specific items together with a sample showing the history and development of the lines we have focused on. Our apologies to researchers whose work we have omitted to mention.

Thanks are due to the Centre de Recerca Matemàtica and its director Manuel Castellet for sponsoring and hosting the Advanced Course, to the CRM staff Maria Julià and Consol Roca for smoothly working out innumerable details, and to Pere Ara for the mathematical organization of the course. We would also like to thank the participants of the course for their attention and their many comments on the material, as well as Jacques Alev, Mauro Costantini, Iain Gordon, and Ed Letzter for helpful comments on the first draft of this book. Errors undoubtedly remain, but one advantage of coauthorship is that each of us can blame the other.

July, 2001

Ken Brown   Ken Goodearl
Department of Mathematics   Department of Mathematics
University of Glasgow   University of California
Glasgow, G12 8QW   Santa Barbara, CA 93106
UK   USA

PART I. BACKGROUND AND BEGINNINGS

CHAPTER I.1

BEGINNINGS AND FIRST EXAMPLES

The origins of the subject of quantum groups lie in mathematical physics, about which we will say nothing except that it is the source of the word 'quantum' in the subject. As in physics, much of the research into quantum groups is aimed at developing 'quantum analogues of classical phenomena', a catch phrase whose imprecision has been exploited to the full. Two decades of research in the area have produced many thousands of papers going in a wide variety of directions. We shall not attempt any overview of this landscape, but instead will focus on some algebraic portions of its terrain.

ORIGINS AND MOTIVATIONS

I.1.1. A bald outline of the origins of quantum groups can be diagrammed as below:

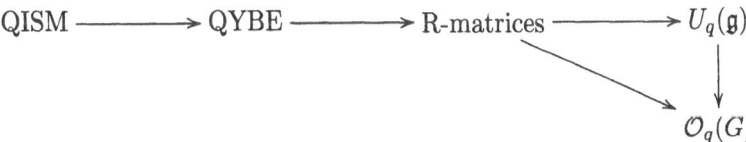

We will only briefly comment on the individual vertices of this diagram; more detailed information can be found in the references given at the end of the chapter. The starting point for the study of quantum groups lies in the Quantum Inverse Scattering Method developed by L. D. Faddeev and the 'Leningrad school' of mathematical physics, with the aim of solving certain 'integrable quantum systems'. A key ingredient in this method is the Quantum Yang-Baxter Equation, which applies to linear transformations on tensor squares of vector spaces (see I.10.2). Solutions to the QYBE were routinely denoted by R and so picked up the name 'R-matrices'. While there is no general method for solving the QYBE, it was discovered in the early 1980s that interesting and useful solutions could be constructed from the representation theory of certain algebras resembling deformations of enveloping algebras of semisimple Lie algebras. The first such algebra, a deformation of $U(\mathfrak{sl}_2(\mathbb{C}))$, arose from a 1981 paper of Kulish and Reshetikhin [**129**]. In the mid-1980s, Drinfel'd and Jimbo independently discovered analogous deformations corresponding to arbitrary semisimple Lie algebras $\mathfrak{g}$ [**50, 51, 110**]. These Hopf algebras were dubbed *quantized enveloping algebras*. The original version, denoted $U_\hbar(\mathfrak{g})$, is a topologically complete $\mathbb{C}[[\hbar]]$-algebra. The later version, a $\mathbb{C}[q^{\pm 1}]$-algebra denoted $U_q(\mathfrak{g})$, is the one on which we shall concentrate here. Thus, the representation theory of $U_q(\mathfrak{g})$, which turned out to possess striking parallels with the well known representation theory of $\mathfrak{g}$, started out as a machine for producing R-matrices.

In the classical setting, a complex semisimple Lie algebra $\mathfrak{g}$ arises as the Lie algebra of a complex semisimple Lie group G, and there is a Hopf algebra duality (a 'perfect Hopf pairing') between the enveloping algebra $U(\mathfrak{g})$ and the coordinate ring $\mathcal{O}(G)$. Via this duality, the representations of $\mathfrak{g}$, in other words the modules over $U(\mathfrak{g})$, correspond to *comodules* over $\mathcal{O}(G)$. Hence, it was natural to look for deformations of the coordinate ring $\mathcal{O}(G)$ which possess Hopf pairings with the quantized enveloping algebra $U_q(\mathfrak{g})$. A deformation of $\mathcal{O}(SL_2(\mathbb{C}))$, which turned out to be paired with $U_q(\mathfrak{sl}_2(\mathbb{C}))$, was introduced in 1987 by Manin [**148**]. A deformation of $\mathcal{O}(SU_2(\mathbb{C}))$, introduced in 1987 by Woronowicz [**211**] and independently by Vaksman-Soibelman [**205**], marks the start of the development of 'compact quantum groups', a branch of the subject which we do not address. Around the same time, Faddeev, Reshetikhin, and Takhtajan [**54, 186**] constructed deformations of $\mathcal{O}(G)$ for the classical simple Lie groups G (i.e., those of types A_n, B_n, C_n, D_n), paired with the corresponding quantized enveloping algebras, using generators and relations culled from a certain 'universal R-matrix' associated with $U_q(\mathfrak{g})$. Each of the four cases required different generators and relations, however. In the meantime, a systematic approach has been developed for the general case (where G is not necessarily simple, and components of exceptional types are allowed), yielding a pairing between $U_q(\mathfrak{g})$ and a Hopf algebra which can be viewed as a deformation of $\mathcal{O}(G)$. This new algebra is called a (single-parameter) *quantized coordinate ring* of G, and we shall label it $\mathcal{O}_q(G)$.

The reader will find explicit descriptions of the algebras $U_q(\mathfrak{g})$ and $\mathcal{O}_q(G)$ in later chapters of the book. The current standard procedure for introducing these algebras is to present $U_q(\mathfrak{g})$ by generators and relations and to construct $\mathcal{O}_q(G)$ as a subalgebra of the Hopf dual of $U_q(\mathfrak{g})$ generated by certain 'coordinate functions' of $U_q(\mathfrak{g})$-modules. In the case $G = SL_n(\mathbb{C})$, convenient generators and relations for the quantized coordinate ring are available, and we will work with $\mathcal{O}_q(SL_n(\mathbb{C}))$ in that form.

I.1.2. A second major thread running through our subject has roots in a philosophy propounded by Grothendieck: One should study objects by means of the functions on them. Let us outline first how this philosophy leads naturally to the concept of a Hopf algebra (although that was not Hopf's approach). Following Grothendieck, if we have a discrete, topological, Lie, or algebraic group G, for instance, we should study the algebra of arbitrary, continuous, C^∞, or polynomial functions, respectively, from G to a base field k. We shall concentrate on the case of an algebraic group; thus G is both a group and an algebraic variety, and we consider its coordinate ring $\mathcal{O}(G)$, the ring of polynomial functions on G. Assuming k algebraically closed, basic algebraic geometry reveals how $\mathcal{O}(G)$ captures the variety G, namely as the maximal ideal space of $\mathcal{O}(G)$.

To see where the group structure of G resides, just consider its effect on functions. View the three main group-theoretic axioms in terms of maps: the multiplication map $\mu : G \times G \to G$, the inclusion map $\{1\} \to G$ where 1 is the identity of G, and the inversion map $i : G \to G$ (that is, $i(g) = g^{-1}$ for $g \in G$). Since G is an algebraic group, these maps are morphisms of algebraic varieties, and composing them with polynomial functions gives other polynomial functions. Thus, we obtain three corresponding algebra homomorphisms:

$$\Delta : \mathcal{O}(G) \xrightarrow{f \mapsto f \circ \mu} \mathcal{O}(G \times G) \xrightarrow{\cong} \mathcal{O}(G) \otimes \mathcal{O}(G)$$

$$\varepsilon : \mathcal{O}(G) \xrightarrow{f \mapsto f(1)} k$$

$$S : \mathcal{O}(G) \xrightarrow{f \mapsto f \circ i} \mathcal{O}(G),$$

called the *comultiplication*, the *counit*, and the *antipode*, respectively. The group axioms for G lead to various relations among Δ, ε, and S, such as coassociativity of Δ, which we do not give here (see (I.9.1), (I.9.7), (I.9.9)). We simply point out that on writing out the obvious axioms satisfied by the system

$$\left(\mathcal{O}(G), \text{multiplication}, \text{unit}, \text{comultiplication}, \text{counit}, \text{antipode}\right),$$

one obtains the axioms defining a Hopf algebra (I.9.9).

I.1.3. Grothendieck's philosophy was extended to quantum groups by Drinfel'd, who proclaimed that one should 'quantize' classical coordinate rings such as $\mathcal{O}(G)$ by deforming them to noncommutative Hopf algebras, and that one should study these new Hopf algebras as if they consisted of 'noncommuting functions' on a nonexistent object, namely a 'quantum group' corresponding to G. Thus, quantum groups *per se* do not exist, only their algebras of functions; for convenient abbreviation, the function algebras themselves are then called quantum groups.

Similarly, one can look for quantizations of general algebraic varieties, obtained by deformations of their coordinate rings. (The word 'deformation' is used here only suggestively, not in its technical sense.) In general, there may not be any Hopf algebra structure, but if a variety supports an action of an algebraic group, that action should have an analogue in any quantization.

INITIAL EXAMPLES

We now present some basic examples of quantized coordinate rings that have been constructed in accordance with the principles outlined above. These constructions make sense over an arbitrary base field k, with respect to an arbitrary parameter $q \in k^\times$.

I.1.4. Example. Since the coordinate ring of the affine plane, k^2, is a commutative polynomial ring in two indeterminates, the natural way to deform it is to pass to a skew polynomial ring in two indeterminates. There are many possibilities, of course; from the perspective of quantum groups the simplest one to choose is *Manin's quantum plane*:

$$\mathcal{O}_q(k^2) = k\langle x, y \mid xy = qyx\rangle$$
$$\text{or} \quad k\langle x_1, x_2 \mid x_1 x_2 = qx_2 x_1\rangle.$$

It should be reasonably clear that this algebra really is a skew polynomial ring, a point to which we shall return below.

A few comments concerning notation may be appropriate here. It has become common in noncommutative algebra to use angle brackets, $\langle\ \rangle$, to denote objects described by generators (with or without relations). For example, the expression $k\langle x, y \mid xy = qyx\rangle$ above refers to the k-algebra given by two generators x and y satisfying the relation $xy = qyx$. Similarly, in the context of groups, the expression $\langle x, y \mid xyx = y^{-1}\rangle$ would mean the group given by two generators x and y satisfying the relation $xyx = x^{-1}$. Returning to algebras, an expression like $k\langle x, y\rangle$ would mean either a free k-algebra (in which x and y satisfy no relations) or an algebra with two generators whose relations are understood from the context (e.g., the subalgebra generated by elements x and y from a given algebra). Finally, angle brackets are also used to describe ideals. For instance, the precise recipe for the algebra $\mathcal{O}_q(k^2)$ is $k\langle X, Y\rangle/\langle XY - qYX\rangle$, where $k\langle X, Y\rangle$ is the free k-algebra on X and Y and $\langle XY - qYX\rangle$ denotes the ideal of $k\langle X, Y\rangle$ generated by $XY - qYX$.

I.1.5. Example. The coordinate ring of $SL_2(k)$ can be expressed as a quotient

$$k[X_{11}, X_{12}, X_{21}, X_{22}]/\langle X_{11}X_{22} - X_{12}X_{21} - 1\rangle$$

where $k[X_{11}, X_{12}, X_{21}, X_{22}]$ is a polynomial ring in four indeterminates that correspond to the functions evaluating the four matrix entries of elements of $SL_2(k)$. It is useful to emphasize that $SL_2(k)$ is the closed subvariety of the set $M_2(k)$ of 2×2 matrices over k determined by the equation $X_{11}X_{22} - X_{12}X_{21} = 1$, that $M_2(k)$ is just a copy of affine 4-space, and that $k[X_{11}, X_{12}, X_{21}, X_{22}]$ should be identified with the coordinate ring of $M_2(k)$.

The comultiplication and counit on $\mathcal{O}(SL_2(k))$ are induced from corresponding maps on $\mathcal{O}(M_2(k))$ arising from multiplication of arbitrary matrices and evaluation at the identity matrix. These are the unique k-algebra homomorphisms

$$\Delta : \mathcal{O}(M_2(k)) \to \mathcal{O}(M_2(k)) \otimes \mathcal{O}(M_2(k)) \qquad \text{and} \qquad \varepsilon : \mathcal{O}(M_2(k)) \to k$$

such that

$$\Delta(X_{ij}) = X_{i1} \otimes X_{1j} + X_{i2} \otimes X_{2j} \qquad \text{and} \qquad \varepsilon(X_{ij}) = \delta_{ij} \qquad (1)$$

for $i, j = 1, 2$; with these maps, $\mathcal{O}(M_2(k))$ becomes a bialgebra (Exercise I.1.A). Writing $\overline{X}_{ij}$ for the image of X_{ij} in $\mathcal{O}(SL_2(k))$, the antipode S on $\mathcal{O}(SL_2(k))$ is the unique k-algebra automorphism such that

$$
\begin{aligned}
S(\overline{X}_{11}) &= \overline{X}_{22} & S(\overline{X}_{12}) &= -\overline{X}_{12} \\
S(\overline{X}_{21}) &= -\overline{X}_{21} & S(\overline{X}_{22}) &= \overline{X}_{11}
\end{aligned}
\qquad (2)
$$

(Exercise I.1.B).

There is some additional structure here that we should also take into account, namely multiplication of row or column vectors by matrices. At the level of coordinate rings, these operations correspond to k-algebra homomorphisms

$$\mathcal{O}(k^2) \to \mathcal{O}(M_2(k)) \otimes \mathcal{O}(k^2) \qquad \text{and} \qquad \mathcal{O}(k^2) \to \mathcal{O}(k^2) \otimes \mathcal{O}(M_2(k))$$

such that

$$x_i \mapsto X_{i1} \otimes x_1 + X_{i2} \otimes x_2 \qquad \text{and} \qquad x_j \mapsto x_1 \otimes X_{1j} + x_2 \otimes X_{2j} \qquad (3)$$

for $i, j = 1, 2$. (For such formulas it is convenient to label the indeterminates in $\mathcal{O}(k^2)$ as x_1 and x_2 rather than as x and y.)

I.1.6. Example. We now turn to the question of quantizing $SL_2(k)$. First, it is natural to start by trying to quantize $M_2(k)$. Following Drinfel'd's philosophy, and taking Manin's quantum plane as our quantization of the affine plane, we should try to construct a bialgebra B, generated as a k-algebra by elements X_{ij} satisfying (1), which supports k-algebra homomorphisms

$$\mathcal{O}_q(k^2) \to B \otimes \mathcal{O}_q(k^2) \qquad \text{and} \qquad \mathcal{O}_q(k^2) \to \mathcal{O}_q(k^2) \otimes B \qquad (4)$$

satisfying (3). The latter requirements imply that we must have

$$
\begin{aligned}
X_{1j} X_{2j} &= q X_{2j} X_{1j} \\
X_{i1} X_{i2} &= q X_{i2} X_{i1} \\
X_{11} X_{22} - X_{22} X_{11} &= q X_{21} X_{12} - q^{-1} X_{12} X_{21} \\
&= q X_{12} X_{21} - q^{-1} X_{21} X_{12}
\end{aligned}
\qquad (5)
$$

for all i, j (Exercise I.1.C). Provided $q^2 \neq -1$, equations (5) imply that

$$X_{12}X_{21} = X_{21}X_{12}$$
$$X_{11}X_{22} - X_{22}X_{11} = (q - q^{-1})X_{12}X_{21}. \tag{6}$$

No further relations are necessary: The k-algebra B given by generators X_{11}, X_{12}, X_{21}, X_{22} and relations (5), (6) does support a bialgebra structure satisfying (1) as well as k-algebra homomorphisms (4) satisfying (3) (Exercise I.1.D).

In case $q^2 = -1$, the relations (5) suffice to construct a bialgebra B and homomorphisms (4) satisfying (1), (3) (Exercise I.1.E). However, this produces an algebra with a different Hilbert series than $\mathcal{O}(M_2(k))$, and the ordered monomials $X_{11}^{\bullet}X_{12}^{\bullet}X_{21}^{\bullet}X_{22}^{\bullet}$ do not span B in this case (Exercise I.1.E). Thus (and also for the sake of uniformity), we impose the relations (5), (6) with all choices of q, as follows. To simplify future discussions, we abbreviate the names of the generators X_{ij} to a, b, c, d.

I.1.7. Definition. The single parameter *quantum 2×2 matrix algebra* (short for the 'quantized coordinate ring of 2×2 matrices') over k is the k-algebra $\mathcal{O}_q(M_2(k))$ given by generators a, b, c, d and relations

$$ab = qba \qquad ac = qca \qquad\qquad bc = cb$$
$$bd = qdb \qquad cd = qdc \qquad ad - da = (q - q^{-1})bc.$$

(When formulas such as (1) or (3) are under discussion, it is helpful to revert to the labelling $X_{11} = a$, $X_{12} = b$, $X_{21} = c$, $X_{22} = d$.) The algebra $\mathcal{O}_q(M_2(k))$ is a bialgebra with comultiplication and counit satisfying

$$\Delta(a) = a \otimes a + b \otimes c \qquad\qquad \Delta(b) = a \otimes b + b \otimes d$$
$$\Delta(c) = d \otimes c + c \otimes a \qquad\qquad \Delta(d) = d \otimes d + c \otimes b$$
$$\varepsilon(a) = \varepsilon(d) = 1 \qquad\qquad\qquad \varepsilon(b) = \varepsilon(c) = 0.$$

I.1.8. Example I.1.6 continued. To obtain a quantization of $SL_2(k)$ from $\mathcal{O}_q(M_2(k))$, we should expect to set some element playing the role of 'quantum determinant' equal to 1. In the classical setting, the determinant can be obtained from the exterior algebra $\Lambda(k^2)$: There is a k-algebra homomorphism $\Lambda(k^2) \rightarrow \mathcal{O}(M_2(k)) \otimes \Lambda(k^2)$ satisfying the left hand formula of (3), and

$$x_1 \wedge x_2 \mapsto \det(X_{ij}) \otimes (x_1 \wedge x_2).$$

Hence, we build a quantum analogue of this situation.

The *quantum exterior algebra* $\Lambda_q(k^2)$ is defined to be the k-algebra given by generators ξ_1 and ξ_2 and relations

$$\xi_1^2 = \xi_2^2 = 0 \qquad\qquad \xi_2\xi_1 = -q\xi_1\xi_2.$$

There is a unique k-algebra homomorphism $\Lambda_q(k^2) \to \mathcal{O}_q(M_2(k)) \otimes \Lambda_q(k^2)$ such that

$$\xi_i \mapsto X_{i1} \otimes \xi_1 + X_{i2} \otimes \xi_2$$

for all i, and

$$\xi_1\xi_2 \mapsto (X_{11}X_{22} - qX_{12}X_{21}) \otimes \xi_1\xi_2$$

(Exercise I.1.F). Hence, we set $D_q = X_{11}X_{22} - qX_{12}X_{21}$. It is easily checked that D_q lies in the centre of $\mathcal{O}_q(M_2(k))$ (Exercise I.1.G), and so the factor algebra $\mathcal{O}_q(M_2(k))/\langle D_q - 1 \rangle$ cannot collapse completely. Finally, we check that this factor algebra supports a k-algebra anti-automorphism S which, together with homomorphisms induced by Δ and ε, yields a Hopf algebra structure (Exercise I.1.H). We record these conclusions using the abbreviated notation of Definition I.1.7.

I.1.9. Definitions. The single parameter 2×2 *quantum determinant* is the element

$$D_q = ad - qbc = da - q^{-1}bc \ \in \ Z(\mathcal{O}_q(M_2(k))).$$

The single parameter *quantum* $SL_2(k)$ is the factor algebra

$$\mathcal{O}_q(SL_2(k)) = \mathcal{O}_q(M_2(k))/\langle D_q - 1 \rangle.$$

This algebra becomes a Hopf algebra with comultiplication, counit, and antipode satisfying

$$\Delta(\bar{a}) = \bar{a} \otimes \bar{a} + \bar{b} \otimes \bar{c} \qquad\qquad \Delta(\bar{b}) = \bar{a} \otimes \bar{b} + \bar{b} \otimes \bar{d}$$
$$\Delta(\bar{c}) = \bar{d} \otimes \bar{c} + \bar{c} \otimes \bar{a} \qquad\qquad \Delta(\bar{d}) = \bar{d} \otimes \bar{d} + \bar{c} \otimes \bar{b}$$
$$\varepsilon(\bar{a}) = \varepsilon(\bar{d}) = 1 \qquad\qquad \varepsilon(\bar{b}) = \varepsilon(\bar{c}) = 0$$
$$S(\bar{a}) = \bar{d} \qquad\qquad S(\bar{b}) = -q^{-1}\bar{b}$$
$$S(\bar{c}) = -q\bar{c} \qquad\qquad S(\bar{d}) = \bar{a}.$$

Note that the classical coordinate ring of $GL_2(k)$ is obtained from that of $M_2(k)$ by simply inverting the determinant function. Since D_q is a central element in $\mathcal{O}_q(M_2(k))$, its powers form a denominator set, and so D_q can be inverted in an Ore localization. This gives us another natural quantization, as follows.

I.1.10. Definition. The single parameter *quantum* $GL_2(k)$ is the localization

$$\mathcal{O}_q(GL_2(k)) = \mathcal{O}_q(M_2(k))[D_q^{-1}].$$

It has a Hopf algebra structure, extending the comultiplication and counit from $\mathcal{O}_q(M_2(k))$, such that

$$
\begin{aligned}
S(a) &= dD_q^{-1} & S(b) &= -q^{-1}bD_q^{-1} \\
S(c) &= -qcD_q^{-1} & S(d) &= aD_q^{-1}
\end{aligned}
\tag{7}
$$

(Exercise I.1.I).

Further examples will be given later. In particular, the algebras we have just introduced all have analogues in higher dimensions, denoted $\mathcal{O}_q(k^n)$, $\mathcal{O}_q(M_n(k))$, $\mathcal{O}_q(SL_n(k))$, and $\mathcal{O}_q(GL_n(k))$, and there are versions with collections of parameters replacing the single parameter q. These algebras will be presented in the following chapter. Discussion of the quantized coordinate rings $\mathcal{O}_q(G)$, for semisimple groups G other then $SL_n(k)$, will be deferred until enough of the representation theory of the corresponding quantized enveloping algebras $U_q(\mathfrak{g})$ has been outlined to allow the $\mathcal{O}_q(G)$s to be defined.

SKEW POLYNOMIAL RINGS

Many quantized algebras have the structure of iterated skew polynomial rings, and we will make considerable use of such structure. Let us put our preferred notation and conventions on record here.

I.1.11. Conventions. The symbol $R[x; \tau, \delta]$ stands for a skew polynomial ring in which $xr = \tau(r)x + \delta(r)$ for $r \in R$. Hence, δ must be a *left τ-derivation* on R, that is, an additive map satisfying $\delta(rs) = \tau(r)\delta(s) + \delta(r)s$ for $r, s \in R$. (There is an analogous definition of a *right τ-derivation*, requiring that $\delta(rs) = \delta(r)\tau(s) + r\delta(s)$ for $r, s \in R$.) We always assume that τ is an automorphism of R (not just an endomorphism), and if R is a k-algebra, we assume that τ and δ are k-linear. Similar conventions will be observed for skew-Laurent rings $R[x^{\pm 1}; \tau]$.

In order to avoid excessive labelling of isomorphisms, it is convenient to define skew polynomial rings as overrings with certain properties rather than as the results of special constructions. Thus, given a ring R, the statement $T = R[x; \tau, \delta]$ means that

    (1) T is an overring of R and x is an element of T.
    (2) T is a free left R-module with basis $\{x^n \mid n = 0, 1, 2, \dots\}$.
    (3) $xr = \tau(r)x + \delta(r)$ for all $r \in R$.
    (4) τ is an automorphism of R and δ is a τ-derivation on R.

Condition (4) is partly redundant, since it follows from (2) and (3) that τ must be a ring endomorphism of R and δ a τ-derivation.

We assume that elements of $R[x; \tau, \delta]$ are preferentially written with left-hand coefficients, that is, in the form

$$u = r_n x^n + r_{n-1} x^{n-1} + \cdots + r_1 x + r_0$$

with $r_n, \ldots, r_0 \in R$. If $u \neq 0$, we may assume that $r_n \neq 0$; then n is called the *degree* of u and r_n is the *leading coefficient* of u. A general formula for the coefficients of products in $R[x; \tau, \delta]$ would be too complicated to be of much use, but an easy induction establishes that

$$x^n r = \tau^n(r) x^n + [\text{lower terms}]$$

for $n > 0$ and $r \in R$. In particular, when the element u above is written with right-hand coefficients, it takes the form

$$u = x^n \tau^{-n}(r_n) + x^{n-1} r'_{n-1} + \cdots + x r'_1 + r'_0$$

for some $r'_{n-1}, \ldots, r'_0 \in R$.

Finally, note that for any elements

$$u = r_m x^m + r_{m-1} x^{m-1} + \cdots + r_1 x_1 + r_0$$
$$v = s_n x^n + s_{n-1} x^{n-1} + \cdots + s_1 x_1 + s_0 \tag{8}$$

in $R[x; \tau, \delta]$, where $r_m, \ldots, r_0, s_n, \ldots, s_0 \in R$, we have

$$uv = r_m \tau^m(s_n) x^{m+n} + [\text{lower terms}].$$

If the leading coefficients r_m and s_n are regular elements (i.e., non-zero-divisors), then $r_m \tau^m(s_n) \neq 0$ (taking advantage of the assumption that τ is an automorphism), and so $uv \neq 0$. This observation yields the first conclusion of the following lemma.

I.1.12. Lemma. (a) *A skew polynomial ring $R[x; \tau, \delta]$ over a domain R is a domain.*

(b) *If R is a prime ring, then so is any skew polynomial ring $R[x; \tau, \delta]$.*

Proof. (b) Let u and v be any two nonzero elements of $R[x; \tau, \delta]$, say with degrees m, n and leading coefficients r_m, s_n as in (8) above. Then r_m and $\tau^m(s_n)$ are nonzero elements of R. Since R is prime, there is some $t \in R$ such that $r_m t \tau^m(s_n) \neq 0$. Hence,

$$u \tau^{-m}(t) v = r_m t \tau^m(s_n) x^{m+n} + [\text{lower terms}] \neq 0.$$

Thus $R[x; \tau, \delta]$ is prime. $\square$

We shall also require the skew polynomial version of the Hilbert Basis Theorem:

I.1.13. Theorem. *A skew polynomial ring $R[x; \tau, \delta]$ over a right (left) noetherian ring R is right (left) noetherian.*

Proof. E.g., [83, Theorem 1.12] or [158, Theorem 1.2.9]. (The conclusion can fail if τ is not assumed to be an automorphism.) $\square$

BASIC STRUCTURE

I.1.14. Now that we have presented some interesting noncommutative algebras, we would like to know something of their structure. According to Drinfel'd's philosophy, much of this structure should parallel that of the corresponding classical coordinate rings, taking suitable account of the noncommutativity. Thus, our first motto is:

Quantized coordinate rings should be affine noetherian domains.

(Recall that an algebra is said to be *affine* provided it is finitely generated as an algebra.) The examples discussed so far are presented with finite sets of generators and relations, and so they are obviously affine (in fact, finitely presented) algebras. We shall see that several can also be presented as iterated skew polynomial algebras over k (that is, with the initial stage equal to k), whence they are noetherian domains by iterated application of Lemma I.1.12 and Theorem I.1.13. With more complicated examples, it is often easier to note that the algebra *resembles* an iterated skew polynomial ring that to prove that it actually *is* one. We shall use the quantum plane and quantum 2×2 matrices to illustrate two methods for dealing with this issue.

The key problem in proving that a given algebra is an iterated skew polynomial algebra is to verify that ordered monomials in the putative indeterminates are linearly independent over the base field. This is needed, in particular, to ensure that the algebra in question is not a proper homomorphic image of a corresponding iterated skew polynomial algebra, as is the case, for instance, with $\Lambda_q(k^2)$. We mention two different methods for verifying the linear independence of these monomials. One can be called the *model approach*, which is to construct an actual iterated skew polynomial algebra modeled on the given algebra, and to show that the two algebras are isomorphic. The *Diamond Lemma approach* involves working directly with the given generators and relations, using the Diamond Lemma (see Appendix I.11) to obtain the desired linear independence.

I.1.15. Example. The easiest example is the quantum plane, $\mathcal{O}_q(k^2)$. It follows quickly from the relation $xy = qyx$ that the monomials $x^i y^j$ span $\mathcal{O}_q(k^2)$.

If one uses the Diamond Lemma (Exercise I.1.J) or any other method to show that the $x^i y^j$ are linearly independent over k, then it is clear that the k-subalgebra of $\mathcal{O}_q(k^2)$ generated by x is a (commutative) polynomial ring $k[x]$, and that $\mathcal{O}_q(k^2)$ is a free left $k[x]$-module with basis $\{y^j \mid j \geq 0\}$. Since $yx^i = q^{-i}x^i y$ for all i, we have $yf = \tau(f)y$ for all $f \in k[x]$, where τ is the k-algebra automorphism of $k[x]$ given by $\tau(f(x)) = f(q^{-1}x)$. Therefore $\mathcal{O}_q(k^2) = k[x][y; \tau]$.

On the other hand, we can simply build a model of $k[x][y; \tau]$ and show that it is isomorphic to $\mathcal{O}_q(k^2)$. Thus, let $k[t]$ be an ordinary polynomial ring, let σ be the k-algebra automorphism of $k[t]$ such that $\sigma(t) = q^{-1}t$, and form the skew polynomial algebra $A = k[t][z; \sigma]$. Since $tz = qzt$, there is a unique k-algebra homomorphism $\phi : \mathcal{O}_q(k^2) \to A$ such that $\phi(x) = t$ and $\phi(y) = z$. Since the elements $\phi(x^i y^j) = t^i z^j$ in A are linearly independent, the monomials $x^i y^j$ must be linearly independent in $\mathcal{O}_q(k^2)$, and therefore ϕ is an isomorphism.

I.1.16. Example. It is easily checked that the monomials $a^\bullet b^\bullet c^\bullet d^\bullet$ span the algebra $\mathcal{O}_q(M_2(k))$. The Diamond Lemma can be used to show that these monomials are linearly independent (Exercise I.1.K). With that fact in hand, we see that $A_1 = k[a]$ is a polynomial ring, and that each of the subalgebras A_2, A_3, and $A_4 = \mathcal{O}_q(M_2(k))$ generated by $\{a, b\}$, $\{a, b, c\}$, and $\{a, b, c, d\}$, respectively, is a free module over the preceding subalgebra, with bases $\{b^m \mid m \geq 0\}$, $\{c^m \mid m \geq 0\}$, and $\{d^m \mid m \geq 0\}$, respectively. Next, observe that $bA_1 = A_1 b$ and $cA_2 = A_2 c$, whence there are automorphisms τ_2 of A_1 and τ_3 of A_2 such that $bx = \tau_2(x)b$ for all $x \in A_1$ and $cy = \tau_3(y)c$ for all $y \in A_2$. It follows that $A_2 = A_1[b; \tau_2]$ and $A_3 = A_2[c; \tau_3]$. Finally, while $dA_3 \neq A_3 d$, we do have $dA_3 + A_3 = A_3 d + A_3$. Since d is regular and $dA_3 \cap A_3 = A_3 d \cap A_3 = 0$ (Exercise I.1.K), there exist maps $\tau_4, \delta_4 : A_3 \to A_3$ such that $dz = \tau_4(z)d + \delta_4(z)$ for all $z \in A_3$. It follows that τ_4 is an automorphism and δ_4 a τ_4-derivation on A_3 (Exercise I.1.L), and so $A_4 = A_3[d; \tau_4, \delta_4]$. Therefore we have expressed $\mathcal{O}_q(M_2(k))$ as an iterated skew polynomial algebra of the form

$$k[a][b; \tau_2][c; \tau_3][d; \tau_4, \delta_4].$$

The model approach to this algebra requires an existence lemma for skew derivations (Exercise I.1.M) at the fourth stage. The first three stages are clear – we construct an iterated skew polynomial algebra

$$B = k[x][y; \sigma_2][z; \sigma_3]$$

where $k[x]$ is a polynomial ring, σ_2 is the k-algebra automorphism of $k[x]$ such that $\sigma_2(x) = q^{-1}x$, and σ_3 is the k-algebra automorphism of $k[x][y; \sigma_2]$ such that $\sigma_3(x) = q^{-1}x$ and $\sigma_3(y) = y$. It is also clear that there is a k-algebra automorphism σ_4 on B such that $\sigma_4(x) = x$ while $\sigma_4(y) = q^{-1}y$ and $\sigma_4(z) = q^{-1}z$. We also need a σ_4-derivation ∂_4 on B such that $\partial_4(x) = (q^{-1} - q)yz$ while $\partial_4(y) = \partial_4(z) = 0$. Such a skew derivation exists by Exercise I.1.N, and so there exists a skew polynomial ring $A = B[w; \sigma_4, \partial_4]$.

Since A has been constructed so that x, y, z, w satisfy the defining relations of a, b, c, d, there is a k-algebra homomorphism $\phi : \mathcal{O}_q(M_2(k)) \to A$ mapping a, b, c, d to x, y, z, w, respectively. The monomials $x^\bullet y^\bullet z^\bullet w^\bullet$ in A are linearly independent, and so it follows that the monomials $a^\bullet b^\bullet c^\bullet d^\bullet$ in $\mathcal{O}_q(M_2(k))$ are linearly independent. Therefore ϕ is an isomorphism.

I.1.17. Theorem. *The algebras $\mathcal{O}_q(k^2)$, $\mathcal{O}_q(M_2(k))$, and $\mathcal{O}_q(GL_2(k))$ are noetherian domains.*

Proof. As discussed in Examples I.1.15 and I.1.16, $\mathcal{O}_q(k^2)$ and $\mathcal{O}_q(M_2(k))$ are iterated skew polynomial algebras over k; hence, these algebras are noetherian domains. Since $\mathcal{O}_q(GL_2(k))$ is the localization of $\mathcal{O}_q(M_2(k))$ with respect to the central regular element D_q, it follows immediately that $\mathcal{O}_q(GL_2(k))$ is a noetherian domain. $\square$

I.1.18. Since $\mathcal{O}_q(SL_2(k))$ is a factor of $\mathcal{O}_q(M_2(k))$, it is clearly noetherian, but further information is needed to show that it is a domain. Going back to the classical situation, recall that the varieties $SL_2(k) \times k^\times$ and $GL_2(k)$ are isomorphic, via the map

$$\left(\begin{pmatrix} a & b \\ c & d \end{pmatrix}, \delta \right) \mapsto \begin{pmatrix} \delta a & \delta b \\ c & d \end{pmatrix}.$$

(Warning: This map is not a group isomorphism!) At the level of coordinate rings, it follows that

$$\mathcal{O}(GL_2(k)) \cong \mathcal{O}(SL_2(k)) \otimes \mathcal{O}(k^\times) \cong \mathcal{O}(SL_2(k))[z^{\pm 1}].$$

There exists an analogous isomorphism $\mathcal{O}_q(GL_2(k)) \cong \mathcal{O}_q(SL_2(k))[z^{\pm 1}]$ (Exercise I.1.O). In particular, $\mathcal{O}_q(SL_2(k))$ embeds in $\mathcal{O}_q(GL_2(k))$, whence $\mathcal{O}_q(SL_2(k))$ must be a domain. Let us record this:

I.1.19. Theorem. $\mathcal{O}_q(SL_2(k))$ *is a noetherian domain.* $\square$

NOTES

A number of discussions of the origins of quantum groups, far more extensive than ours, can be found in the literature. Let us just mention the papers [51] and [186] and the books [34], [125], [146]. For some background on the QYBE, see, for example, [34, §7.5,], [112, §8.5], [130, Introduction], [146, §4.4]. Manin's quantum plane appeared in [148, 150], and he promulgated the viewpoint that quantizations of $M_2(k)$, $SL_2(k)$, and $GL_2(k)$ should support quantizations of k^2 as comodule algebras [149, 151].

EXERCISES

Exercise I.1.A. Show that there exist k-algebra homomorphisms

$$\Delta : \mathcal{O}(M_2(k)) \to \mathcal{O}(M_2(k)) \otimes \mathcal{O}(M_2(k)) \qquad \text{and} \qquad \varepsilon : \mathcal{O}(M_2(k)) \to k$$

satisfying equations (1), and that $(\mathcal{O}(M_2(k)), \Delta, \varepsilon)$ is a bialgebra. $\square$

Exercise I.1.B. Show that the comultiplication and counit on $\mathcal{O}(M_2(k))$ induce k-algebra homomorphisms

$$\Delta : \mathcal{O}(SL_2(k)) \to \mathcal{O}(SL_2(k)) \otimes \mathcal{O}(SL_2(k)) \qquad \text{and} \qquad \varepsilon : \mathcal{O}(SL_2(k)) \to k,$$

and that $(\mathcal{O}(SL_2(k)), \Delta, \varepsilon)$ is a bialgebra. Then show that there is a k-algebra automorphism S on $\mathcal{O}(SL_2(k))$ satisfying equations (2), and that the system $(\mathcal{O}(SL_2(k)), \Delta, \varepsilon, S)$ is a Hopf algebra. $\square$

Exercise I.1.C. Verify that if there exist k-algebra homomorphisms (4) satisfying conditions (3), then equations (5) must hold. In case $q^2 \neq -1$, derive equations (6). $\square$

Exercise I.1.D. Let B be the k-algebra given by generators X_{11}, X_{12}, X_{21}, X_{22} and relations (5), (6). Show that there exist k-algebra homomorphisms $\Delta : B \to B \otimes B$ and $\varepsilon : B \to k$ satisfying (1), and that (B, Δ, ε) is a bialgebra. Show also that there exist k-algebra homomorphisms (4) satisfying equations (3). $\square$

Exercise I.1.E. Let B be the k-algebra given by generators X_{11}, X_{12}, X_{21}, X_{22} and relations (5) only. Verify all the conclusions of Exercise I.1.D in this setting. Show that the ordered monomials $X_{11}^\bullet X_{12}^\bullet X_{21}^\bullet X_{22}^\bullet$ do not span B. $\square$

Exercise I.1.F. Verify the existence of a k-algebra homomorphism

$$\phi : \Lambda_q(k^2) \to \mathcal{O}_q(M_2(k)) \otimes \Lambda_q(k^2)$$

such that $\phi(\xi_i) = X_{i1} \otimes \xi_1 + X_{i2} \otimes \xi_2$ for $i = 1, 2$, and show that

$$\phi(\xi_1\xi_2) = (X_{11}X_{22} - qX_{12}X_{21}) \otimes \xi_1\xi_2. \quad \square$$

Exercise I.1.G. Show that D_q is central in $\mathcal{O}_q(M_2(k))$. $\square$

Exercise I.1.H. Set $A = \mathcal{O}_q(M_2(k))/\langle D_q - 1 \rangle$, and show that the comultiplication and counit on $\mathcal{O}_q(M_2(k))$ induce k-algebra homomorphisms $\Delta : A \to A \otimes A$ and $\varepsilon : A \to k$. Next, show that there is a k-algebra <u>anti</u>-automorphism S on A such that

$$S(\overline{X}_{11}) = \overline{X}_{22} \qquad\qquad S(\overline{X}_{12}) = -q^{-1}\overline{X}_{12}$$
$$S(\overline{X}_{21}) = -q\overline{X}_{21} \qquad\qquad S(\overline{X}_{22}) = \overline{X}_{11}.$$

Finally, show that $(A, \Delta, \varepsilon, S)$ is a Hopf algebra. $\square$

Exercise I.1.I. Set $A = \mathcal{O}_q(GL_2(k))$, and show that the comultiplication and counit on $\mathcal{O}_q(M_2(k))$ induce k-algebra homomorphisms $\Delta : A \to A \otimes A$ and $\varepsilon : A \to k$. [Hint: Show that $\Delta(D_q) = D_q \otimes D_q$ and $\varepsilon(D_q) = 1$.] Next, show that there is a k-algebra anti-automorphism S on A satisfying equations (7). Finally, show that $(A, \Delta, \varepsilon, S)$ is a Hopf algebra. $\square$

Exercise I.1.J. Use the Diamond Lemma to show that the monomials $x^\bullet y^\bullet$ in $\mathcal{O}_q(k^2)$ are linearly independent over k. $\square$

Exercise I.1.K. Use the Diamond Lemma to show that the monomials $a^\bullet b^\bullet c^\bullet d^\bullet$ in $\mathcal{O}_q(M_2(k))$ are linearly independent over k (cf. Example I.11.7). Then show that d is a regular element of $\mathcal{O}_q(M_2(k))$ satisfying $dA_3 \cap A_3 = A_3 d \cap A_3 = 0$, where A_3 is the k-subalgebra of $\mathcal{O}_q(M_2(k))$ generated by a, b, c. $\square$

Exercise I.1.L. Let $R \subseteq S$ be rings, and suppose that there is a regular element $d \in S$ such that $dR + R = Rd + R$ and $dR \cap R = Rd \cap R = 0$. Then there are unique maps $\tau, \delta : R \to R$ such that $dr = \tau(r)d + \delta(r)$ for all $r \in R$. Show that τ is an automorphism of R and that δ is a τ-derivation on R. □

Exercise I.1.M. Let $F = k\langle X_1, \ldots, X_t \rangle$ be the free algebra over k on letters $X_1, \ldots, X_t$, and let τ be a k-algebra endomorphism of F. Given any $f_1, \ldots, f_t \in F$, show that there exists a unique k-linear τ-derivation δ on F such that $\delta(X_i) = f_i$ for all i. [Hint: Construct a k-algebra homomorphism $\phi : F \to M_2(F)$ such that
$$\phi(X_i) = \begin{pmatrix} \sigma(X_i) & f_i \\ 0 & X_i \end{pmatrix} \text{ for all } i.]$$
    Now let $I = \langle G \rangle$ be the ideal of F generated by some subset $G \subseteq F$. If $\tau(g), \delta(g) \in I$ for all $g \in G$, show that I is stable under τ and δ. Conclude that τ and δ induce an endomorphism $\overline{\tau}$ and a $\overline{\tau}$-derivation $\overline{\delta}$ on F/I. □

Exercise I.1.N. If B and σ_4 are as in Example I.1.16, show that there exists a σ_4-derivation ∂_4 on B such that $\partial_4(x) = (q^{-1} - q)yz$ while $\partial_4(y) = \partial_4(z) = 0$. □

Exercise I.1.O. Construct a k-algebra isomorphism of $\mathcal{O}_q(GL_2(k))$ onto the Laurent polynomial ring $\mathcal{O}_q(SL_2(k))[z^{\pm 1}]$ mapping a, b, c, d to $\overline{a}z$, $\overline{b}z$, $\overline{c}$, $\overline{d}$, respectively. □

FURTHER QUANTIZED COORDINATE RINGS

Many other quantizations of classical algebras, in addition to the basic examples discussed in the previous chapter, have been constructed. We present a selection here, from those given in terms of generators and relations.

GENERATORS AND RELATIONS

I.2.1. Definitions. Single parameter quantized coordinate rings of both affine spaces and algebraic tori are constructed in the same manner as Manin's quantum plane. Thus, *quantum affine n-space* is the k-algebra $\mathcal{O}_q(k^n)$ with generators $x_1, \ldots, x_n$ and relations $x_i x_j = q x_j x_i$ for all $i < j$. Similarly, the *quantum torus of rank n* is the k-algebra $\mathcal{O}_q((k^\times)^n)$ with generators $x_1^{\pm 1}, \ldots, x_n^{\pm 1}$ and the same relations as $\mathcal{O}_q(k^n)$.

It is not difficult to build multiparameter versions of these algebras using relations of the form $x_i x_j = q_{ij} x_j x_i$ for some scalars q_{ij}. If such relations are given for all pairs of indices i and j, then $x_i^2 = q_{ii} x_i^2$ for all i and $x_i x_j = q_{ij} q_{ji} x_i x_j$ for all i, j. Since we are quantizing a polynomial ring, we would like to obtain something like a skew polynomial ring; in particular, we do not want to have $x_i x_j = 0$ for any i, j. Hence, we restrict the matrix of scalars $\boldsymbol{q} = (q_{ij})$ to be *multiplicatively antisymmetric*, that is, $q_{ii} = 1$ and $q_{ji} = q_{ij}^{-1}$ for all i, j.

Given a multiplicatively antisymmetric matrix $\boldsymbol{q} \in M_n(k^\times)$, the corresponding *multiparameter quantum affine space* is the k-algebra $\mathcal{O}_q(k^n)$ with generators $x_1, \ldots, x_n$ and relations $x_i x_j = q_{ij} x_j x_i$ for all i, j, and the *multiparameter quantum torus* $\mathcal{O}_q((k^\times)^n)$ is given by generators $x_1^{\pm 1}, \ldots, x_n^{\pm 1}$ and the same relations. The single parameter versions are recovered in case $q_{ij} = q$ for all $i < j$.

We may also express $\mathcal{O}_q((k^\times)^n)$ as a localization of $\mathcal{O}_q(k^n)$ with respect to the multiplicative set generated by $x_1, \ldots, x_n$ (Exercise I.2.A).

I.2.2. Definitions. The single parameter *quantum $n \times n$ matrix algebra*, denoted $\mathcal{O}_q(M_n(k))$, is given by generators X_{ij} for $i, j = 1, \ldots, n$ and relations chosen so that each 'square' of four generators, i.e., those at positions

$$\begin{matrix} (i,j) & (i,j') \\ (i',j) & (i',j') \end{matrix}$$

(where $i < i'$ and $j < j'$), generate a copy of $\mathcal{O}_q(M_2(k))$. More precisely,

$$X_{ij}X_{lm} = \begin{cases} qX_{lm}X_{ij} & (i < l, \ j = m) \\ qX_{lm}X_{ij} & (i = l, \ j < m) \\ X_{lm}X_{ij} & (i < l, \ j > m) \\ X_{lm}X_{ij} + (q - q^{-1})X_{im}X_{lj} & (i < l, \ j < m). \end{cases}$$

We can also define a *multiparameter quantum $n \times n$ matrix algebra*, denoted $\mathcal{O}_{\lambda,p}(M_n(k))$, where $\lambda \in k^\times$ and $\boldsymbol{p} = (p_{ij}) \in M_n(k^\times)$ is multiplicatively antisymmetric. This algebra has generators X_{ij} for $i, j = 1, \ldots, n$ and relations

$$X_{lm}X_{ij} = \begin{cases} p_{li}p_{jm}X_{ij}X_{lm} + (\lambda - 1)p_{li}X_{im}X_{lj} & (l > i, \ m > j) \\ \lambda p_{li}p_{jm}X_{ij}X_{lm} & (l > i, \ m \leq j) \\ p_{jm}X_{ij}X_{lm} & (l = i, \ m > j). \end{cases}$$

The single parameter case is recovered when $\lambda = q^{-2}$ and $p_{ij} = q$ for all $i > j$.

These algebras become bialgebras using comultiplications and counits such that

$$\Delta(X_{ij}) = \sum_{l=1}^{n} X_{il} \otimes X_{lj} \qquad \text{and} \qquad \varepsilon(X_{ij}) = \delta_{ij} \tag{1}$$

for all i, j (Exercise I.2.B).

I.2.3. Definitions. As in (I.1.8), one can construct quantum exterior algebras of arbitrary degree and use them to recover a quantum determinant element (see Exercise I.2.C). In the single parameter case, the *quantum determinant* in $\mathcal{O}_q(M_n(k))$ can be expressed as

$$D_q = \sum_{\pi \in S_n} (-q)^{\ell(\pi)} X_{1,\pi(1)} X_{2,\pi(2)} \cdots X_{n,\pi(n)}, \tag{2}$$

where $\ell(\pi)$ denotes the *length* of the permutation π, that is, the minimum length of an expression for π as a product of adjacent transpositions $(i, i+1)$. The *multiparameter quantum determinant* in $\mathcal{O}_{\lambda,p}(M_n(k))$ is

$$D_{\lambda,p} = \sum_{\pi \in S_n} \left(\prod_{\substack{1 \leq i < j \leq n \\ \pi(i) > \pi(j)}} (-p_{\pi(i),\pi(j)}) \right) X_{1,\pi(1)} X_{2,\pi(2)} \cdots X_{n,\pi(n)}. \tag{3}$$

I.2.4. Definitions. The quantum determinant D_q is central in $\mathcal{O}_q(M_n(k))$ (Exercise I.2.E), and so we can form the single parameter *quantum $SL_n(k)$* and *quantum $GL_n(k)$* just as in Definitions I.1.9 and I.1.10:

$$\mathcal{O}_q(SL_n(k)) = \mathcal{O}_q(M_n(k))/\langle D_q - 1\rangle$$
$$\mathcal{O}_q(GL_n(k)) = \mathcal{O}_q(M_n(k))[D_q^{-1}].$$

In the multiparameter case, $D_{\lambda,p}$ is not always central, but it still generates a denominator set (Exercise I.2.F). Hence, we form the *multiparameter quantum $GL_n(k)$*,

$$\mathcal{O}_{\lambda,p}(GL_n(k)) = \mathcal{O}_{\lambda,p}(M_n(k))[D_{\lambda,p}^{-1}],$$

for all λ and p, but the *multiparameter quantum $SL_n(k)$*,

$$\mathcal{O}_{\lambda,p}(SL_n(k)) = \mathcal{O}_{\lambda,p}(M_n(k))\langle D_{\lambda,p} - 1\rangle,$$

only for those choices of λ and p for which $D_{\lambda,p}$ is central (see Exercise I.2.F).

   All of the algebras just introduced have Hopf algebra structures, with comultiplications and counits induced from those on the corresponding quantum matrix algebras (Exercises I.2.G and I.2.H).

I.2.5. For each connected, semisimple, complex algebraic group G, there are quantizations $\mathcal{O}_q(G)_k$ which can be defined over most base fields k containing a non-root of unity q. As mentioned above, the definition of $\mathcal{O}_q(G)_k$ requires some knowledge of the representation theory of $U_q(\mathfrak{g})_k$ (where $\mathfrak{g}$ is the Lie algebra of G), and so we postpone further discussion of $\mathcal{O}_q(G)_k$ to Chapter I.7. Multiparameter quantizations of G can be obtained by twisting $\mathcal{O}_q(G)_k$ with Hopf cocycles, as described in [**96**].

I.2.6. Definition. Finally, let us mention quantizations of the Weyl algebras. Recall that the classical Weyl algebra $A_n(k)$ can be viewed as an algebra of differential operators on the polynomial ring $\mathcal{O}(k^n)$. Quantized Weyl algebras have been constructed as algebras of 'q-difference operators' on $\mathcal{O}_q(k^n)$ or $\mathcal{O}_q(k^n)$ (cf. [**147**] and [**113**, §2.9]). In particular, the simplest quantization of $A_1(k)$ is the k-algebra $A_1^q(k)$ given by generators x, y satisfying the relation $xy - qyx = 1$.

   Now let $Q = (q_1, \ldots, q_n)$ be a vector in $(k^\times)^n$ and $\Gamma = (\gamma_{ij})$ a multiplicatively antisymmetric $n \times n$ matrix over k. The corresponding *multiparameter quantized Weyl algebra of degree n over k* is the k-algebra $A_n^{Q,\Gamma}(k)$ with generators $x_1, y_1, \ldots, x_n, y_n$ satisfying the following relations:

$$
\begin{aligned}
y_i y_j &= \gamma_{ij} y_j y_i && \text{(all } i, j) \\
x_i x_j &= q_i \gamma_{ij} x_j x_i && (i < j) \\
x_i y_j &= \gamma_{ji} y_j x_i && (i < j) \\
x_i y_j &= q_j \gamma_{ji} y_j x_i && (i > j) \\
x_j y_j &= 1 + q_j y_j x_j + \sum_{l<j}(q_l - 1) y_l x_l && \text{(all } j).
\end{aligned}
$$

While these algebras are not quantized coordinate rings, they share many properties with the latter algebras, and many of the results we will prove apply equally well to quantized coordinate rings and quantized Weyl algebras.

BASIC STRUCTURE

I.2.7. Theorem. *Each of the algebras*

$$\mathcal{O}_q(k^n), \qquad \mathcal{O}_q(k^n), \qquad \mathcal{O}_q(M_n(k)), \qquad \mathcal{O}_{\lambda,p}(M_n(k)), \qquad A_n^{Q,\Gamma}(k)$$

is an iterated skew polynomial algebra over k. Consequently, these algebras are noetherian domains.

Proof. Exercise I.2.I. □

I.2.8. Corollary. *The algebras*

$$\mathcal{O}_q((k^\times)^n), \qquad \mathcal{O}_q((k^\times)^n), \qquad \mathcal{O}_q(GL_n(k)), \qquad \mathcal{O}_{\lambda,p}(GL_n(k))$$

are noetherian domains.

Proof. Each of these algebras is an Ore localization of a noetherian domain with respect to a denominator set consisting of nonzero elements. □

In order to deal with $\mathcal{O}_q(SL_n(k))$, we need an isomorphism similar to that constucted in Exercise I.1.O, as follows.

I.2.9. Lemma. $\mathcal{O}_q(GL_n(k)) \cong \mathcal{O}_q(SL_n(k))[z^{\pm 1}]$.

Proof. Set $Y_{1j} = D_q^{-1} X_{1j} \in \mathcal{O}_q(GL_n(k))$ for all j, and $Y_{ij} = X_{ij} \in \mathcal{O}_q(GL_n(k))$ for $i > 1$ and all j. Since D_q is central in $\mathcal{O}_q(GL_n(k))$, it is easily checked that that the Y_{ij} satisfy the defining relations of the X_{ij}. Hence, there is a unique k-algebra homomorphism

$$\theta_0 : \mathcal{O}_q(M_n(k)) \longrightarrow \mathcal{O}_q(GL_n(k))$$

such that $\theta_0(X_{ij}) = Y_{ij}$ for all i, j. Observe that $\theta_0(D_q) = 1$. Hence, θ_0 induces a k-algebra homomorphism $\theta_1 : \mathcal{O}_q(SL_n(k)) \to \mathcal{O}_q(GL_n(k))$ such that $\theta_1(\overline{X}_{ij}) = Y_{ij}$ for all i, j. Then θ_1 extends to a k-algebra homomorphism $\theta : \mathcal{O}_q(SL_n(k))[z^{\pm 1}] \to \mathcal{O}_q(GL_n(k))$ such that $\theta(z) = D_q$.

Similarly, one constructs a k-algebra homomorphism $\xi : \mathcal{O}_q(GL_n(k)) \to \mathcal{O}_q(SL_n(k))[z^{\pm 1}]$ such that $\xi(X_{1j}) = z\overline{X}_{1j}$ for all j while $\xi(X_{ij}) = \overline{X}_{ij}$ for $i > 1$ and all j. Finally, one checks that θ and ξ are inverses of each other (Exercise I.2.J). □

I.2.10. Theorem. *The algebras $\mathcal{O}_q(SL_n(k))$ and $\mathcal{O}_{\lambda,p}(SL_n(k))$ are noetherian domains.*

Proof. Since $\mathcal{O}_q(M_n(k))$ and $\mathcal{O}_{\lambda,p}(M_n(k))$ are noetherian (Theorem I.2.7), so are $\mathcal{O}_q(SL_n(k))$ and $\mathcal{O}_{\lambda,p}(SL_n(k))$. In view of Lemma I.2.9, there is an embedding of $\mathcal{O}_q(SL_n(k))$ in $\mathcal{O}_q(GL_n(k))$. Since the latter is a domain by Corollary I.2.8, it follows that $\mathcal{O}_q(SL_n(k))$ is a domain. The case of $\mathcal{O}_{\lambda,p}(SL_n(k))$ is handled in the same manner (Exercise I.2.K). □

QUOTIENT DIVISION RINGS

I.2.11. We have now introduced several families of noetherian domains. Each has (by Ore's Theorem) a division ring of fractions, and we can ask what these division rings look like, the easiest to describe being the quotient division rings of quantum affine spaces. We use the notation Fract A for the quotient division ring of a noetherian domain A.

Since, typically, quotient division rings are simpler than the underlying algebras (recall the concept of birational equivalence in algebraic geometry), one asks whether the quotient division rings for a given class of algebras fall into a short list of prototypical examples. A model situation occurs in Lie theory, where the *Gel'fand-Kirillov conjecture* [59] asserted that for any finite dimensional algebraic Lie algebra $\mathfrak{g}$ over an algebraically closed field k of characteristic zero, the quotient division ring Fract $U(\mathfrak{g})$ should be isomorphic to the quotient division ring of some Weyl algebra over a purely transcendental field extension of k. The conjecture has been confirmed for $\mathfrak{g} = \mathfrak{sl}_n(k)$ and $\mathfrak{g} = \mathfrak{gl}_n(k)$ [60] and for $\mathfrak{g}$ solvable [16, 114, 155], but counterexamples have been constructed for the general case [4]. In the semisimple case, aside from type A, the conjecture remains open, but a weakened version (with $U(\mathfrak{g})$ replaced by $U(\mathfrak{g}) \otimes_{Z(U(\mathfrak{g}))} \widetilde{Z}$ for a suitable extension $\widetilde{Z}$ of the centre) has been proved [61, 62].

Taking inspiration from the above classical work, Alev-Dumas [3] and Iohara-Malikov [103] proposed quantum analogues of the conjecture. In particular, Alev and Dumas asked for a direct parallel with the classical case: With generic parameters, the quotient division rings of $U_q(\mathfrak{g})$ and analogous algebras should be isomorphic to quotient division rings of quantized Weyl algebras over transcendental extensions of the base field. They verified this conjecture for $U_q^+(\mathfrak{sl}_n(k))$, $U_q(\mathfrak{sl}_2(k))$, and $\breve{U}_q(\mathfrak{sl}_3(k))$ when q is not a root of unity [3, Théorème 2.15, Proposition 4.3, Théorème 4.6]. Further, they proved that for the cases of quantized Weyl algebras $A_n^{Q,\Gamma}(k)$ appearing in their results, Fract $A_n^{Q,\Gamma}(k) \cong$ Fract $\mathcal{O}_q(k^{2n})$ for suitable matrices q [3, Théorème 3.5]. (The latter isomorphisms were shown by Jordan to hold for arbitrary $A_n^{Q,\Gamma}(k)$ [113, §3.1].) This result justifies formulating the conjecture in terms of quotient division rings of quantum affine spaces, which is how Iohara and Malikov presented and proved it for $U_q^+(\mathfrak{sl}_n(k))$ with q not a root of unity [103, Theorem 3.5].

In its current, somewhat expanded form, the *quantum Gel'fand-Kirillov conjecture* asserts that quotient division rings of quantized coordinate rings and quantized enveloping algebras should be isomorphic to quotient division rings of quantum affine spaces over purely transcendental extensions of the base field. We conclude the chapter by discussing a few cases in which the conjecture has been confirmed. For further discussion and references, see (II.10.4).

I.2.12. Example. Let $A = \mathcal{O}_q(SL_2(k))$ and $D = $ Fract A. Since $\bar{a}\bar{d} - q\bar{b}\bar{c} = 1$ in A, we have $\bar{d} = \bar{a}^{-1}(1 + q\bar{b}\bar{c})$ in D, and so D is generated as a division algebra by

$\bar{a}, \bar{b}, \bar{c}$. These three elements satisfy the defining relations of $\mathcal{O}_{\boldsymbol{q}}(k^3)$ where

$$\boldsymbol{q} = \begin{pmatrix} 1 & q & q \\ q^{-1} & 1 & 1 \\ q^{-1} & 1 & 1 \end{pmatrix},$$

and the corresponding homomorphism $\mathcal{O}_{\boldsymbol{q}}(k^3) \to A$ is an embedding (Exercise I.2.L). Hence, $D \cong \operatorname{Fract} \mathcal{O}_{\boldsymbol{q}}(k^3)$.

Since A is a single parameter quantization, one would like to tighten up the result above, so that D becomes isomorphic to the quotient division ring of some single parameter quantum affine space. This can be done as follows.

First observe that the element $\bar{b}\bar{c}^{-1}$ is central in D, and that D is generated as a division algebra by $\bar{a}, \bar{b}, \bar{b}\bar{c}^{-1}$. Since $\bar{b}\bar{c}^{-1}$ lies in the field $Z(D)$, that field contains the quotient field of $k[\bar{b}\bar{c}^{-1}]$, which is a rational function field because $\bar{b}\bar{c}^{-1}$ is transcendental over k (Exercise I.2.M). Hence, if we build D by first forming $k(\bar{b}\bar{c}^{-1})$ and then adjoining $\bar{a}$ and $\bar{b}$ to obtain a quantum plane over this field, we see that $D \cong \operatorname{Fract} \mathcal{O}_q(K^2)$ where $K = k(\bar{b}\bar{c}^{-1})$ is a purely transcendental extension of k.

I.2.13. Example. Let $D = \operatorname{Fract} \mathcal{O}_q(GL_2(k)) = \operatorname{Fract} \mathcal{O}_q(M_2(k))$. As in the example above, we can generate D by either a, b, c, D_q or a, b, bc^{-1}, D_q. The second set of generators has the advantage that two of the generators, namely bc^{-1} and D_q, are central. It follows that $D \cong \operatorname{Fract} \mathcal{O}_{\boldsymbol{q}}(k^4) \cong \operatorname{Fract} \mathcal{O}_q(K^2)$ where $\boldsymbol{q}$ is a suitable 4×4 multiplicatively antisymmetric matrix over k while $K = k(bc^{-1}, D_q)$ is a purely transcendental extension field of k (Exercise I.2.N).

I.2.14. Theorem. *There exist multiplicatively antisymmetric matrices $\boldsymbol{q}, \boldsymbol{r}$, and $\boldsymbol{s}$ over k of appropriate sizes such that*

$$\operatorname{Fract} \mathcal{O}_q(M_n(k)) \cong \operatorname{Fract} \mathcal{O}_{\boldsymbol{q}}(k^{n^2})$$
$$\operatorname{Fract} \mathcal{O}_{\lambda,\boldsymbol{p}}(M_n(k)) \cong \operatorname{Fract} \mathcal{O}_{\boldsymbol{r}}(k^{n^2})$$
$$\operatorname{Fract} A_n^{Q,\Gamma}(k) \cong \operatorname{Fract} \mathcal{O}_{\boldsymbol{s}}(k^{2n}).$$

Proof. See [**36**, Propositions 3–5], [**177**, Theorem 3.8], [**164**, Theorem 1.24], and [**113**, §3.1]. The cited results give details about the entries of $\boldsymbol{q}, \boldsymbol{r}, \boldsymbol{s}$. Panov's calculations also provide an isomorphism $\operatorname{Fract} \mathcal{O}_q(M_n(k)) \cong \operatorname{Fract} \mathcal{O}_t(K^{n(n-1)})$ for an explicit t, where $K = k(z_1, \dots, z_n)$ is a purely transcendental extension field of k. □

NOTES

We shall not attribute the single-parameter quantized coordinate rings $\mathcal{O}_q(k^n)$, $\mathcal{O}_q(M_n(k))$, $\mathcal{O}_q(SL_n(k))$, and $\mathcal{O}_q(GL_n(k))$ to any particular source, since they appeared in a number of places. The multiparameter algebras $\mathcal{O}_{\lambda,\boldsymbol{p}}(M_n(k))$, together with $\mathcal{O}_{\lambda,\boldsymbol{p}}(SL_n(k))$ and $\mathcal{O}_{\lambda,\boldsymbol{p}}(GL_n(k))$, were constructed by Artin-Schelter-Tate [8] and Sudbury [200], while the quantized Weyl algebras $A_n^{Q,\Gamma}(k)$ appeared in work of Demidov [46] (cf. [47]) and Maltsiniotis [147]. There are also quantized coordinate rings of symplectic and Euclidean spaces, which were defined by Faddeev-Reshetikhin-Takhtadjan [54, 186]; simpler sets of generators and relations for these algebras were described by Musson [167] and Oh [174, 175]. Further discussion of and references for quantized coordinate rings can be found in the survey part of [70].

EXERCISES

Exercise I.2.A. Show that the multiplicative set X in $\mathcal{O}_q(k^n)$ generated by $x_1, \ldots, x_n$ is a denominator set, and that $\mathcal{O}_q(k^n)[X^{-1}] \cong \mathcal{O}_q((k^\times)^n)$. □

Exercise I.2.B. Let A be either $\mathcal{O}_q(M_n(k))$ or $\mathcal{O}_{\lambda,\boldsymbol{p}}(M_n(k))$, and show that there exist k-algebra homomorphisms $\Delta : A \to A \otimes A$ and $\varepsilon : A \to k$ satisfying equations (1). Then show that (A, Δ, ε) is a bialgebra. □

Exercise I.2.C. Define the single parameter *quantum exterior algebra* $\Lambda_q(k^n)$ to be the k-algebra with generators $\xi_1, \ldots, \xi_n$ and relations

$$\xi_i^2 = 0 \qquad\qquad \xi_i \xi_j = -q \xi_j \xi_i$$

for all $i > j$. Show that there is a k-algebra homomorphism

$$\phi : \Lambda_q(k^n) \to \mathcal{O}_q(M_n(k)) \otimes \Lambda_q(k^n)$$

such that $\phi(\xi_i) = \sum_{j=1}^n X_{ij} \otimes \xi_j$ for all i, and that

$$\phi(\xi_1 \xi_2 \cdots \xi_n) = D_q \otimes (\xi_1 \xi_2 \cdots \xi_n)$$

for some element $D_q \in \mathcal{O}_q(M_n(k))$. Then verify formula (2) for D_q. Show also that there is a k-algebra homomorphism $\psi : \Lambda_q(k^n) \to \Lambda_q(k^n) \otimes \mathcal{O}_q(M_n(k))$ such that $\psi(\xi_j) = \sum_{i=1}^n \xi_i \otimes X_{ij}$ for all j, and that $\psi(\xi_1 \xi_2 \cdots \xi_n) = (\xi_1 \xi_2 \cdots \xi_n) \otimes D_q$.

Similarly, define $\Lambda_{\boldsymbol{p}}(k^n)$ to be the k-algebra with generators $\xi_1, \ldots, \xi_n$ and relations

$$\xi_i^2 = 0 \qquad\qquad \xi_i \xi_j = -p_{ij} \xi_j \xi_i$$

for all i, j. Show that there is a k-algebra homomorphism

$$\phi : \Lambda_{\boldsymbol{p}}(k^n) \to \mathcal{O}_{\lambda,\boldsymbol{p}}(M_n(k)) \otimes \Lambda_{\boldsymbol{p}}(k^n)$$

such that $\phi(\xi_i) = \sum_{j=1}^n X_{ij} \otimes \xi_j$ for all i, and that

$$\phi(\xi_1 \xi_2 \cdots \xi_n) = D_{\lambda,\boldsymbol{p}} \otimes (\xi_1 \xi_2 \cdots \xi_n)$$

where $D_{\lambda,\boldsymbol{p}}$ is given by formula (3). □

Exercise I.2.D. The purpose of this exercise is to establish the *quantum Laplace expansions* for D_q, using the maps ϕ and ψ defined in the previous exercise. First set

$$\widehat{\xi_j} = \xi_1 \cdots \xi_{j-1}\xi_{j+1} \cdots \xi_n \in \Lambda_q(k^n)$$

for $j = 1, \ldots, n$, and show that $\xi_i \widehat{\xi_j} = \delta_{ij}(-q)^{j-1}\xi_1\xi_2 \cdots \xi_n$ for all i, j. Next, show that there are (unique) elements $A_{ij} \in \mathcal{O}_q(M_n(k))$ such that

$$\phi(\widehat{\xi_i}) = \sum_{j=1}^{n} A_{ij} \otimes \widehat{\xi_j}$$

for all i. (The element A_{ij} is called the i,j-th *quantum minor* of $\mathcal{O}_q(M_n(k))$, because it is the image of the quantum determinant of $\mathcal{O}_q(M_{n-1}(k))$ under the natural isomorphism of that algebra with the subalgebra of $\mathcal{O}_q(M_n(k))$ generated by the X_{lm} with $l \neq i$ and $m \neq j$.) Now show that

$$\sum_{l=1}^{n}(-q)^{l-j}X_{il}A_{jl} = \sum_{l=1}^{n}(-q)^{i-l}A_{li}X_{lj} = \delta_{ij}D_q \tag{†}$$

for all i, j. [Hints: Apply ϕ to the equation $\xi_i\widehat{\xi_j} = \delta_{ij}(-q)^{j-1}\xi_1\xi_2 \cdots \xi_n$ to show that the first sum equals $\delta_{ij}D_q$. To deal with the second sum, apply ψ to an equation involving $\widehat{\xi_i}\xi_j$.] $\square$

Exercise I.2.E. Show that D_q is central in $\mathcal{O}_q(M_n(k))$, as follows. Set $X = (X_{ij})$ and $Y = (Y_{ij}) = ((-q)^{i-j}A_{ji})$; equations (†) from the previous exercise imply that $XY = YX = D_qI_n$, where I_n is the $n \times n$ identity matrix. Then consider the product XYX. $\square$

Exercise I.2.F. It is known that

$$D_{\lambda,\boldsymbol{p}}X_{ij} = \lambda^{j-i}\left(\prod_{l=1}^{n}p_{jl}p_{li}\right)X_{ij}D_{\lambda,\boldsymbol{p}}$$

for all i, j [**8**, Theorem 3]. Use these relations to show that the powers of $D_{\lambda,\boldsymbol{p}}$ form a denominator set in $\mathcal{O}_{\lambda,\boldsymbol{p}}(M_n(k))$. Also, conclude that $D_{\lambda,\boldsymbol{p}}$ is central in $\mathcal{O}_{\lambda,\boldsymbol{p}}(M_n(k))$ if and only if

$$\lambda^i \prod_{l=1}^{n}p_{il} = \lambda^j \prod_{l=1}^{n}p_{jl}$$

for all i, j. $\square$

Exercise I.2.G. It is known that there exists a k-algebra anti-automorphism S on $\mathcal{O}_q(GL_n(k))$ such that $S(X_{ij}) = (-q)^{i-j}A_{ji}D_q^{-1}$ for all i, j [**180**, Theorem 5.3.2]. Show that $\left(\mathcal{O}_q(GL_n(k)), \Delta, \varepsilon, S\right)$ is a Hopf algebra.

Show that the antipode on $\mathcal{O}_q(GL_n(k))$ induces a k-algebra anti-automorphism S on $\mathcal{O}_q(SL_n(k))$ such that $S(\overline{X}_{ij}) = (-q)^{i-j}\overline{A}_{ji}$ for all i, j, and that $\left(\mathcal{O}_q(SL_n(k)), \Delta, \varepsilon, S\right)$ is a Hopf algebra. □

Exercise I.2.H. It is known that the $n \times n$ matrix (X_{ij}) over $\mathcal{O}_{\lambda,\boldsymbol{p}}(M_n(k))$ has an inverse (Y_{ij}) over $\mathcal{O}_{\lambda,\boldsymbol{p}}(GL_n(k))$, and that there is a k-algebra anti-automorphism S on $\mathcal{O}_{\lambda,\boldsymbol{p}}(GL_n(k))$ such that $S(X_{ij}) = Y_{ij}$ for all i, j [**8**, Theorem 3]. Show that $\mathcal{O}_{\lambda,\boldsymbol{p}}(GL_n(k))$ and $\mathcal{O}_{\lambda,\boldsymbol{p}}(SL_n(k))$ are Hopf algebras. □

Exercise I.2.I. Prove Theorem I.2.7. □

Exercise I.2.J. Check the details of the proof of Lemma I.2.9. □

Exercise I.2.K. Prove that $\mathcal{O}_{\lambda,\boldsymbol{p}}(GL_n(k)) \cong \mathcal{O}_{\lambda,\boldsymbol{p}}(SL_n(k))[z^{\pm 1}]$ when $D_{\lambda,\boldsymbol{p}}$ is central in $\mathcal{O}_{\lambda,\boldsymbol{p}}(M_n(k))$. Conclude that $\mathcal{O}_{\lambda,\boldsymbol{p}}(SL_n(k))$ is a domain. □

Exercise I.2.L. Let $B = \mathcal{O}_q(k^3)$ where $q = \begin{pmatrix} 1 & q & q \\ q^{-1} & 1 & 1 \\ q^{-1} & 1 & 1 \end{pmatrix}$. Show that there is a k-algebra embedding $B \to \mathcal{O}_q(SL_2(k))$ that sends x_1, x_2, x_3 to $\bar{a}$, $\bar{b}$, $\bar{c}$, respectively. □

Exercise I.2.M. Show that the element $\bar{b}\bar{c}^{-1} \in \operatorname{Fract}\mathcal{O}_q(SL_2(k))$ is transcendental over k. □

Exercise I.2.N. Show that the elements $bc^{-1}, D_q \in \operatorname{Fract}\mathcal{O}_q(GL_2(k))$ are algebraically independent over k. □

THE QUANTIZED ENVELOPING ALGEBRA OF $\mathfrak{sl}_2(k)$

In this chapter, we introduce the definition of the simplest quantized enveloping algebra and derive some of its most basic properties. The first version of this algebra was defined by Kulish and Reshetikhin [129] in 1981, but it is fair to say that – like many of the best ideas in mathematics – its beauty and importance relative to an amazingly wide range of topics in mathematics and physics only became gradually apparent over the succeeding years.

I.3.1. Let k be a field and let $q \in k$, with $q \neq 0, 1, -1$. The *quantized enveloping algebra* of $\mathfrak{sl}_2(k)$ is the k-algebra $U_q(\mathfrak{sl}_2(k))$ with generators E, F, K, K^{-1} subject to the relations

$$KEK^{-1} = q^2 E \qquad KFK^{-1} = q^{-2}F \qquad EF - FE = \frac{K - K^{-1}}{q - q^{-1}}$$

as well, of course, as the relations $KK^{-1} = 1 = K^{-1}K$.

The *positive Borel subalgebra* of $U_q(\mathfrak{sl}_2(k))$, denoted $U_q^{\geq 0}(\mathfrak{sl}_2(k))$, is the subalgebra of $U_q(\mathfrak{sl}_2(k))$ generated by E and by K, K^{-1}. Similarly, the *negative Borel subalgebra*, $U_q^{\leq 0}(\mathfrak{sl}_2(k))$, is generated by F and K, K^{-1}. These two algebras are clearly localizations of the coordinate ring of quantum 2-space $\mathcal{O}_{q^2}(k^2)$ as defined in (I.1.4). The *positive nilpotent subalgebra* of $U_q(\mathfrak{sl}_2(k))$ is the subalgebra $U^+(\mathfrak{sl}_2(k))$ generated by E; similarly, the *negative nilpotent subalgebra* is the subalgebra $U^-(\mathfrak{sl}_2(k))$ generated by F. The *Cartan subalgebra* of $U_q(\mathfrak{sl}_2(k))$ is $U_0 := k[K, K^{-1}]$.

Proposition. 1. $U_q(\mathfrak{sl}_2(k))$ is an *iterated skew polynomial ring* over $k[K, K^{-1}]$ of the form $k[K, K^{-1}][E; \tau_1][F; \tau_2, \delta_2]$.

2. $U_q(\mathfrak{sl}_2(k))$ is a noetherian domain.

3. Similar conclusions apply to the subalgebras defined before the proposition.

Proof. 1. Let τ_1 be the k-algebra automorphism of $k[K, K^{-1}]$ defined by $\tau_1(K) = q^{-2}K$. Then there is a k-algebra automorphism of $k[K, K^{-1}][E; \tau_1]$ defined by $\tau_2(E) = E$ and $\tau_2(K) = q^2 K$, (as is easily confirmed, since τ_2 preserves the relations defining $k[K, K^{-1}][E; \tau_1])$. Moreover there is a unique left τ_2-derivation δ_2 of $k[K, K^{-1}][E; \tau_1]$ such that $\delta_2(K) = 0$ and

$$\delta_2(E) = \frac{K^{-1} - K}{q - q^{-1}},$$

(Exercise I.3.A). It's easy to check that the algebra $k[K, K^{-1}][E; \tau_1][F; \tau_2, \delta_2]$ is isomorphic to $U_q(\mathfrak{sl}_2(k))$.

2. Lemma I.1.12 and Theorem I.1.13.

3. This is clear. $\square$

Notice that the proof of 1 is an example of the *model approach* to proving that an algebra is an iterated skew polynomial ring outlined in (I.1.14). An immediate consequence of the proposition is the

Corollary. PBW Theorem for $U_q(\mathfrak{sl}_2(k))$. *The set*

$$\{E^i K^j F^t : i, j, t \in \mathbb{Z}, \ i, t \geq 0\}$$

is a k-vector space basis for $U_q(\mathfrak{sl}_2(k))$. $\square$

When it's clear from the context that we're dealing with subalgebras of $U_q(\mathfrak{sl}_2(k))$, we'll simply write $U^{\geq 0}$, $U^{\leq 0}$, U^+, U^- and U_0 for the appropriate algebras. One major difference from the classical theory of semisimple Lie algebras and their enveloping algebras is already apparent -- there are *many* positive Borel subalgebras of, for example, $\mathfrak{sl}_2(k)$, namely the (infinitely many) conjugates of the space $\mathfrak{b}^+$ of upper triangular matrices in $\mathfrak{sl}_2(k)$; and so there are many "Borel subalgebras" of $U(\mathfrak{sl}_2(k))$ [98, Section 16]. Similar comments apply to the other subalgebras. This is a special case of a general phenomenon: quantum groups are "less symmetric" and correspondingly "more rigid" than their classical counterparts.

I.3.2. Lemma. 1. *There is a unique automorphism ω of $U_q(\mathfrak{sl}_2(k))$ with $\omega(E) = F$, $\omega(F) = E$, and $\omega(K) = K^{-1}$. It satisfies $\omega^2 = 1$.*

2. *There is a unique antiautomorphism τ of $U_q(\mathfrak{sl}_2(k))$ with $\tau(E) = E$, $\tau(F) = F$, and $\tau(K) = K^{-1}$. It satisfies $\tau^2 = 1$.*

Proof. This is a routine check, (Exercise I.3.B). $\square$

I.3.3. Grading of $U_q(\mathfrak{sl}_2(k))$. (See Appendix I.12.7.) Assign degrees 1 to E, -1 to F, and 0 to K. Then the relations in (I.3.1) are all homogeneous, so that $U_q(\mathfrak{sl}_2(k))$ is a $\mathbb{Z}$-graded algebra in which, for $n \in \mathbb{Z}$, the nth homogeneous component is the k-vector space spanned by $\{E^i K^j F^t : i, j, t \in \mathbb{Z}, \ i, t \geq 0, \ i - t = n\}$. Since

$$K(E^i K^j F^t)K^{-1} = q^{2(i-t)} E^i K^j F^t,$$

it's clear that, *provided q is not a root of unity*, this grading is the same as the eigenspace decomposition for the action of K on $U_q(\mathfrak{sl}_2(k))$ by conjugation. Of course if q *is* a root of unity then the grading is finer than the eigenspace decomposition.

I.3.4. Hopf algebra structure on $U_q(\mathfrak{sl}_2(k))$. We claim that the data

$$\Delta(K) = K \otimes K \qquad\qquad \varepsilon(K) = 1 \qquad\qquad S(K) = K^{-1}$$
$$\Delta(E) = E \otimes 1 + K \otimes E \qquad \varepsilon(E) = 0 \qquad\qquad S(E) = -K^{-1}E$$
$$\Delta(F) = F \otimes K^{-1} + 1 \otimes F \qquad \varepsilon(F) = 0 \qquad\qquad S(F) = -FK$$

determine a structure of Hopf k-algebra on $U_q(\mathfrak{sl}_2(k))$. Here is a summary of the steps which have to be carried through in order to justify this claim.

Lemma 1. *There exist unique k-algebra homomorphisms*

$$\Delta : U_q(\mathfrak{sl}_2(k)) \longrightarrow U_q(\mathfrak{sl}_2(k)) \otimes U_q(\mathfrak{sl}_2(k)) \qquad \text{and} \qquad \varepsilon : U_q(\mathfrak{sl}_2(k)) \longrightarrow k$$

extending the definitions above.

Proof. It's trivial that ε extends to a homomorphism. For Δ one has to check that the relations in (I.3.1) are preserved. This is a straightforward exercise. □

Lemma 2. Δ *is coassociative.*

Proof. Given Lemma 1, to confirm that the first diagram in (I.9.1) is commutative one only has to check this with E, F and K in the argument; and this is easy. □

Lemma 3. *The second diagram in (I.9.1) is commutative.*

Proof. Again, only the generators need to be checked, and this is trivial. □

Lemma 4. S *is an antiautomorphism of $U_q(\mathfrak{sl}_2(k))$.*

Proof. One first has to check that the relations in (I.3.1) hold in $U_q(\mathfrak{sl}_2(k))^{\mathrm{op}}$ when E, F, K, K^{-1} are replaced by $-K^{-1}E$, $-FK$, K^{-1}, and K respectively. This is routine. Then observe that the endomorphism S^2 of $U_q(\mathfrak{sl}_2(k))$ satisfies

$$S^2(u) = K^{-1}uK$$

for all $u \in U_q(\mathfrak{sl}_2(k))$. (Of course it's enough to check this when u is a generator.) So S^2 is bijective and hence so is S. □

Lemma 5. *The relations (I.9.9) are satisfied by S.*

Proof. One first checks that, with m denoting the multiplication map,

$$m \circ (\mathrm{id} \otimes S) \circ \Delta(u) = \varepsilon(u) \tag{1}$$

for $u = E, F, K, K^{-1}$, and similarly for $m \circ (S \otimes \mathrm{id}) \circ \Delta$. This is easy. Then confirm, by a straightforward calculation using Lemmas 1 and 4, that, if $a, b \in U_q(\mathfrak{sl}_2(k))$ and (1) holds for a and b, then it holds for ab. And similarly for the other relation. □

Gathering the above together, we conclude:

Theorem. $U_q(\mathfrak{sl}_2(k))$ *is a Hopf algebra with the definitions of* Δ, S *and* ε *given at the beginning of this paragraph.* $\square$

I.3.5. Limit as $q \longrightarrow 1$. We could have presented $U_q(\mathfrak{sl}_2(k))$ as the algebra generated by E, F, $K^{\pm 1}$, L, subject to the relations

$$KEK^{-1} = q^2 E \qquad KFK^{-1} = q^{-2}F \qquad EF - FE = L$$
$$KK^{-1} = 1 \qquad K^{-1}K = 1 \qquad (q - q^{-1})L = K - K^{-1}.$$

Moreover these relations imply, when $q \neq \pm 1$,

$$[L, E] = q(EK + K^{-1}E) \qquad\qquad [L, F] = -q(KF + FK^{-1}).$$

The presentation given by these 8 relations has the advantage over (I.3.1) that it makes sense when $q = 1$. Let's label by U_1 the algebra we get when we take $q = 1$ in the above presentation.

Lemma. $U_1 \cong U(\mathfrak{sl}_2(k))[K]/\langle K^2 - 1\rangle$, *(and so)*

$$U(\mathfrak{sl}_2(k)) \quad\cong\quad U_1/\langle K - 1\rangle.$$

Proof. See Exercise I.3.D.

NOTES

The material in this chapter is standard and can be found in many places. Our main sources have been [**107**] and [**122**].

EXERCISES

Exercise I.3.A. Let α be an endomorphism of a ring R, let δ be any map from R to R, and define a map $\phi : R \longrightarrow M_2(R)$ by the rule $\phi(r) = \begin{pmatrix} \alpha(r) & \delta(r) \\ 0 & r \end{pmatrix}$. Show that δ is an α-derivation of R if and only if ϕ is a ring homomorphism.

Exercise I.3.B. Prove Lemma I.3.2.

Exercise I.3.C. Check the details of the proof (I.3.4) that $U_q(\mathfrak{sl}_2(k))$ is a Hopf algebra with the given definitions of Δ, ε and S.

Exercise I.3.D. Prove all the claims in (I.3.5).

CHAPTER I.4

THE FINITE DIMENSIONAL
REPRESENTATIONS OF $U_q(\mathfrak{sl}_2(k))$

I.4.1. Recall that in the analogous classical setting of the finite dimensional representation theory of $U(\mathfrak{sl}_2(k))$ (when k has characteristic zero), there is a unique irreducible $U(\mathfrak{sl}_2(k))$-module of dimension n for each non-negative integer n, [**98**, Chapter 7]; and every finite dimensional $U(\mathfrak{sl}_2(k))$-module is semisimple [**98**, Theorem 6.3]. We shall see that there are analogous results for $U_q(\mathfrak{sl}_2(k))$ when q is not a root of unity. For convenience we'll write U for $U_q(\mathfrak{sl}_2(k))$ in this chapter, with corresponding notation for the subalgebras defined in (I.3.1).

I.4.2. Verma modules. For $0 \neq \lambda \in k$, there is a one-dimensional $U^{\geq 0}$-module $k(\lambda)$ with

$$Ev = 0 \qquad\qquad\qquad Kv = \lambda v$$

for all $v \in V$. (In fact, if k is algebraically closed, these are the only finite dimensional irreducible $U^{\geq 0}$-modules, (Exercise I.4.B).) Define the *Verma module* $M(\lambda)$ by

$$M(\lambda) \quad := \quad U \otimes_{U^{\geq 0}} k(\lambda). \tag{1}$$

Thus $M(\lambda)$ is the cyclic U-module $U/I(\lambda)$, where

$$I(\lambda) \quad = \quad UE + U(K - \lambda),$$

and, since U is a free right $U^{\geq 0}$-module with basis $\{F^t : t \in \mathbb{Z}_{\geq 0}\}$, by Proposition I.3.1, $M(\lambda)$ has vector space basis

$$m_i \quad := \quad F^i + I(\lambda), \qquad i \geq 0.$$

We use in (4) below notation from Exercise I.4.C. It's easy to check (Exercise I.4.E) that

$$Km_i \quad = \quad \lambda q^{-2i} m_i, \tag{2}$$

$$Fm_i \quad = \quad m_{i+1}, \tag{3}$$

$$Em_i \quad = \quad \begin{cases} 0, & i = 0 \\ [i]\dfrac{\lambda q^{1-i} - \lambda^{-1}q^{i-1}}{q - q^{-1}} m_{i-1}, & i > 0. \end{cases} \tag{4}$$

I.4.3. Just as in the classical case of semisimple Lie algebras, a powerful tool in our analysis of a finite dimensional U-module M will be the eigenspaces of M under the action of the Cartan subalgebra $U_0 = k[K, K^{-1}]$ of U. (In fact this will be the case also for bigger quantized enveloping algebras than $U_q(\mathfrak{sl}_2(k))$.) If A is a commutative k-algebra, V is an A-module and $f : A \to k$ is an algebra homomorphism, we'll write

$$V_f \quad := \quad \{v \in V : av = f(a)v \text{ for all } a \in A\}.$$

Note that V_f is an A-submodule of V, called the A-*eigenspace* of V with *eigenvalue* f. Often, V_f is called the f-*weightspace* of V, and f is called a *weight* of V. (When the algebra A in question is clear from the context we'll omit mention of it from this terminology.) When $A = k[K, K^{-1}]$ each such homomorphism f is determined by $\lambda = f(K) \in k^\times$, and we'll write V_λ rather than V_f in this case, and refer to λ as a weight.

Theorem. *Suppose that q is not a root of unity. Let $\lambda \in k^\times$.*

1. If $\lambda \neq \pm q^n$ for any non-negative integer n then $M(\lambda)$ is irreducible.

2. Suppose that $\lambda = \pm q^n$ where $n \in \mathbb{Z}_{\geq 0}$, and set $N(\lambda) = \sum_{i \geq n+1} km_i$. Then $N(\lambda)$ is a U-submodule of $M(\lambda)$ with

$$N(\lambda) \quad \cong \quad M(\pm q^{-(n+2)}).$$

Moreover $N(\lambda)$ is the only proper submodule of $M(\lambda)$.

Proof. From (I.4.2)(2) and our assumption on q we see that the distinct eigenspaces of $M(\lambda)$ under the action of U_0 are

$$M(\lambda)_{\lambda q^{-2i}} \quad = \quad km_i.$$

Now suppose that A is any non-zero submodule of $M(\lambda)$, so that A must be spanned by those m_i which it contains. Now (I.4.2)(3) shows that

$$A \quad = \quad \sum_{i \geq j} km_i$$

for some j. Suppose $A \neq M(\lambda)$, so that $j > 0$. By (I.4.2)(4), Em_j is a scalar multiple of m_{j-1}, so that $Em_j = 0$ and it follows from (I.4.2)(4) that

$$\lambda q^{1-j} - \lambda^{-1} q^{j-1} \quad = \quad 0.$$

That is, $\lambda^2 = q^{2(j-1)}$, so

$$\lambda \quad = \quad \pm q^{j-1}$$

and 1 is proved.

Conversely, suppose that $\lambda = \pm q^n$ where $n \in \mathbb{Z}_{\geq 0}$. Reversing the above argument shows that in this case $N(\lambda)$ is a proper non-zero submodule of $M(\lambda)$, and $N(\lambda)$ is cyclic with generator m_{n+1}. Now

$$E m_{n+1} = 0$$

and

$$K m_{n+1} \quad = \quad \pm q^n q^{-2(n+1)} m_{n+1} \quad = \quad \pm q^{-(n+2)} m_{n+1}.$$

Thus $N(\lambda)$ is an image of $M(\pm q^{-(n+2)})$. But 1 shows that the latter module is irreducible, so 2 is proved.   □

Notation. For each non-negative integer n, we'll denote the irreducible U-module $M(\pm q^n)/N(\pm q^n)$ by $V(n, \pm)$. Thus it follows from the above theorem and its proof that $V(n, +)$ has basis

$$\overline{m}_i \quad := \quad m_i + N(q^n), \quad 0 \leq i \leq n,$$

so that

$$\dim_k V(n, +) \quad = \quad n + 1.$$

Moreover, for $i = 0, \ldots, n$,

$$K \overline{m}_i \quad = \quad q^{n-2i} \overline{m}_i, \tag{5}$$

$$F \overline{m}_i \quad = \quad \begin{cases} \overline{m}_{i+1}, & i < n \\ 0, & i = n \end{cases} \tag{6}$$

$$E \overline{m}_i \quad = \quad \begin{cases} 0, & i = 0 \\ [i][n + 1 - i] \overline{m}_{i-1}, & i > 0. \end{cases} \tag{7}$$

There is a similar description of the $(n + 1)$-dimensional irreducible U-module $V(n, -)$.

I.4.4. Converse of Theorem I.4.3. We begin with the

Lemma. *Suppose that q is not a root of unity. Let M be a finite dimensional U-module. Then there are positive integers r and s such that, for all $m \in M$,*

$$E^r m \quad = \quad F^s m \quad = \quad 0.$$

Proof. For each irreducible polynomial $f \in k[X]$, write

$$M_{(f)} = \{ m \in M : f(K)^n m = 0 \text{ for some } n > 0 \}.$$

By the theory of finitely generated modules over a PID,

$$M \;=\; \bigoplus_{f \in \mathcal{F}} M_{(f)} \tag{8}$$

for a finite set $\mathcal{F}$ of non-associate irreducible polynomials with $M_{(f)} \neq 0$ for all $f \in \mathcal{F}$. By the defining relations for U, $f(K)E = Ef(q^2 K)$ for $f \in \mathcal{F}$, so that, for $r \geq 0$,

$$f(K)E^r \;=\; E^r f(q^{2r} K).$$

Hence

$$E^r M_{(f)} \subseteq M_{(f(q^{-2r} X))}. \tag{9}$$

We claim that, for large enough r,

$$M_{(f(q^{-2r} X))} \;=\; 0. \tag{10}$$

If not, then by (8) there exist distinct positive integers r and s with $f(q^{-2r} X)$ and $f(q^{-2s} X)$ associates. But since $M_{(X)} = 0$, (as K is a unit), f has non-zero constant term. Since the constant terms of $f(q^{-2r} X)$ and $f(q^{-2s} X)$ are thus equal and non-zero, these polynomials must in fact be equal. Comparing their terms of highest degree now yields

$$q^{2(r-s)t} \;=\; 1$$

for some $t > 0$, contradicting our initial hypothesis. Thus (10) is proved. The result for E follows from (8), (9) and (10), and a similar argument deals with F. $\square$

We can now state the desired

Theorem. *Suppose that q is not a root of unity. Let M be a finite dimensional irreducible U-module. Then either $M \cong V(n, +)$ or $M \cong V(n, -)$, where $n = \dim_k(M) - 1$.*

Proof. (For k algebraically closed.) By the lemma there exists $0 \neq m' \in M$ with $Em' = 0$. The defining relations show that the set of such elements is a U_0-submodule M_0 of M. Since k is algebraically closed, M_0 contains a U_0-eigenvector. Thus we can find $0 \neq m \in M$ and $\lambda \in k^\times$ with

$$Em = 0 \quad \text{and} \quad Km = \lambda m.$$

But $M = Um$ since M is irreducible, so that M is a factor of $M(\lambda)$. We can therefore complete the proof by appealing to Theorem I.4.3. $\square$

Remark. The proof of the above result for arbitrary fields of characteristic not 2 is slightly more involved [**107**, Chapter 2]. It proceeds by way of a subsidiary result which we state separately below since we'll need it later. Notice that the algebraic closure of k is used above only to ensure the existence of a U_0-eigenvector, so this proposition can be used instead.

Proposition. *Suppose that q is not a root of unity and that k does not have characteristic 2. Let M be a finite dimensional U-module. Then M is the direct sum of its U_0-weightspaces. The weights of M have the form $\pm q^r$ for $r \in \mathbb{Z}$.*

Proof. By Exercise I.4.I, we have to show that the minimal polynomial of K acting on M has the form $\prod_i (X - \lambda_i)$ with the λ_i distinct and of the form $\pm q^r$ with $r \in \mathbb{Z}$. By Lemma I.4.4 there is a positive integer s with $F^s M = 0$. For $t \in \mathbb{Z}$, $t \geq 0$ and using the notation of Exercise I.4.D, define

$$h_t \quad = \quad \prod_{j=-(t-1)}^{t-1} [K; t - s + j]. \tag{11}$$

In particular $h_0 = 1$. We claim that, for $0 \leq t \leq s$,

$$F^{s-t} h_t M \quad = \quad 0. \tag{12}$$

This follows by induction; for $t = 0$ it is the definition of s, and the induction step is Exercise I.4.J. Taking $t = s$ in (12) yields

$$0 = h_s M = \left(\prod_{j=-(s-1)}^{s-1} (q - q^{-1})^{-1} q^j K^{-1} (K^2 - q^{-2j}) \right) M.$$

Multiplying this expression by a suitable non-zero scalar and by a suitable power of K we deduce

$$0 = \left(\prod_{j=-(s-1)}^{s-1} (K^2 - q^{-2j}) \right) M = \left(\prod_{j=-(s-1)}^{s-1} (K - q^{-j})(K + q^{-j}) \right) M.$$

The hypotheses on q and on k ensure that this forces the minimal polynomial of K acting on M to have the desired form. $\square$

I.4.5. Complete reducibility of finite dimensional U-modules. Again following the classical case, the proof of complete reducibility makes use of the centre of the algebra. It's easy to confirm that

$$C_q \quad := \quad FE + \frac{qK + q^{-1}K^{-1}}{(q - q^{-1})^2}$$

is in the centre of U. (See Exercise I.4.F.) We could deduce from general theory that C_q must act by multiplication by a scalar on any finite dimensional irreducible U-module when k is algebraically closed. But in fact we can easily prove this directly, and even specify the scalar when M is finite dimensional. Moreover, the argument doesn't depend on the algebraic closure of k, and applies to Verma modules as well.

Lemma. *Let $\lambda \in k^\times$. Then C_q acts on $M(\lambda)$ by scalar multiplication by*

$$\frac{q\lambda + q^{-1}\lambda^{-1}}{(q - q^{-1})^2}.$$

Hence C_q acts on $M(\lambda)$ and $M(\mu)$ by the same scalar if and only if $\lambda = \mu$ or $\lambda\mu = q^{-2}$.

Proof. Clearly, from (I.4.2), C_q acts on m_0 by the stated scalar. Since $M(\lambda) = Um_0$, the first claim follows. The second is an elementary calculation (Exercise I.4.F). $\square$

Corollary. *Suppose that q is not a root of unity. Then C_q acts by multiplication by the same scalar on finite dimensional irreducible U-modules V and W if and only if $V \cong W$.*

Proof. By Theorem I.4.4, $V \cong V(n, \epsilon)$ and $W \cong V(m, \epsilon')$ for some $n, m \geq 0$ and $\epsilon, \epsilon' \in \{\pm\}$. If C_q acts by the same scalar on these modules then it has to act by the same scalar on $M(\epsilon q^n)$ and $M(\epsilon' q^m)$. Thus, by the lemma, either $\epsilon q^n = \epsilon' q^m$, so that $V \cong W$, or

$$\epsilon\epsilon' q^{n+m} = q^{-2}.$$

The second case forces $q^{n+m+2} = \pm 1$, contradicting our hypothesis on q. $\square$

Theorem. *Suppose that q is not a root of unity and that k does not have characteristic 2. Let M be a finite dimensional U-module. Then M is completely reducible.*

Proof. Fix a composition series

$$0 = M_0 \subseteq M_1 \subseteq \ldots \subseteq M_d = M \tag{13}$$

of M. Each factor M_i/M_{i-1} is isomorphic to $V(n, \epsilon)$ for some $n \geq 0$ and $\epsilon \in \{\pm\}$, by Theorem I.4.4.

Step 1: We claim that, without loss of generality, all irreducible factors in (13) are isomorphic to a fixed $V(n_0, \epsilon_0)$. By the lemma, for each $i = 1, \ldots, d$, C_q acts on M_i/M_{i-1} by multiplication by a scalar λ_i. Thus

$$\left(\prod_{i=1}^{d} (C_q - \lambda_i) \right) M = 0. \tag{14}$$

For $\lambda \in k$, set

$$M_{(\lambda)} := \{m \in M : (C_q - \lambda)^n m = 0 \text{ for some } n \geq 0\},$$

a U-submodule. Then M is the direct sum of the non-zero $M_{(\lambda)}$, so we may assume that $M = M_{(\lambda)}$ for some $\lambda \in k$. So each λ_i in (14) equals λ. The corollary now yields our claim.

Step 2: For each U_0-eigenvalue $\mu \in k^\times$,

$$\dim_k M_\mu \;\; = \;\; d \cdot \dim_k V(n_0, \epsilon_0)_\mu.$$

For, by Proposition I.4.4, M is the direct sum of its U_0-eigenspaces. So Step 2 follows from Step 1.

Step 3: We show now that

$$M \;\; \cong \;\; V(n_0, \epsilon_0)^{(\oplus d)}. \tag{15}$$

Write $\lambda_0 = \epsilon_0 q^{n_0}$. By Step 2,

$$\dim_k M_{\lambda_0} \;\; = \;\; d \tag{16}$$

and

$$M_{q^2 \lambda_0} \;\; = \;\; 0. \tag{17}$$

Let $0 \neq v \in M_{\lambda_0}$. By (17), $Ev = 0$, so Uv is isomorphic to a factor of $M(\lambda_0)$. Being finite dimensional, $Uv \cong V(n_0, \epsilon_0)$ by Theorem I.4.3. Let $m_1, \ldots, m_d$ be a basis of M_{λ_0}. Then

$$M \;\; = \;\; \sum_{i=1}^{d} U m_i,$$

since the λ_0-weightspace of $M / \sum_{i=1}^{d} U m_i$ has dimension 0, but all composition factors of this factor are isomorphic to $V(n_0, \epsilon_0)$. Thus M is a sum of irreducible modules and hence is completely reducible. $\square$

Remarks. 1. The theorem is false when k has characteristic 2 – see Exercise I.4.G. However, even in this case the result can be retrieved if we impose the extra hypothesis that M is the direct sum of its U_0-eigenspaces [**107**, Theorem 2.9].

2. When q is not a root of unity, $Z(U_q(\mathfrak{sl}_2(k))) = k[C_q]$, a polynomial k-algebra in one variable, [**107**, Proposition 2.18].

NOTES

The material in this chapter is standard. Our treatment follows [**107**, Chapter 2].

EXERCISES

Exercise I.4.A. Let k be an algebraically closed field, let $0 \neq q \in k$ and let $\Lambda = \mathcal{O}_q(k^2)$ be the quantum plane $k\langle X, Y : XY = qYX \rangle$. Determine the finite dimensional irreducible representations of Λ for (i) q not a root of 1 and (ii) q a root of 1.

Exercise I.4.B. Use Exercise I.4.A to determine the irreducible representations of $U_q^+(\mathfrak{sl}_2(k))$, for q generic and q a root of unity.

Exercise I.4.C. (Essentially just notation.) For $n \in \mathbb{Z}$, define $[n]_t \in \mathbb{Q}(t)$ by

$$[n]_t \quad = \quad \frac{t^n - t^{-n}}{t - t^{-1}}.$$

Show that $[n]_t \in \mathbb{Z}[t, t^{-1}]$ for all $n \in \mathbb{Z}$. For $q \in k^\times$ let $[n]_q$ be the image of $[n]_t$ under the homomorphism $\mathbb{Z}[t, t^{-1}] \longrightarrow k$ got by sending t to q. We'll often omit the suffix q when this is clear from the context.

Exercise I.4.D. In $U_q(\mathfrak{sl}_2(k))$, write, for $a \in \mathbb{Z}$,

$$[K; a] \quad = \quad \frac{Kq^a - K^{-1}q^{-a}}{q - q^{-1}}.$$

Prove that (i) for all $a, b, c \in \mathbb{Z}$,

$$[b + c][K; a] \quad = \quad [b][K; a + c] + [c][K; a - b];$$

(ii) for all $a \in \mathbb{Z}$,

$$[K; a]E \quad = \quad E[K; a + 2] \qquad \text{and} \qquad [K; a]F \quad = \quad F[K; a - 2];$$

(iii) in the notation of (I.3.2),

$$\omega([K; a]) \quad = \quad -[K; -a];$$

(iv) for all positive integers s,

$$EF^s \quad = \quad F^s E + [s]F^{s-1}[K; 1 - s];$$

(v) (by applying ω to (iv)), for all positive integers r,

$$FE^r \quad = \quad E^r F - [r]E^{r-1}[K; r - 1].$$

Exercise I.4.E. Use Exercise I.4.D to check (I.4.2)(2),(3),(4).

Exercise I.4.F. (Notation I.4.5.) Show that C_q is in the centre of $U_q(\mathfrak{sl}_2(k))$. Complete the proof of Lemma I.4.5. Show that

$$C_q = EF + \frac{q^{-1}K + qK^{-1}}{(q - q^{-1})^2},$$

so that $\omega(C_q) = C_q = \tau(C_q)$, in the notation of (I.3.2).

Exercise I.4.G. Show that Theorem I.4.5 is false when k has characteristic 2: Take $M = K^{(2)}$ with both E and F acting as 0 and K as $\begin{pmatrix} 1 & 1 \\ 0 & 1 \end{pmatrix}$.

Exercise I.4.H. (Notation I.4.3.) Show that $V(n, \epsilon) \cong V(m, \epsilon') \iff n = m$ and $\epsilon = \epsilon'$.

Exercise I.4.I. Prove that an endomorphism of a finite dimensional vector space V is diagonalizable if and only if its minimal polynomial splits into linear factors, and each factor occurs with multiplicity one.

Exercise I.4.J. Prove (I.4.4)(12). (See [**107**, Chapter 2, Appendix].)

PRIMER ON SEMISIMPLE LIE ALGEBRAS

Most of the standard ingredients from the structure theory of semisimple Lie algebras also play a role in the development of quantized enveloping algebras. We summarize those ingredients in this chapter, in order to have them conveniently available, and to establish notation.

I.5.1. Notation and definitions. A finite dimensional complex Lie algebra $\mathfrak{g}$ is *semisimple* if it has no non-zero abelian ideals. In this case $\mathfrak{g}$ contains a *Cartan subalgebra* $\mathfrak{h}$. (That is, $\mathfrak{h}$ is a maximal abelian subalgebra of $\mathfrak{g}$ consisting of semisimple elements; here, an element x of $\mathfrak{g}$ is *semisimple* if the adjoint endomorphism $\operatorname{ad} x : y \mapsto [xy]$ of $\mathfrak{g}$ is semisimple.) Under the action of $\operatorname{ad}\mathfrak{h}$, the Lie algebra $\mathfrak{g}$ decomposes as a direct sum of *root spaces*, these being the eigenspaces under this action in the sense of (I.4.3),

$$\mathfrak{g} \;=\; \mathfrak{h} \oplus \bigoplus_{\alpha \in \Phi} \mathfrak{g}_\alpha \;=\; \mathfrak{g}_0 \oplus \bigoplus_{\alpha \in \Phi} \mathfrak{g}_\alpha,$$

where Φ, the set of *roots*, is $\{0 \neq \alpha \in \mathfrak{h}^* : \mathfrak{g}_\alpha \neq 0\}$. In this case Φ forms a *(reduced) root system* in $\mathfrak{h}^*$ in the sense of [**194**, p. 25], simply called a *root system* in [**98**, p. 42]. If $\alpha \in \Phi$ then $\dim_{\mathbb{C}} \mathfrak{g}_\alpha = 1$ and $\mathfrak{g}_{-\alpha} \neq 0$. From the Jacobi identity,

$$[\mathfrak{g}_\alpha \mathfrak{g}_\beta] \subseteq \mathfrak{g}_{\alpha+\beta}. \tag{1}$$

¿From the properties of root systems, Φ contains a set Φ^+ of *positive roots* and corresponding basis of $\mathfrak{h}^*$ consisting of *simple roots*

$$\Delta = \{\alpha_1, \dots, \alpha_n\} \subseteq \Phi^+.$$

Here, $n = \dim_{\mathbb{C}} \mathfrak{h}$ is the *rank* of $\mathfrak{g}$. We have $\Phi = \Phi^+ \sqcup \Phi^-$, where $\Phi^- = -\Phi^+$, and a *triangular decomposition*

$$\mathfrak{g} = \mathfrak{n}^+ \oplus \mathfrak{h} \oplus \mathfrak{n}^-,$$

where $\mathfrak{n}^+ = \sum_{\alpha \in \Phi^+} \mathfrak{g}_\alpha$ and $\mathfrak{n}^- = \sum_{\alpha \in \Phi^-} \mathfrak{g}_\alpha$. Thanks to (1), $\mathfrak{n}^+$ and $\mathfrak{n}^-$ are nilpotent subalgebras of $\mathfrak{g}$. There is a corresponding decomposition

$$U(\mathfrak{g}) \;=\; U(\mathfrak{n}^+) \otimes U(\mathfrak{h}) \otimes U(\mathfrak{n}^-),$$

and $U(\mathfrak{g})$ has a PBW basis with generating set consisting of the elements $\{e_1, \ldots, e_N, h_1, \ldots, h_n, f_1, \ldots, f_N\}$, where $\{h_1, \ldots, h_n\}$ is a basis for $\mathfrak{h}$ and $\{e_1, \ldots, e_N\}$ [resp. $\{f_1, \ldots, f_N\}$] is a set of positive [resp. negative] root vectors. Clearly, $\mathfrak{n}^+$ is an ideal of $\mathfrak{h} \oplus \mathfrak{n}^+ := \mathfrak{b}^+$ and $\mathfrak{n}^-$ is an ideal of $\mathfrak{h} \oplus \mathfrak{n}^- := \mathfrak{b}^-$, so that $\mathfrak{b}^+$ and $\mathfrak{b}^-$, respectively the *positive* and *negative Borel subalgebras* of $\mathfrak{g}$ associated to our choice of $\mathfrak{h}$ and Φ^+, are solvable subalgebras of $\mathfrak{g}$.

The *Killing form* $(\ , \ ) : \mathfrak{g} \times \mathfrak{g} \longrightarrow \mathbb{C}$ given by $(x, y) := \mathrm{Tr}(\mathrm{ad}\, x \circ \mathrm{ad}\, y)$ is $\mathfrak{g}$-invariant, (meaning that $([xy], z) = (x, [yz])$ for $x, y, z \in \mathfrak{g}$), and nondegenerate. In fact, $(\mathfrak{g}_\alpha, \mathfrak{g}_\beta) \neq 0$ only if $\beta = -\alpha$, so that $(\ , \ )$ is non-degenerate even when restricted to $\mathfrak{h}$. Thus we can (and shall) use $(\ , \ )$ to identify $\mathfrak{g}$ with $\mathfrak{g}^*$ and $\mathfrak{h}$ with $\mathfrak{h}^*$.

Associated with $\mathfrak{g}$ is its *Weyl group* W, the finite group of automorphisms of $\mathfrak{h}$ (or equivalently of $\mathfrak{h}^*$) generated by the *simple reflections*, these being the reflections s_α in the hyperplanes orthogonal to $\alpha \in \Delta$. The Killing form is invariant under the action of W, and $w(\Phi) = \Phi$ for each $w \in W$, [**98**, Lemma 9.2]. If (as we may) we take $\mathfrak{g}$ to be the Lie algebra of a connected complex semisimple algebraic group G, then $\mathfrak{h}$ is the Lie algebra of a maximal torus T of G, and then

$$W = N_G(T)/T,$$

where $N_X(Y)$ denotes the normalizer of a subgroup Y of a group X.

The classification of the finite dimensional semisimple Lie algebras $\mathfrak{g}$ consists of the following components:

(1) There is a bijective correspondence between isomorphism classes of such Lie algebras $\mathfrak{g}$ and reduced root systems Φ.

(2) Reduced root systems are *sums* of *irreducible* root systems. The latter correspond to the *simple* Lie algebras $\mathfrak{g}$, these being the ones with no proper ideals. An arbitrary semisimple Lie algebra is a finite direct sum of simple Lie algebras.

(3) (Isomorphism classes of) irreducible root systems can be bijectively associated to connected (nonempty) Dynkin diagrams.

(4) Connected Dynkin diagrams are classified: they fall into four infinite families, A_n $(n \geq 1)$, B_n $(n \geq 2)$, C_n $(n \geq 3)$, D_n $(n \geq 4)$, together with G_2, F_4, E_6, E_7, and E_8.

The *Cartan matrix* of $\mathfrak{g}$ is the $n \times n$ matrix $C = (a_{ij})$ with

$$a_{ij} = 2(\alpha_i, \alpha_j)/(\alpha_i, \alpha_i). \qquad (2)$$

When $\mathfrak{g}$ is simple, the simple roots of $\mathfrak{g}$ are either all of the same length, the *simply laced* case, (types A_n, D_n, E_6, E_7, E_8), or have two lengths, *long* and *short*. Usually, the form is normalised so that $(\alpha, \alpha) = 2$ for short roots. Notice that the Cartan matrix of a simple Lie algebra $\mathfrak{g}$ is symmetric if and only if $\mathfrak{g}$ is simply laced. For a general simple $\mathfrak{g}$ (with the normalisation mentioned above), the integers

$$d_i := (\alpha_i, \alpha_i)/2, \quad (1 \leq i \leq n), \tag{3}$$

belong to $\{1, 2, 3\}$. If $D := \mathrm{diag}(d_i)$, then DC is symmetric.

Writing s_i for s_{α_i}, the Weyl group W has presentation

$$W = \langle s_1, \dots, s_n : s_i^2 = (s_i s_j)^{m_{ij}} = 1, \text{ for } 1 \leq i \neq j \leq n \rangle, \tag{4}$$

where $m_{ij} = 2, 3, 4, 6$ according as $a_{ij} a_{ji} = 0, 1, 2, 3$. Thanks to its definition as a reflection, the action of s_i on $\mathfrak{h}$ is given by

$$s_i(\lambda) \quad = \quad \lambda - \frac{2(\alpha_i, \lambda)}{(\alpha_i, \alpha_i)} \alpha_i. \tag{5}$$

The *length* $\ell(w)$ of an element w of W is r, where

$$w = s_{i_1} s_{i_2} \cdots s_{i_r} \tag{6}$$

is an expression for w of shortest length as a product of simple reflections s_{i_j}. There is a unique element of longest length, denoted w_0, and

$$\ell(w_0) = N,$$

where N is the number of positive roots of $\mathfrak{g}$. Indeed, suppose we write $w_0 = s_{i_1} \cdots s_{i_N}$ and define

$$\beta_1 := \alpha_{i_1}, \quad \beta_2 := s_{i_1}(\alpha_{i_2}), \quad \dots, \quad \beta_N := s_{i_1} \cdots s_{i_{N-1}}(\alpha_{i_N}).$$

Then $\{\beta_1, \dots, \beta_N\}$ is the set of positive roots of $\mathfrak{g}$ determined by the set $\Delta = \{\alpha_1, \dots, \alpha_n\}$ of simple roots, (and in particular contains Δ as a subset).

I.5.2. Example: $\mathfrak{g} = \mathfrak{sl}_{n+1}(\mathbb{C})$. In this case we can take

$$\mathfrak{h} \quad = \quad \{\Lambda = \mathrm{diag}(\lambda_i) : \sum_{i=1}^{n+1} \lambda_i = 0\},$$

with $\mathfrak{n}^+$ [resp. $\mathfrak{n}^-$] the upper [resp. lower] triangular matrices in $\mathfrak{sl}_{n+1}(\mathbb{C})$. So $\mathfrak{g}$ has rank n. The roots Φ are the functionals

$$\begin{aligned} \alpha_{i,j} : \mathfrak{h} &\longrightarrow \mathbb{C} \\ \Lambda &\longmapsto \lambda_i - \lambda_j, \end{aligned} \tag{7}$$

for $1 \leq i, j \leq n+1$, $i \neq j$, so that $\mathfrak{g}_{\alpha_{i,j}} = \mathbb{C} E_{ij}$ $(i \neq j)$, and we can take $\Delta = \{\alpha_i := \alpha_{i,i+1} : 1 \leq i \leq n\}$. Here, E_{ij} is the $(n+1) \times (n+1)$ matrix with

1 in the (ij)th entry and 0 elsewhere. One gets $(\alpha_i, \alpha_i) = 2$ for $1 \leq i \leq n$, and $(\alpha_i, \alpha_{i\pm1}) = -1$, with $(\alpha_i, \alpha_j) = 0$ otherwise, so that

$$
C \quad = \quad
\begin{pmatrix}
2 & -1 & 0 & 0 & \cdots & 0 & 0 & 0 \\
-1 & 2 & -1 & 0 & \cdots & 0 & 0 & 0 \\
0 & -1 & 2 & -1 & \cdots & 0 & 0 & 0 \\
\vdots & & & & \ddots & & & \vdots \\
0 & 0 & 0 & 0 & \cdots & 2 & -1 & 0 \\
0 & 0 & 0 & 0 & \cdots & -1 & 2 & -1 \\
0 & 0 & 0 & 0 & \cdots & 0 & -1 & 2
\end{pmatrix} .
\tag{8}
$$

The Weyl group of $\mathfrak{sl}_{n+1}(\mathbb{C})$ is the symmetric group S_{n+1}. For $i = 1, \dots, n$, we calculate (for instance from (I.5.1)(5)) that the reflection s_{α_i} replaces all occurrences of the subscript i by $i+1$, and vice versa, in the $\alpha_{i,j}$, and leaves all other subscripts fixed. This yields a homomorphism from W to S_{n+1}, and since S_{n+1} is generated by $\{s_i := (i, i+1) : 1 \leq i \leq n\}$, we get an isomorphism as claimed.

We have $\ell(w_0) = N = \frac{1}{2}n(n+1)$. For example, two reduced expressions for w_0 in S_{n+1} are

$$
w_0 = s_1 s_3 s_5 \cdots s_2 s_4 s_6 \cdots s_1 s_3 \cdots \qquad \text{and} \qquad w_0 = s_2 s_4 s_6 \cdots s_1 s_3 s_5 \cdots s_2 s_4 \cdots ,
$$

where in each case the pattern is repeated until $\frac{1}{2}n(n+1)$ simple reflections have been used. For $n > 2$ there are many more expressions than these for w_0; in general the number is unknown, and is of considerable interest, for instance in the theory of crystal bases.

I.5.3. Finite dimensional representation theory of $\mathfrak{g}$. Keep the notation of (I.5.1). The *root lattice* of $\mathfrak{g}$ is

$$
Q := \mathbb{Z}\Phi = \bigoplus_{i=1}^{n} \mathbb{Z}\alpha_i \subseteq \mathfrak{h}^* .
$$

The *fundamental weights* $\varpi_1, \dots, \varpi_n \in \mathfrak{h}^*$ are defined by the conditions

$$
(\varpi_i, \alpha_j) = d_j \delta_{ij}, \quad 1 \leq i, j \leq n.
\tag{9}
$$

The *weight lattice* is

$$
P := \bigoplus_{i=1}^{n} \mathbb{Z}\varpi_i \subseteq \mathfrak{h}^* .
$$

One easily calculates from (1), (I.5.1)(2) and (I.5.1)(3) that

$$
\alpha_j = \sum_{i=1}^{n} a_{ij} \varpi_i
$$

for $j = 1, \dots, n$, so that $Q \subseteq P$ and $[P : Q] < \infty$. We provide P with a partial order $\leq$, in which, for weights μ and λ in P, we set

$$\lambda \leq \mu \quad \Longleftrightarrow \quad \mu - \lambda \in Q^+ := \sum_{i=1}^{n} \mathbb{Z}_{\geq 0} \alpha_i.$$

Since the Weyl group W permutes Φ, W acts as automorphisms of Q, and hence of P. Notice from (I.5.1)(5) that

$$W(\lambda) \subseteq \lambda + Q \tag{10}$$

for $\lambda \in \mathfrak{h}$. In particular, it follows that any lattice between Q and P (that is, any additive subgroup of P containing Q) is invariant under W.

The starting points for the finite dimensional representation theory of $\mathfrak{g}$ are:

(1) Every finite dimensional $\mathfrak{g}$-module is completely reducible.
(2) There is a bijective correspondence between finite dimensional irreducible $\mathfrak{g}$-modules and the set of *dominant integral weights* $P^+ = \sum_{i=1}^{n} \mathbb{Z}_{\geq 0} \varpi_i$; the correspondence assigning to a module V its *highest weight*, namely the eigenvalue corresponding to a *highest weight vector* v, which is a $U(\mathfrak{h})$-eigenvector killed by $\mathfrak{n}^+$, with eigenvalue an element of P^+.
(3) The finite dimensional irreducible $\mathfrak{g}$-module $V(\lambda)$ with highest weight $\lambda \in P^+$ is a factor of the Verma module

$$M(\lambda) \quad := \quad U(\mathfrak{g}) \otimes_{U(\mathfrak{b}^+)} \mathbb{C}_\lambda,$$

where $\mathbb{C}_\lambda$ is the one-dimensional $U(\mathfrak{b}^+)$-module with weight λ. Hence the weights of $V(\lambda)$ are of the form $\lambda - \alpha$, for $\alpha \in Q^+ := \sum_{i=1}^{n} \mathbb{Z}_{\geq 0} \alpha_i$. Thus, in terms of the partial ordering of weights introduced above, the weights μ of $V(\lambda)$ all satisfy $\mu \leq \lambda$.

I.5.4. Example: $\mathfrak{g} = \mathfrak{sl}_{n+1}(\mathbb{C})$. Keep the notation of (I.5.2). The fundamental weights of $\mathfrak{sl}_{n+1}(\mathbb{C})$ are given by

$$\varpi_i(\Lambda) \quad = \quad \lambda_1 + \dots + \lambda_i$$

for $\Lambda = \operatorname{diag}(\lambda_j) \in \mathfrak{h}$ and $i = 1, \dots, n$. Thus, from (I.5.2)(7) and (I.5.2)(8) or by direct calculation, for $j = 1, \dots, n$, (and writing $\varpi_0 = \varpi_{n+1} = 0$),

$$\alpha_j = -\varpi_{j-1} + 2\varpi_j - \varpi_{j+1}.$$

It follows that $[P : Q] = n + 1$; indeed the *fundamental group* P/Q is cyclic of order $n + 1$, [**98**, Exercise 4, p. 72].

The natural representation of $\mathfrak{sl}_{n+1}(\mathbb{C})$ on $V = \mathbb{C}_{n+1}$ has highest weight ϖ_1; ϖ_i is the highest weight of the ith exterior power of V.

I.5.5. Serre's presentation of semisimple Lie algebras. Since the construction of quantized enveloping algebras which we'll outline in the next chapter is modelled on Serre's presentation of ordinary enveloping algebras, we first indicate here the classical result. We keep the notation of paragraphs (I.5.1) and (I.5.3). Serre proved that $\mathfrak{g}$ can be presented as the complex Lie algebra with generators $e_1, \ldots, e_n, f_1, \ldots, f_n, h_1, \ldots, h_n$ and relations

$$[e_i f_j] = \delta_{ij} h_i \qquad\qquad [h_i h_j] = 0$$
$$[h_i e_j] = a_{ij} e_j \qquad\qquad [h_i f_j] = -a_{ij} f_j$$
$$\mathrm{ad}(e_i)^{1-a_{ij}} e_j = \mathrm{ad}(f_i)^{1-a_{ij}} f_j = 0 \qquad (i \neq j).$$

A corresponding presentation of $U(\mathfrak{g})$ is obtained by turning the Lie products into commutators and expanding the adjoint expressions. Thus, $U(\mathfrak{g})$ is the $\mathbb{C}$-algebra with generators $e_1, \ldots, e_n, f_1, \ldots, f_n, h_1, \ldots, h_n$ and relations

$$[e_i, f_j] = \delta_{ij} h_i \qquad\qquad [h_i, h_j] = 0$$
$$[h_i, e_j] = a_{ij} e_j \qquad\qquad [h_i, f_j] = -a_{ij} f_j$$
$$\sum_{l=0}^{1-a_{ij}} (-1)^l \binom{1-a_{ij}}{l} e_i^{1-a_{ij}-l} e_j e_i^l = 0 \qquad (i \neq j)$$
$$\sum_{l=0}^{1-a_{ij}} (-1)^l \binom{1-a_{ij}}{l} f_i^{1-a_{ij}-l} f_j f_i^l = 0 \qquad (i \neq j).$$

NOTES

Good references for the material of this chapter are [98] and [194].

STRUCTURE AND REPRESENTATION THEORY
OF $U_q(\mathfrak{g})$ WITH q GENERIC

In this chapter we introduce the quantized enveloping algebras $U_q(\mathfrak{g})$ for arbitrary semisimple Lie algebras $\mathfrak{g}$, obtaining results which generalise those we proved for $\mathfrak{g} = \mathfrak{sl}_2(k)$ in Chapter I.4. The role of binomial coefficients in the presentations of these algebras is taken by a q-analogue of binomial coefficients, which we begin by defining. As before, q denotes a nonzero scalar in our base field k.

I.6.1. q-binomial coefficients. The definition of quantized enveloping algebras requires q-analogues of binomial coefficients which we now introduce. Let $n \geq i$ be nonnegative integers. First choose an indeterminate t, and define the following rational function in $k(t)$:

$$\binom{n}{i}_t = \frac{(t^n - 1)(t^{n-1} - 1) \cdots (t - 1)}{(t^i - 1)(t^{i-1} - 1) \cdots (t - 1)(t^{n-i} - 1)(t^{n-i-1} - 1) \cdots (t - 1)} .$$

Observe that $\binom{n}{i}_t$ is actually a polynomial in $k[t]$, and so it can be evaluated at any scalar. Its evaluation at $t = q$ is the *Gaussian q-binomial coefficient* $\binom{n}{i}_q$. When q is not a root of unity, $\binom{n}{i}_q \neq 0$ for $n \geq i \geq 0$.

In particular, in $\mathcal{O}_q(k^2)$ we have the identity

$$(x + y)^n = \sum_{i=0}^{n} \binom{n}{i}_q y^i x^{n-i},$$

the *q-binomial theorem*, for any positive integer n. This is a convenient aid in developing other identities, such as the *q-Pascal identity*

$$\binom{n}{i}_q = \binom{n-1}{i}_q + q^{n-i} \binom{n-1}{i-1}_q = \binom{n-1}{i-1}_q + q^i \binom{n-1}{i}_q$$

for $n > i > 0$.

In the development of quantized enveloping algebras, it is more convenient to use an alternative type of q-binomial coefficients. These, denoted $\begin{bmatrix} n \\ i \end{bmatrix}_q$, are the evaluations at $t = q$ of the polynomials

$$\begin{bmatrix} n \\ i \end{bmatrix}_t = \frac{(t^n - t^{-n})(t^{n-1} - t^{1-n}) \cdots (t - t^{-1})}{(t^i - t^{-i}) \cdots (t - t^{-1})(t^{n-i} - t^{i-n}) \cdots (t - t^{-1})} .$$

We can also write, in the notation of Exercise I.4.C,

$$\begin{bmatrix} n \\ i \end{bmatrix}_t = \frac{[n]_t[n-1]_t \cdots [1]_t}{[i]_t[i-1]_t \cdots [1]_t[n-i]_t[n-i-1]_t \cdots [1]_t} \ .$$

The two types of q-binomial coefficients are related by the equation

$$\begin{bmatrix} n \\ i \end{bmatrix}_q = q^{i(i-n)} \binom{n}{i}_{q^2} \ . \tag{1}$$

I.6.2. The quantized enveloping algebra of $\mathfrak{sl}_n(k)$. This is the k-algebra $U_q(\mathfrak{sl}_n(k))$ with generators $E_1, \ldots, E_{n-1}$, $F_1, \ldots, F_{n-1}$, $K_1^{\pm 1}, \ldots, K_{n-1}^{\pm 1}$ and relations

$$
\begin{array}{llll}
K_i E_j K_i^{-1} = q^2 E_j & \text{and} & K_i F_j K_i^{-1} = q^{-2} F_j & (i = j) \\[4pt]
K_i E_j K_i^{-1} = q^{-1} E_j & \text{and} & K_i F_j K_i^{-1} = q F_j & (|i-j| = 1) \\[4pt]
K_i E_j K_i^{-1} = E_j & \text{and} & K_i F_j K_i^{-1} = F_j & (|i-j| \geq 2) \\[4pt]
K_i K_j = K_j K_i & \text{and} & E_i F_j - F_j E_i = \delta_{ij} \dfrac{K_i - K_i^{-1}}{q - q^{-1}} & \\[8pt]
E_i E_j = E_j E_i & \text{and} & F_i F_j = F_j F_i & (|i-j| \geq 2) \\[4pt]
E_i^2 E_j - (q+q^{-1}) E_i E_j E_i + E_j E_i^2 = 0 & & & (|i-j| = 1) \\[4pt]
F_i^2 F_j - (q+q^{-1}) F_i F_j F_i + F_j F_i^2 = 0 & & & (|i-j| = 1).
\end{array}
$$

Certain subalgebras of $U_q(\mathfrak{sl}_n(k))$ are viewed as quantizations of the enveloping algebras of positive and negative Borel and nilpotent Lie subalgebras of $\mathfrak{sl}_n(k)$. Thus, the subalgebra generated by the $K_i^{\pm 1}$ and the E_i (respectively, the $K_i^{\pm 1}$ and the F_i) is denoted $U_q(\mathfrak{b}^+)$ or $U_q^{\geq 0}(\mathfrak{sl}_n(k))$ (respectively, $U_q(\mathfrak{b}^-)$ or $U_q^{\leq 0}(\mathfrak{sl}_n(k))$), while the subalgebra generated by the E_i (respectively, the F_i) is denoted $U_q(\mathfrak{n}^+)$ or $U_q^+(\mathfrak{sl}_n(k))$ (respectively, $U_q(\mathfrak{n}^-)$ or $U_q^-(\mathfrak{sl}_n(k))$).

As we shall outline in (I.6.5), there is a Hopf algebra structure on $U_q(\mathfrak{sl}_n(k))$ determined by the following data:

$$
\begin{array}{lll}
\Delta(K_i) = K_i \otimes K_i & \varepsilon(K_i) = 1 & S(K_i) = K_i^{-1} \\[4pt]
\Delta(E_i) = E_i \otimes 1 + K_i \otimes E_i & \varepsilon(E_i) = 0 & S(E_i) = -K_i^{-1} E_i \\[4pt]
\Delta(F_i) = F_i \otimes K_i^{-1} + 1 \otimes F_i & \varepsilon(F_i) = 0 & S(F_i) = -F_i K_i.
\end{array}
$$

Note that $U_q(\mathfrak{b}^+)$ and $U_q(\mathfrak{b}^-)$ are sub-Hopf algebras of $U_q(\mathfrak{sl}_n(k))$.

I.6.3. Quantized enveloping algebras of complex semisimple Lie algebras. We now describe the quantized enveloping algebra corresponding to any (finite dimensional) complex semisimple Lie algebra $\mathfrak{g}$. Although $\mathfrak{g}$ and its classical enveloping algebra are only defined over $\mathbb{C}$, quantized enveloping algebras can be defined over arbitrary fields. This is because $\mathfrak{g}$ plays only a symbolic role in the construction, which is based on the Cartan matrix of $\mathfrak{g}$.

Let n be the rank of $\mathfrak{g}$, and let $C = (a_{ij})$ be the Cartan matrix of $\mathfrak{g}$ corresponding to some choice of Cartan subalgebra $\mathfrak{h}$ and simple roots $\alpha_1, \ldots, \alpha_n$. We continue here and throughout to use the standard notation introduced in (I.5.1) and (I.5.3). Let $q \in k^\times$, set

$$q_i = q^{d_i} \tag{2}$$

for $i = 1, \ldots, n$, and assume that $q_i \neq \pm 1$. The *quantized enveloping algebra of $\mathfrak{g}$ over k* associated with the above choices is the k-algebra with generators $E_1, \ldots, E_n$, $F_1, \ldots, F_n$, $K_1^{\pm 1}, \ldots, K_n^{\pm 1}$ satisfying the following relations:

$$K_i E_j K_i^{-1} = q_i^{a_{ij}} E_j \qquad\qquad K_i F_j K_i^{-1} = q_i^{-a_{ij}} F_j$$

$$K_i K_j = K_j K_i \qquad\qquad E_i F_j - F_j E_i = \delta_{ij} \frac{K_i - K_i^{-1}}{q_i - q_i^{-1}}$$

$$\sum_{l=0}^{1-a_{ij}} (-1)^l \begin{bmatrix} 1-a_{ij} \\ l \end{bmatrix}_{q_i} E_i^{1-a_{ij}-l} E_j E_i^l = 0 \qquad (i \neq j)$$

$$\sum_{l=0}^{1-a_{ij}} (-1)^l \begin{bmatrix} 1-a_{ij} \\ l \end{bmatrix}_{q_i} F_i^{1-a_{ij}-l} F_j F_i^l = 0 \qquad (i \neq j).$$

Since the definition of this algebra depends only on C, k, and q, an appropriate notation might be $U_q(C; k)$. However, most authors prefer to refer to $\mathfrak{g}$ rather than to C, and to suppress the mention of k; typical notations are $U_q(\mathfrak{g}_\mathbb{C})$ and, most commonly, simply $U_q(\mathfrak{g})$. When we wish to emphasize the field k, we shall refer to the *k-form* of the quantized enveloping algebra and denote it by $U_q(\mathfrak{g})_k$. Subalgebras $U_q^{\geq 0}(\mathfrak{g})$, $U_q^{\leq 0}(\mathfrak{g})$, $U_q^+(\mathfrak{g})$, $U_q^-(\mathfrak{g})$ are defined just as in (I.6.2).

In the case $\mathfrak{g} = \mathfrak{sl}_n(\mathbb{C})$ (which has rank $n-1$), the algebra $U_q(\mathfrak{g})_k$ as just defined coincides with the algebra $U_q(\mathfrak{sl}_n(k))$ defined in (I.6.2). The latter notation is preferred over $U_q(\mathfrak{sl}_n(\mathbb{C}))_k$, for obvious reasons.

One can also define slightly larger versions of $U_q(\mathfrak{g})$, where in place of the generators $K_1^{\pm 1}, \ldots, K_n^{\pm 1}$ one uses a group of units $\{K_\lambda \mid \lambda \in M\}$ isomorphic to some fixed lattice M lying between Q and P (that is, $K_0 = 1$ and $K_\lambda K_\mu = K_{\lambda + \mu}$ for $\lambda, \mu \in M$). Then $K_i = K_{\alpha_i}$ and the first pair of relations for $U_q(\mathfrak{g})$ is replaced by

$$K_\lambda E_j K_{-\lambda} = q^{(\lambda, \alpha_j)} E_j \qquad \text{and} \qquad K_\lambda F_j K_{-\lambda} = q^{-(\lambda, \alpha_j)} F_j.$$

We'll denote this algebra by $U_q(\mathfrak{g}; M)$. The case given first, where $M = Q$, is called the *adjoint form* of the quantized enveloping algebra; when no lattice M is mentioned, it will be understood that this is the case being considered. At the other extreme, when $M = P$, the resulting quantized enveloping algebra is called the *simply connected form* and is denoted $\breve{U}_q(\mathfrak{g})$.

I.6.4. Cartan automorphism. The following lemma generalises Lemma I.3.2 and is proved in the same way. See Exercise I.6.A.

Lemma. 1. *There is a unique automorphism ω of $U_q(\mathfrak{g}; M)$ with $\omega(E_i) = F_i$, $\omega(F_i) = E_i$ and $\omega(K_\lambda) = K_\lambda^{-1}$ for $i = 1, \dots, n$ and $\lambda \in M$. Further, $\omega^2 = 1$.*

2. *There is a unique antiautomorphism τ of $U_q(\mathfrak{g}; M)$ with $\tau(E_i) = E_i$, $\tau(F_i) = F_i$ and $\tau(K_\lambda) = K_\lambda^{-1}$ for $i = 1, \dots, n$ and $\lambda \in M$. Further, $\tau^2 = 1$.* $\square$

I.6.5. Hopf structure. $U_q(\mathfrak{g}; M)$ is a Hopf algebra – for $\mathfrak{g} = \mathfrak{sl}_n(k)$ the data are in (I.6.2). For general $\mathfrak{g}$, we set

$$\Delta(K_\lambda) = K_\lambda \otimes K_\lambda \qquad \varepsilon(K_\lambda) = 1 \qquad S(K_\lambda) = K_\lambda^{-1}$$
$$\Delta(E_i) = E_i \otimes 1 + K_{\alpha_i} \otimes E_i \qquad \varepsilon(E_i) = 0 \qquad S(E_i) = -K_{\alpha_i}^{-1} E_i$$
$$\Delta(F_i) = F_i \otimes K_{\alpha_i}^{-1} + 1 \otimes F_i \qquad \varepsilon(F_i) = 0 \qquad S(F_i) = -F_i K_{\alpha_i}$$

for $i = 1, \dots, n$ and $\lambda \in M$. Note that, as with $\mathfrak{sl}_n(k)$, $U_q(\mathfrak{b}^+)$ and $U_q(\mathfrak{b}^-)$ are sub-Hopf algebras of $U_q(\mathfrak{sl}_n(k))$.

Theorem. *With these definitions, $U_q(\mathfrak{g}; M)$ is a Hopf algebra.*

Proof. Here is a sketch of the proof of the theorem. Details can be found in [**107**, Chapter 4]. Take $M = Q$ for simplicity.

Step 1. Write $q_i = q^{d_i}$ (notation I.6.3(2)) and notice that

$$K_i E_i K_i^{-1} = q_i^2 E_i \qquad\qquad K_i F_i K_i^{-1} = q_i^{-2} F_i.$$

Combined with the other relevant relations this means that for $i = 1, \dots, n$ there are homomorphisms (which turn out to be injective – see (I.6.6) below)

$$\psi_i : U_{q_i}(\mathfrak{sl}_2(k)) \longrightarrow U_q(\mathfrak{g}) \tag{3}$$

for $i = 1, \dots, n$. So versions of the commutator formulae (Exercise I.4.D) hold in $U_q(\mathfrak{g})$.

Step 2. Define $\widetilde{U}_q(\mathfrak{g})$ to be the algebra generated by the same elements as $U_q(\mathfrak{g})$, and with the same relations, except that the *quantized Serre relations* – those of the last two types from (I.6.2) and (I.6.3) – are omitted. There is thus an epimorphism $\widetilde{U}_q(\mathfrak{g}) \longrightarrow U_q(\mathfrak{g})$.

Step 3. $U_q(\mathfrak{g})$ and $\widetilde{U}_q(\mathfrak{g})$ are Q-graded; (see (I.12.7) for the definition). We simply set

$$\deg E_i = \alpha_i \qquad \deg F_i = -\alpha_i \qquad \deg K_i = \deg K_i^{-1} = 0,$$

and note that the relations are all homogeneous.

Step 4. $\widetilde{U}_q(\mathfrak{g})$ is a Hopf algebra under the corresponding definitions to those given for $U_q(\mathfrak{g})$ at the start of paragraph (I.6.5). We have first to check that the images of the generators under Δ, S and ε satisfy the relations for $\widetilde{U}_q(\mathfrak{g})$. It follows easily that the conjugation relations are preserved, and the $[E_i, F_i]$ relations are straightforward to check. Having proved that each of the maps is an algebra homomorphism (or anti-homomorphism, in the case of S), it's enough to check the diagram commutativity of (I.9.1), and the formulae of (I.9.9), on generators. For this, we can simply quote the result for $U_q(\mathfrak{sl}_2(k))$, Theorem I.3.4.

Step 5. The ideal of $\widetilde{U}_q(\mathfrak{g})$ generated by the left sides of the quantized Serre relations is a Hopf ideal, and hence $U_q(\mathfrak{g})$ is a Hopf algebra. This is a messy calculation – see [**107**, 4.10 and Chapter 4, Appendix A]. □

For simplicity, we shall from now on state results only for the case $U_q(\mathfrak{g})$. Since $U_q(\mathfrak{g})$ is a subalgebra of $U_q(\mathfrak{g}, M)$ for any other choice of the lattice M, and indeed the latter algebra is a crossed product of the former by a finite abelian group, many results (such as, for example, the PBW theorem and the action of the braid group) can easily be extended to the general case. The reader should be warned, however, that not all such translations from one choice of the lattice M to another are trivial; in fact, some aspects of the theory are more straightforward for $M = P$ (that is, for $\check{U}_q(\mathfrak{g})$) than for $M = Q$. In particular, the centre of $\check{U}_q(\mathfrak{g})$ is a polynomial ring over k [**116**, Section 7.3, p. 218], whereas the center of $U_q(\mathfrak{g})$ need not be a polynomial ring even for $\mathfrak{g} = \mathfrak{sl}_3(\mathbb{C})$ [**118**, Example 8.8].

I.6.6. Triangular decompositions for $\widetilde{U}_q(\mathfrak{g})$ and for $U_q(\mathfrak{g})$. Consider first $\widetilde{U}_q(\mathfrak{g})$. One shows easily from the defining relations that, in an obvious notation,

$$\{F_I K_\lambda E_J : I, J \text{ finite sequences of simple roots, } \lambda \in Q\}$$

spans $\widetilde{U}_q(\mathfrak{g})$ as a vector space. Then, by considering the action of $\widetilde{U}_q(\mathfrak{g})$ on a suitable module, one finds that this set is actually a basis for $\widetilde{U}_q(\mathfrak{g})$ [**107**, Proposition 4.16]. Let $\widetilde{U}^+$ [resp., $\widetilde{U}^-$] denote the free k-algebra generated by $\{E_i : 1 \leq i \leq n\}$ [resp. $\{F_i : 1 \leq i \leq n\}$], and let $U_0 = k\langle K_\lambda : \lambda \in Q\rangle \subseteq \widetilde{U}_q(\mathfrak{g})$. The vector space basis described above shows that there is a triangular decomposition, as vector spaces,

$$\widetilde{U}^- \otimes U_0 \otimes \widetilde{U}^+ \;\cong\; \widetilde{U}_q(\mathfrak{g}) \tag{4}$$

with the isomorphism given by multiplication $\mu : u \otimes v \otimes w \mapsto uvw$. Thus U_0 is a group algebra of a free abelian group of rank n, n being the rank of $\mathfrak{g}$.

Now we deal with $U_q(\mathfrak{g})$. Let I^- [resp. I^+] be the ideal of $\widetilde{U}^-$ [resp. $\widetilde{U}^+$] generated by the left hand sides of the quantized Serre relations. It's not hard to see [**107**, Lemma 4.20, 4.21] that $\mu(I^- \otimes U_0 \otimes \widetilde{U}^+ + \widetilde{U}^- \otimes U_0 \otimes I^+)$ is the kernel of the canonical epimorphism from $\widetilde{U}_q(\mathfrak{g})$ to $U_q(\mathfrak{g})$, so that there is a triangular decomposition as vector spaces

$$U_q^-(\mathfrak{g}) \otimes U_0 \otimes U_q^+(\mathfrak{g}) \quad \cong \quad U_q(\mathfrak{g}) \tag{5}$$

induced from (4) by μ. Here, U_0 is as above, the group algebra of $\langle K_\lambda : \lambda \in Q \rangle \cong Q$. Moreover, grading considerations show that E_i^r does not belong to I^+ and F_i^s does not belong to I^- for all $i = 1, \dots, n$. From this together with (5) we can show that the maps ψ_i of (I.6.5)(3) are embeddings.

Our next task is to get a better understanding of $U_q^\pm(\mathfrak{g})$, for which at the moment we lack a vector space basis.

I.6.7. Braid group actions. Assume in this paragraph that q is not a root of 1. We continue to study $U_q(\mathfrak{g})$. The *braid group* B_W associated to the Weyl group W is got by omitting the involution relations from the presentation (I.5.1)(4) of W:

$$B_W \quad := \quad \langle T_1, \dots, T_n : T_i T_j T_i \cdots = T_j T_i T_j \cdots \ \text{for}\ i \neq j \rangle,$$

where there are m_{ij} generators on each side of the relations. Clearly from (I.5.1)(4) there is an epimorphism $B_W \longrightarrow W : T_i \longmapsto s_i$.

Recall the definition (I.6.3)(2) of $q_i = q^{d_i}$, and recall also that, for $m \in \mathbb{Z}$ and $v \in k \setminus \{0, \pm 1\}$, we define $[m]_v := \frac{v^m - v^{-m}}{v - v^{-1}}$ (Exercise I.4.C). Define the divided powers, for $s \geq 0$ and $i = 1, \dots, n$,

$$E_i^{(s)} \quad := \quad \frac{E_i^s}{[s]_{q_i}!} \qquad \text{and} \qquad F_i^{(s)} \quad := \quad \frac{F_i^s}{[s]_{q_i}!},$$

where $[s]_{q_i}! = [s]_{q_i}[s-1]_{q_i} \cdots [1]_{q_i}$. Lusztig (and independently Levendorskii and Soibelman) discovered an action of B_W as algebra – but not Hopf algebra – automorphisms of $U_q(\mathfrak{g})$ given for $i = 1, \dots, n$ by

$$T_i E_i = -F_i K_i$$

$$T_i F_i = -K_i^{-1} E_i$$

$$T_i E_j = \sum_{s=0}^{-a_{ij}} (-1)^{s-a_{ij}} q_i^{-s} E_i^{(-a_{ij}-s)} E_j E_i^{(s)} \qquad (j \neq i) \tag{6}$$

$$T_i F_j = \sum_{s=0}^{-a_{ij}} (-1)^{s-a_{ij}} q_i^{s} F_i^{(s)} F_j F_i^{(-a_{ij}-s)} \qquad (j \neq i)$$

$$T_i K_\alpha = K_{s_i(\alpha)} \qquad (\alpha \in Q).$$

The proof that (6) does indeed define an action of B_W on $U_q(\mathfrak{g})$ is long and technical; the details are set out in [**107**, Chapter 8], for example, (although even here the case of $\mathfrak{g}$ of type G_2 is omitted). Notice that (6) extends the action of W by reflections on $\langle K_i : 1 \le i \le n \rangle = Q$.

The action (6) is a quantization of the classical action of B_W on $\mathfrak{g}$ and on $U(\mathfrak{g})$, which is given by defining, for $u \in U(\mathfrak{g})$ and using notation (I.5.5),

$$T_i(u) \quad := \quad \exp(\mathrm{ad}\, e_i)\exp(-\,\mathrm{ad}\, f_i)\exp(\mathrm{ad}\, e_i)(u);$$

this makes sense because $U(\mathfrak{g})$ is ad-locally finite. It is no longer true that the (right or left) adjoint action of $U_q(\mathfrak{g})$ on itself is locally finite; nevertheless, the quantization can be achieved. The trick is to define appropriate operators T_w on $U_q(\mathfrak{g})$, for $w \in W$, such that the canonical epimorphism $B_W \to W$ sends $T_w \mapsto w$. In the classical setting the analogous operators permute (up to sign) the root vectors of $\mathfrak{g}$. This gives us the idea of how to find the missing generators of $U_q(\mathfrak{g})$, corresponding to the root vectors which are not in $\pm\Delta$ – namely, we apply suitable $T_w \in B_W$ to $\{E_i, F_i : 1 \le i \le n\}$.

I.6.8. Generators for non-simple roots. In this paragraph we sketch how the procedure described in the previous sentence is done. Then we study the simplest non-trivial example, namely $\mathfrak{sl}_3(\mathbb{C})$.

Recall first some facts from (I.5.1), (see [**116**, Appendix A]). The *length* $\ell(w)$ of an element w of W is r, where

$$w = s_{i_1} s_{i_2} \cdots s_{i_r} \tag{7}$$

is an expression for w of shortest length as a product of simple reflections s_{i_j}. There is a unique element of longest length, denoted w_0, and

$$\ell(w_0) = N,$$

where N is the number of positive roots of $\mathfrak{g}$. Indeed, suppose we write $w_0 = s_{i_1} \cdots s_{i_N}$ and define

$$\beta_1 := \alpha_{i_1}, \quad \beta_2 := s_{i_1}(\alpha_{i_2}), \quad \ldots, \quad \beta_N := s_{i_1} \cdots s_{i_{N-1}}(\alpha_{i_N}).$$

Then $\{\beta_1, \ldots, \beta_N\}$ is the set of positive roots of $\mathfrak{g}$ determined by the set $\Delta = \{\alpha_1, \ldots, \alpha_n\}$ of simple roots, (and in particular contains Δ as a subset). Analogously, we define elements $E_{\beta_1}, E_{\beta_2}, \cdots, E_{\beta_N}$ of $U_q(\mathfrak{g})$ by

$$E_{\beta_j} \quad := \quad T_{i_1} \cdots T_{i_{j-1}}(E_{i_j}). \tag{8}$$

Theorem. (Lusztig, Levendorskii-Soibelman) *Retain the above notation.*

     1. $\{E_{\beta_1}, E_{\beta_2}, \dots, E_{\beta_N}\} \subseteq U_q^+(\mathfrak{g})$.

     2. $\{E_1, \dots, E_n\} \subseteq \{E_{\beta_1}, E_{\beta_2}, \dots, E_{\beta_N}\}$. *More precisely, suppose that* $\alpha_t = \beta_j$. *Then*

$$E_{\beta_j} = E_t.$$

     3. *Fix* r, $1 \leq r \leq N$, *and write* $w = s_{i_1} \cdots s_{i_r} \in W$ *with* $\ell(w) = r$. *Define*

$$U^+(w) := k\langle E_{\beta_1}, \dots, E_{\beta_r}\rangle \subseteq U_q^+(\mathfrak{g}).$$

Then the monomials

$$E_{\beta_1}^{m_1} E_{\beta_2}^{m_2} \cdots E_{\beta_r}^{m_r}, \quad (m_1, \dots, m_r \in \mathbb{Z}_{\geq 0}),$$

are a basis for $U^+(w)$.

     4. $U^+(w_0) = U_q^+(\mathfrak{g})$.

Proof. See [**107**, Chapter 8], [**44**, Chapter 9]. $\square$

Notes. 1. The notation $U^+(w)$ in 3 is appropriate – this algebra depends only on w. If $w \in W$, we can find $w' \in W$ with $ww' = w_0$ and $\ell(w) + \ell(w') = N$; and $U^+(w)$ doesn't depend on the choice of shortest length expression for w.

     2. Notwithstanding Note 1, the *elements* E_{β_j}, $(1 \leq j \leq N)$, depend heavily on the choice of reduced expression for w_0, as we'll see when $\mathfrak{g} = \mathfrak{sl}_3(\mathbb{C})$ in (I.6.9).

Consider part 3 of the theorem for the particular case $r = N$, $w = w_0$. First note the analogous statements for $U_q^-(\mathfrak{g})$, which we can get by applying the automorphism ω of Lemma I.6.4 to $U_q^+(\mathfrak{g})$. Second, recall the triangular decomposition (I.6.6)(5) and the fact that U_0 is the group algebra of Q. Together, these yield at once the PBW theorem for $U_q(\mathfrak{g})$, which, in view of its importance, we record as a separate corollary. We denote $\omega(E_{\beta_j})$ by F_{β_j} for $j = 1, \dots, N$.

Corollary. PBW Theorem for $U_q(\mathfrak{g})$. *The ordered monomials in the generators* $F_{\beta_1}, F_{\beta_2}, \dots, F_{\beta_N}$, K_λ $(\lambda \in Q)$, $E_{\beta_1}, E_{\beta_2}, \dots, E_{\beta_N}$, *that is, the elements*

$$F_{\beta_1}^{n_1} F_{\beta_2}^{n_2} \cdots F_{\beta_N}^{n_N} K_\lambda E_{\beta_1}^{m_1} E_{\beta_2}^{m_2} \cdots E_{\beta_N}^{m_N}$$

for $m_i, n_j \in \mathbb{Z}_{\geq 0}$ *and* $\lambda \in Q$, *form a vector space basis for* $U_q(\mathfrak{g})$. $\square$

I.6.9. Example: $\mathfrak{g} = \mathfrak{sl}_3(\mathbb{C})$. In this case $n = 2$,

$$W = S_3 = \langle s_1, s_2 : s_1^2 = s_2^2 = 1, \ s_1 s_2 s_1 = s_2 s_1 s_2\rangle,$$

and

$$B_W = \langle T_1, T_2 : T_1 T_2 T_1 = T_2 T_1 T_2\rangle.$$

We have
$$U_q(\mathfrak{g}) \quad = \quad k\langle E_1, E_2, F_1, F_2, K_1^{\pm 1}, K_2^{\pm 1}\rangle.$$

There are two expressions for the longest word of W, namely $w_0 = s_1 s_2 s_1 = s_2 s_1 s_2$. Following the recipe in (I.6.8) using (I.6.7)(6), this gives respectively the two non-simple roots
$$T_1 E_2 \quad = \quad -E_1 E_2 + q^{-1} E_2 E_1$$
and
$$T_2 E_1 \quad = \quad -E_2 E_1 + q^{-1} E_1 E_2.$$

This gives for PBW bases of $U_q^+(\mathfrak{g})$ the ordered monomials in
$$\{E_1,\ E_1 E_2 - q^{-1} E_2 E_1,\ E_2\}$$
and
$$\{E_1,\ E_2 E_1 - q^{-1} E_1 E_2,\ E_2\}$$

respectively. It's easy to check that the two non-simple root vectors are normal elements of $U_q^+(\mathfrak{g})$, and that in each case the factor by the ideal generated by the element is a quantum plane. Compare this with the classical case, where $U^+(\mathfrak{sl}_3(\mathbb{C}))$ is the enveloping algebra of the three-dimensional Heisenberg Lie algebra.

I.6.10. Skew polynomial structure. Keep the notation we developed in (I.6.8). Additionally, for $k = (k_1, \ldots, k_N) \in \mathbb{Z}_{\geq 0}^N$, the symbol E^k will be used to denote $E_{\beta_1}^{k_1} E_{\beta_2}^{k_2} \cdots E_{\beta_N}^{k_N} \in U_q^+(\mathfrak{g})$. Versions of the following result were stated and proved in [**140**, Proposition 5.5.2], [**44**, Theorem 9.3], and [**188**, Theorem 2]; see also [**126**, Chapter 4, Theorem 3.2.3]. None of these authors states the result in precisely the form below. However, one can check (most easily from the arguments using Hall algebras in [**188**]) that the necessary calculations can be reduced to the case where $\mathfrak{g}$ has rank 2 and can then be carried out in a $\mathbb{Z}[q^{\pm 1}]$-subalgebra of $U_q(\mathfrak{g})$ with generators which involve the simple roots E_1 and E_2. This allows one to make the claims on the coefficients z_k as given below. The reduction to the rank 2 case is given, for example, in [**44**, Proof of Theorem 9.3(iv)], and the identities needed for that case are then set out in [**44**, Appendix, (A4)–(A8)].

Proposition. (Levendorskii-Soibelman) *Assume that k has characteristic zero. For $1 \leq i < j \leq N$,*
$$E_{\beta_i} E_{\beta_j} - q^{(\beta_i, \beta_j)} E_{\beta_j} E_{\beta_i} = \sum z_k E^k, \tag{9}$$

where $z_k \in \mathbb{Q}[q^{\pm 1}]$ and $z_k = 0$ unless $k_r = 0$ for $r \leq i$ and $r \geq j$. $\quad\square$

From the proposition and Theorem I.6.8(3) we readily deduce that the algebras $U^+(w)$, for $w \in W$, and in particular $U_q^+(\mathfrak{g})$ itself, are iterated skew polynomial algebras.

Corollary. *Assume that k has characteristic zero. Let $w \in W$ and suppose that w has shortest length expression of the form (I.6.8)(7). Then there exist automorphisms $\sigma_2, \ldots, \sigma_r$ of the appropriate subalgebras, and σ_i-derivations $\delta_2, \ldots, \delta_r$, such that*

$$U^+(w) \quad = \quad k[E_{\beta_1}][E_{\beta_2}; \sigma_2, \delta_2] \ldots [E_{\beta_r}; \sigma_r, \delta_r].$$

Proof. Exercise I.6.E.   □

Of course, thanks to an application of the automorphism ω, the conclusion of the corollary applies to $U_q^-(\mathfrak{g})$. And in view of the conjugation relations, similar results are true for the positive and negative Borel algebras.

I.6.11. Filtered structure. From Corollary I.6.8 we have a PBW basis for $U_q(\mathfrak{g})$ given by the monomials

$$M_{\boldsymbol{r}, \boldsymbol{k}, \alpha} \quad := \quad F^{\boldsymbol{r}} K_\alpha E^{\boldsymbol{k}}$$

where $\boldsymbol{r}, \boldsymbol{k} \in \mathbb{Z}_{\geq 0}^N$ and $\alpha \in Q$. Define the *height* of $\beta = \sum_{i=1}^n a_i \alpha_i \in Q$ by

$$\mathrm{ht}(\beta) \quad := \quad \sum_{i=1}^n a_i,$$

and the *height* of the monomial $M_{\boldsymbol{r}, \boldsymbol{k}, \alpha}$ to be

$$\mathrm{ht}(M_{\boldsymbol{r}, \boldsymbol{k}, \alpha}) \quad := \quad \sum_{j=1}^N (k_j + r_j) \, \mathrm{ht}(\beta_j).$$

Now define the *total degree* of $M_{\boldsymbol{r}, \boldsymbol{k}, \alpha}$ to be

$$d(M_{\boldsymbol{r}, \boldsymbol{k}, \alpha}) \quad := \quad (k_N, \ldots, k_1, r_1, \ldots, r_N, \mathrm{ht}(M_{\boldsymbol{r}, \boldsymbol{k}, \alpha})) \in \mathbb{Z}_{\geq 0}^{(2N+1)}.$$

We impose a total ordering on the semigroup $\mathbb{Z}_{\geq 0}^{(2N+1)}$ by using the lexicographical order with

$$u_1 < u_2 < \ldots < u_{2N+1}$$

where $u_i = (\delta_{i,1}, \delta_{i,2}, \ldots, \delta_{i,2N+1})$. We claim that, with these definitions,

$$U_q(\mathfrak{g}) \quad \text{is a } \mathbb{Z}_{\geq 0}^{(2N+1)}\text{-filtered algebra} \tag{10}$$

in the sense of Appendix I.12.5. To check this, write $U_{\boldsymbol{d}}$, for $\boldsymbol{d} \in \mathbb{Z}_{\geq 0}^{(2N+1)}$, for the subspace of $U_q(\mathfrak{g})$ spanned by the monomials $M_{\boldsymbol{r}, \boldsymbol{k}, \alpha}$ of total degree at most $\boldsymbol{d}$. Note that $U_{\boldsymbol{0}} = U_0 = k\langle K_\alpha : \alpha \in Q \rangle$. We have to show that if M and M' are monomials in $U_{\boldsymbol{d}}$ and $U_{\boldsymbol{d}'}$ respectively, then

$$M M' \in U_{\boldsymbol{d} + \boldsymbol{d}'}. \tag{11}$$

The defining relations (I.6.3) for $U_q(\mathfrak{g})$ mean that this is mostly pretty clear. When using the relations to rewrite MM' as a sum of basis monomials, the only points we need to highlight arise when we are ordering the E_{β_j}s and the F_{β_j}s. Here we use (I.6.10)(9) and the analogous relations for the F_{β_j}; note that conjugation by K_α $(\alpha \in Q)$ shows that the right hand side of (I.6.10)(9) is a sum of monomials with the *same* height as $E_{\beta_i} E_{\beta_j}$, so Proposition I.6.10 and the definition of total degree yields (11) and hence (10).

From the defining relations (I.6.3), Proposition I.6.10 and Corollary I.6.8, we deduce the

Proposition. (De Concini-Kac) *Assume that k has characteristic zero. Equip $U_q(\mathfrak{g})$ with the filtration defined above. The associated graded algebra $\mathrm{gr}(U)$ is the k-algebra on generators*

$$E_{\beta_1}, \ldots, E_{\beta_N}, \ F_{\beta_1}, \ldots, F_{\beta_N}, \ K_\alpha \ (\alpha \in Q),$$

subject to the relations

$$K_\alpha K_\beta = K_{\alpha+\beta} \qquad\qquad K_0 = 1$$
$$K_\alpha E_{\beta_i} = q^{(\alpha,\beta_i)} E_{\beta_i} K_\alpha \qquad K_\alpha F_{\beta_i} = q^{-(\alpha,\beta_i)} F_{\beta_i} K_\alpha$$
$$E_{\beta_i} F_{\beta_j} = F_{\beta_j} E_{\beta_i}$$
$$E_{\beta_i} E_{\beta_j} = q^{(\beta_i,\beta_j)} E_{\beta_j} E_{\beta_i} \qquad F_{\beta_i} F_{\beta_j} = q^{(\beta_i,\beta_j)} F_{\beta_j} F_{\beta_i}$$

for $\alpha, \beta \in Q$ and $1 \leq i, j \leq N$.

Proof. See [**40**, Proposition 1.7] or [**44**, Proposition 10.1]. $\square$

Thus $\mathrm{gr}(U)$ is got from a quantum affine space of dimension $(2N+n)$ by inverting n of the generators. As such, it is of course a noetherian domain. From Lemma I.12.12 and Theorems I.12.13 and I.12.14 we can immediately deduce

Corollary. *Assume that k has characteristic zero. Then $U_q(\mathfrak{g})$ is a noetherian domain of finite global dimension.* $\square$

I.6.12. Finite dimensional representation theory. Assume that k has characteristic not equal to 2, and that the characteristic is not 3 if $\mathfrak{g}$ involves an irreducible component of type G_2. Assume also that

$$q \text{ is not a root of unity.}$$

(The arguments are harder if q is not transcendental over $\mathbb{Q}$, although the conclusions are the same.)

Consider first the 1-dimensional $U_q(\mathfrak{g})$-modules. If V is such a module, relations (I.6.3) show that $E_i V = F_i V = 0$ for $i = 1, \ldots, n$, and then that $(K_i^2 - 1)V =$

0, so that K_i acts on V by multiplication by ± 1 for $i = 1, \ldots, n$. Conversely, any homomorphism $\sigma : \langle K_\lambda : \lambda \in Q \rangle \longrightarrow \{\pm 1\}$ yields a 1-dimensional representation k_σ, where $K_i.1 = \sigma(K_i)1$ for $i = 1, \ldots, n$. Thus, recalling $n = \operatorname{rank} \mathfrak{g}$,

there are 2^n isomorphism classes of 1-dimensional $U_q(\mathfrak{g})$-modules.

Writing $H := \operatorname{Hom}(Q, \mathbb{Z}/2\mathbb{Z})$, we label these 1-dimensional modules k_σ, $(\sigma \in H)$. If V is any finite dimensional $U_q(\mathfrak{g})$-module, the theory for $U_q(\mathfrak{sl}_2(k))$ (Proposition I.4.4) shows that V is the direct sum of its K_i-weight spaces, for each i. Since the K_i commute, V is the direct sum of its U_0-weight spaces. The $U_q(\mathfrak{sl}_2(k))$ theory shows also that each non-zero weight space has the form

$$V_{\lambda,\sigma} \quad := \quad \{v \in V : K_\mu v = \sigma(\mu) q^{(\mu,\lambda)} v \text{ for all } \mu \in Q\}$$

for some $\sigma \in H$ and $\lambda \in P$. We call λ the *weight* of $V_{\lambda,\sigma}$ and say that $v \in V_{\lambda,\sigma}$ is a *highest weight vector* if $E_i v = 0$ for $i = 1, \ldots, n$. It's easy to see that

$$V \quad = \quad \bigoplus_{\sigma \in H} V_\sigma$$

where V_σ is a $U_q(\mathfrak{g})$-module, the sum of the weight spaces of *type σ*. In particular, every finite dimensional irreducible $U_q(\mathfrak{g})$-module V has a well-defined *type*, which is an element σ of H. If V has type τ then $V \otimes k_\sigma$ has type $\tau\sigma$. Thus, the class of (finite dimensional) modules of type $\mathbf{1}$ is closed under $\oplus$, $\otimes$ and ( )* (duals of left $U_q(\mathfrak{g})$-modules are made into left $U_q(\mathfrak{g})$-modules as in (I.9.10)). We denote this category of modules by $\mathcal{C}_q(\mathfrak{g})$, and note that

it is enough to study finite dimensional modules in $\mathcal{C}_q(\mathfrak{g})$.

It turns out that the theory of finite dimensional irreducible $U_q(\mathfrak{g})$-modules of type $\mathbf{1}$ is exactly parallel to the classical theory of finite dimensional irreducible $U(\mathfrak{g})$-modules. We collect the key facts in the following

Theorem. *Continue with the above notation.*

1. *There is a bijection $\lambda \longleftrightarrow V(\lambda)$ between the set of dominant integral weights (that is, elements of $P^+ = \sum_{i=1}^n \mathbb{Z}_{\geq 0}\varpi_i$) and finite dimensional irreducible $U_q(\mathfrak{g})$-modules of type $\mathbf{1}$.*

2. *For $\lambda \in P^+$, $V(\lambda)$ is the unique irreducible image of the corresponding Verma module $M(\lambda) = U_q(\mathfrak{g}) \otimes_{U_q^{\geq 0}(\mathfrak{g})} k(\lambda)$.*

3. *$V(\lambda)$ has a highest weight vector of weight λ, and the weights of $V(\lambda)$ are given by Weyl's character formula.*

4. *Every finite dimensional $U_q(\mathfrak{g})$-module is completely reducible.* $\square$

When we come to study quantized coordinate rings in Chapter I.7 we'll also need some information about tensor products and duals of $U_q(\mathfrak{g})$-modules, which we collect in the

Proposition. 1. If $\lambda = \lambda_1 + \cdots + \lambda_r$ for some $\lambda_i \in P^+$, then $V(\lambda)$ is isomorphic to a direct summand of $V(\lambda_1) \otimes \cdots \otimes V(\lambda_r)$. In particular, every $V(\lambda)$ for $\lambda \in P^+$ is isomorphic to a direct summand of some tensor product of copies of the $V(\varpi_i)$.

2. Let $\lambda \in P^+$. Then

$$\dim_k V(\lambda)_{w\lambda} = \dim_k V(\lambda)^*_{-w\lambda} = 1$$

for all $w \in W$. The weight $w_0\lambda$ is the unique lowest weight of $V(\lambda)$, that is, all weights μ of $V(\lambda)$ satisfy $w_0\lambda \le \mu \le \lambda$. Further, $V(\lambda)^* \cong V(-w_0\lambda)$, and the unique lowest weight of $V(\lambda)^*$ is $-\lambda$.

Proof. 1. Exercise I.6.I.

2. For the one-dimensionality of the weight spaces $V(\lambda)_{w\lambda}$ and the isomorphism $V(\lambda)^* \cong V(-w_0\lambda)$, see [**107**, §5.9, Remark 1, and Proposition 5.16]. (Cf. [**94**, §1.4(3)] and [**116**, §4.4.1].) The remaining statements follow from the facts that the weights of $V(\lambda)^*$ are the negatives of the weights of $V(\lambda)$, and the weights of $V(\lambda)$ all lie in $\lambda - Q^+$ (Exercises I.6.G and I.6.H). □

NOTES

Detailed proofs of most of the results in (I.6.4)–(I.6.11) can be found in [**107**, Chapter 4]. For the most part the proofs of the results in (I.6.12) are routine adaptations of the corresponding arguments for the classical theory. One exception is the complete reducibility result, whose proof uses a description of $V(\lambda)$ in terms of generators and relations which is *deduced* from the classical theory by reduction modulo $(q - 1)$ in a suitable $\mathbb{Q}[q^{\pm 1}]$-lattice. Full details of this and of the other results stated in (I.6.12) can be found in [**107**, Chapter 5], for example, except that for parts 3 and 4 of Theorem I.6.12 with the minimal hypotheses on k and q that we have imposed one should use [**7**, Corollary 7.7] or else take note of the discussion in [**107**, 6.26].

EXERCISES

Exercise I.6.A. Prove Lemma I.6.4.

Exercise I.6.B. Check the details of Step 4 of the proof of Theorem I.6.5.

Exercise I.6.C. Determine the kernel of the canonical homomorphism from $\widetilde{U}_q(\mathfrak{g})$ to $U_q(\mathfrak{g})$, (I.6.6).

Exercise I.6.D. Check the details of Example I.6.9. First confirm the expressions given for the two non-simple root vectors. Then show that these are normal elements of $U_q(\mathfrak{sl}_3(\mathbb{C}))$, and describe the factors by the ideals these elements generate. Show that these two ideals are distinct and prime.

Exercise I.6.E. Prove Corollary I.6.10.

Exercise I.6.F. Prove (I.6.11)(11).

Exercise I.6.G. Given $M, N \in \mathcal{C}_q(\mathfrak{g})$, show that

$$M_\lambda \oplus N_\lambda = (M \oplus N)_\lambda \qquad \text{and} \qquad M_\lambda \otimes N_\mu \subseteq (M \otimes N)_{\lambda+\mu}$$

for all $\lambda, \mu \in P$. Conclude that $M \oplus N$ and $M \otimes N$ lie in $\mathcal{C}_q(\mathfrak{g})$. Conclude also that the weight set of $M \oplus N$ is the union of the weight sets of M and N, and that the weights of $M \otimes N$ are precisely the sums $\lambda + \mu$ where λ is a weight of M and μ is a weight of N.

Choose a basis $\{m_j\}$ for M where each $m_j \in M_{\lambda_j}$ for some $\lambda_j \in P$, and let $\{f_j\}$ be the corresponding dual basis for M^*. Show that each $f_j \in M^*_{-\lambda_j}$. [Hint: Observe that $(K_i f)(m) = f(K_i^{-1} m)$ for $i = 1, \ldots, n$, all $f \in M^*$, and all $m \in M$.] Conclude that the weights of M^* are exactly the negatives of the weights of M, and that $M^* \in \mathcal{C}_q(\mathfrak{g})$.

Exercise I.6.H. If $M \in \mathcal{C}_q(\mathfrak{g})$ is a highest weight module with highest weight $\lambda \in P$, show that all weights μ of M are of the form $\lambda - \beta$ with $\beta \in Q^+$, whence $\mu \leq \lambda$.

Exercise I.6.I. Prove Proposition I.6.12(1).

GENERIC QUANTIZED COORDINATE RINGS
OF SEMISIMPLE GROUPS

Quantized coordinate rings have been constructed for all connected, complex, semisimple algebraic groups G. We describe here the *generic* case, that is, the single-parameter quantized coordinate rings $\mathcal{O}_q(G)$ with q not a root of unity; the root of unity case will be discussed in Chapter III.7. As with quantized enveloping algebras, G is just a suggestive label, and quantized coordinate rings for G can be defined over (almost) arbitrary fields. Since $\mathcal{O}_q(G)$ is built from coordinate functions of certain representations of the quantized enveloping algebra of its Lie algebra, we rely heavily on the results of the previous chapter. Throughout the chapter, q denotes a fixed nonzero scalar, which is not a root of unity, in our base field k.

I.7.1. Fix a connected, complex, semisimple algebraic group G, and let $\mathfrak{g}$ be its Lie algebra. Assume that k has characteristic different from 2, and also different from 3 in case G has a component of type G_2. The notation and choices of data associated with $\mathfrak{g}$ given in the previous chapter will remain in force; in particular, $\mathfrak{h}$ denotes the chosen Cartan subalgebra of $\mathfrak{g}$, and $Q \subseteq P$ denote the root and weight lattices. Fix a lattice L, with $Q \subseteq L \subseteq P$, corresponding to the character group of a maximal torus of G with Lie algebra $\mathfrak{h}$, and set $L^+ = L \cap P^+$. (When G is simply connected, $L = P$.) Note that since $W(\lambda) \subseteq \lambda + Q$ for $\lambda \in P$ (see (I.5.3)(10)), L is stable under the Weyl group W.

Let $\mathcal{C}_q(\mathfrak{g}) = \mathcal{C}_q(\mathfrak{g})_k$ denote the class of type **1** finite dimensional modules over $U_q(\mathfrak{g}) = U_q(\mathfrak{g})_k$, and recall that $\mathcal{C}_q(\mathfrak{g})$ is closed under finite direct sums, tensor products, and duals (I.6.12). Similarly, let $\mathcal{C}_q(\mathfrak{g}, L)$ be the subclass of $\mathcal{C}_q(\mathfrak{g})$ consisting of those modules which are direct sums of copies of $V(\lambda)$s for $\lambda \in L^+$. Alternatively, $\mathcal{C}_q(\mathfrak{g}, L)$ can be described as the class of those modules in $\mathcal{C}_q(\mathfrak{g})$ all of whose weights lie in L, whence it is clear that $\mathcal{C}_q(\mathfrak{g}, L)$ is closed under finite direct sums, tensor products, and duals (Exercise I.7.A).

The quantized coordinate ring of G is built from coordinate functions of the modules in $\mathcal{C}_q(\mathfrak{g}, L)$, defined as follows.

COORDINATE FUNCTIONS

I.7.2. Definitions. Let M be a (left) module over a Hopf algebra H. For any $f \in M^*$ and $v \in M$, define the *coordinate function* $c_{f,v}^M \in H^*$ by the rule

$$c_{f,v}^M(x) = f(xv) \qquad \text{for } x \in H.$$

The annihilator of M is contained in the kernel of $c_{f,v}^M$, so if M is finite dimensional, we have $c_{f,v}^M \in H^\circ$, the Hopf dual of H (I.9.5). In fact, every element of H° can be expressed as a coordinate function of a finite dimensional H-module (Exercise I.7.B). The *coordinate space* of M, denoted $C(M)$, is the linear subspace of H^* spanned by the coordinate functions $c_{f,v}^M$ as f runs over M^* and v over M. If M is finite dimensional, then so is $C(M)$ (Exercise I.7.B).

I.7.3. Lemma. *Let M and N be finite dimensional modules over a Hopf algebra H. Let $f \in M^*$, $g \in N^*$, $v \in M$, and $w \in N$. Then*

$$c_{f,v}^M + c_{g,w}^N = c_{(f,g),(v,w)}^{M \oplus N} \qquad\qquad c_{f,v}^M c_{g,w}^N = c_{f \otimes g, v \otimes w}^{M \otimes N}$$

$$\varepsilon(c_{f,v}^M) = f(v) \qquad\qquad\qquad S(c_{f,v}^M) = c_{v,f}^{M^*}.$$

If $\{v_i\}$ and $\{f_j\}$ are dual bases for M and M^*, then

$$\Delta(c_{f,v}^M) = \sum_i c_{f,v_i}^M \otimes c_{f_i,v}^M.$$

Proof. Exercise I.7.C. □

In particular, the first formula of the lemma yields $c_{f,v}^M = c_{(f,g),(v,0)}^{M \oplus N}$ in the case $w = 0$. Hence, $C(M) \subseteq C(M \oplus N)$.

I.7.4. Corollary. *Let H be a Hopf algebra, and let A be the subalgebra of H° generated by all the coordinate functions of some family C of finite dimensional H-modules.*

(a) *A is a sub-bialgebra of H°.*

(b) *If C is closed under duals, then A is a sub-Hopf algebra of H°.*

(c) *Let $\hat{C}$ denote the closure of C under finite direct sums and tensor products. Then A is the directed union of the coordinate spaces $C(V)$ for $V \in \hat{C}$.* □

DEFINITION OF $\mathcal{O}_q(G)$

I.7.5. Definition. The *(k-form of) the quantized coordinate ring of G* is the k-subalgebra of $U_q(\mathfrak{g})^\circ$ generated by the coordinate functions of the highest weight modules $V(\lambda)$ for $\lambda \in L^+$. We denote this algebra by $\mathcal{O}_q(G)$, or by $\mathcal{O}_q(G)_k$ when we need to emphasize the field k. It is also possible to describe $\mathcal{O}_q(G)$ without reference to coordinate functions, as noted in the following chapter (Exercise I.8.B). It is clear from Lemma I.7.3 that $\mathcal{O}_q(G)$ is also generated by the coordinate functions of the modules in $C_q(\mathfrak{g}, L)$; in fact, we have the following

I.7.6. Lemma. *The algebra $\mathcal{O}_q(G)$ is the directed union of the coordinate spaces $C(V)$ for $V \in \mathcal{C}_q(\mathfrak{g}, L)$.*

Proof. Corollary I.7.4(c). $\square$

I.7.7. There may be many semisimple groups G with the same Lie algebra $\mathfrak{g}$, in which case $U_q(\mathfrak{g})^\circ$ contains a number of subalgebras of the form $\mathcal{O}_q(G)$. For example, $SL_2(\mathbb{C})$ and $PSL_2(\mathbb{C})$ have the same Lie algebra, namely $\mathfrak{sl}_2(\mathbb{C})$. In this case, there is only a single fundamental weight ϖ_1, so that $P = \mathbb{Z}\varpi_1$, while $Q = 2\mathbb{Z}\varpi_1$. The lattices L corresponding to the cases $G = SL_2(\mathbb{C})$ and $G = PSL_2(\mathbb{C})$ are $L = P$ and $L = Q$, respectively. Consequently, $\mathcal{O}_q(PSL_2(\mathbb{C}))$ is generated by the coordinate functions of a smaller family of highest weight modules than $\mathcal{O}_q(SL_2(\mathbb{C}))$ is, whence $\mathcal{O}_q(PSL_2(\mathbb{C}))$ is a subalgebra of $\mathcal{O}_q(SL_2(\mathbb{C}))$. Explicit descriptions of these algebras are given in Exercise I.7.M and Theorem I.7.16, respectively.

I.7.8. Proposition. *$\mathcal{O}_q(G)$ is an affine k-algebra and a sub-Hopf algebra of $U_q(\mathfrak{g})^\circ$. In particular, L^+ is a finitely generated semigroup, and if $\lambda_1, \ldots, \lambda_r$ generate L^+, then the coordinate functions of the modules $V(\lambda_1), \ldots, V(\lambda_r)$ generate $\mathcal{O}_q(G)$ as a k-algebra.*

Proof. Since $\mathcal{O}_q(G)$ is generated by the coordinate functions of the modules in $\mathcal{C}_q(\mathfrak{g}, L)$, it follows from Corollary I.7.4 that $\mathcal{O}_q(G)$ is a sub-Hopf algebra of $U_q(\mathfrak{g})^\circ$.

We next show that L^+ is a finitely generated semigroup. Since P and Q are free abelian groups of rank n, so is L, whence P/L is finite. Hence, there exists a positive integer m such that $m\varpi_i \in L$ for all i. The set

$$L_0 = L^+ \cap \left\{ \sum_{i=1}^n j_i \varpi_i \;\middle|\; j_1, \ldots, j_n \in \{0, 1, \ldots, m-1\} \right\}$$

is finite, and $L_0 \cup \{m\varpi_1, \ldots, m\varpi_n\}$ generates L^+, establishing the claim. Let $\lambda_1, \ldots, \lambda_r$ be a finite set of generators for L^+.

It follows from Proposition I.6.12 that for any $\lambda \in L^+$, the module $V(\lambda)$ is isomorphic to a direct summand of some tensor product of copies of the $V(\lambda_i)$. Hence, Lemma I.7.3 implies that $\mathcal{O}_q(G)$ is generated (as a k-algebra) by the coordinate functions of the $V(\lambda_i)$. If $\{v_{is}\}$ and $\{f_{it}\}$ are bases for $V(\lambda_i)$ and $V(\lambda_i)^*$, then any coordinate function of $V(\lambda_i)$ is a linear combination of the $c_{f_{it}, v_{is}}^{V(\lambda_i)}$ (Exercise I.7.B). Therefore $\mathcal{O}_q(G)$ is an affine k-algebra. $\square$

There is a potential ambiguity in the notation $\mathcal{O}_q(SL_n(\mathbb{C}))$, which is resolved by the following theorem. We discuss the case of SL_2 in more detail below.

I.7.9. Theorem. *If $G = SL_n(\mathbb{C})$, then the Hopf algebra $\mathcal{O}_q(G)_k$ defined in (I.7.5) is isomorphic to Hopf algebra $\mathcal{O}_q(SL_n(k))$ defined in (I.2.4).*

Proof. See [94, Theorem 1.4.1], which is stated for the case $k = \mathbb{C}$; however, the methods extend to the general case. See also [7, Appendix]. $\square$

The SL_2 case

I.7.10. Let us concentrate on the case $G = SL_2(\mathbb{C})$, with $\mathfrak{g} = \mathfrak{sl}_2(\mathbb{C})$. In this case, $L = P$ and ϖ_1 generates L^+, whence Proposition I.7.8 shows that the coordinate functions of the $U_q(\mathfrak{g})$-module $V = V(\varpi_1)$ generate $\mathcal{O}_q(G)$. Recall from (I.4.3) that $V = L(1, +)$ has a basis $\{v_1, v_2\}$ such that

$$
\begin{array}{lll}
Kv_1 = qv_1 & Ev_1 = 0 & Fv_1 = v_2 \\
Kv_2 = q^{-1}v_2 & Ev_2 = v_1 & Fv_2 = 0.
\end{array}
$$

Let $\{f_1, f_2\}$ be the corresponding dual basis for V^*. Thus, $\mathcal{O}_q(G)$ is generated by the elements $x_{ij} = c^V_{f_i, v_j}$ for $i, j = 1, 2$. Note from Lemma I.7.3 that

$$
\varepsilon(x_{ij}) = \delta_{ij} \qquad\qquad \Delta(x_{ij}) = x_{i1} \otimes x_{1j} + x_{i2} \otimes x_{2j}
$$

for all i, j. For a computation of $S(x_{ij})$, see Exercise I.7.D.

In order to understand the algebra structure of $\mathcal{O}_q(G)$, we need to find relations among the x_{ij}. In general, one way to look for relations among coordinate functions of modules over a Hopf algebra is to find homomorphisms among tensor products of these modules, as follows.

I.7.11. Lemma. *Let V and W be finite dimensional modules over a Hopf algebra H, and let $R : V \otimes W \to W \otimes V$ be an H-module homomorphism. Let $\{v_i\}$ and $\{w_j\}$ be bases for V and W, respectively, and let $\{f_i\}$ and $\{g_j\}$ be the corresponding dual bases for V^* and W^*, respectively. There are scalars $R^{st}_{ij} \in k$ such that $R(v_i \otimes w_j) = \sum_{s,t} R^{st}_{ij} w_s \otimes v_t$ for all i, j. Then*

$$
\sum_{s,t} R^{st}_{ij} c^W_{g_l, w_s} c^V_{f_m, v_t} = \sum_{s,t} R^{lm}_{st} c^V_{f_s, v_i} c^W_{g_t, w_j}
$$

for all i, j, l, m.

Proof. Exercise I.7.E. □

I.7.12. Returning to the case at hand, let $V = V(\varpi_1)$ with basis $\{v_1, v_2\}$ as above, and define a linear transformation $R : V \otimes V \to V \otimes V$ so that

$$
\begin{array}{ll}
R(v_1 \otimes v_1) = qv_1 \otimes v_1 & R(v_1 \otimes v_2) = v_2 \otimes v_1 + (q - q^{-1})v_1 \otimes v_2 \\
R(v_2 \otimes v_1) = v_1 \otimes v_2 & R(v_2 \otimes v_2) = qv_2 \otimes v_2.
\end{array}
$$

Then R is an isomorphism of $U_q(\mathfrak{g})$-modules (Exercise I.7.F). Applying Lemma I.7.11 and simplifying, we find that the x_{ij} satisfy the six defining relations of the standard generators of $\mathcal{O}_q(M_2(k))$:

$$
\begin{array}{ll}
x_{11}x_{12} = qx_{12}x_{11} & x_{11}x_{21} = qx_{21}x_{11} \\
x_{12}x_{21} = x_{21}x_{12} & x_{12}x_{22} = qx_{22}x_{12} \\
x_{21}x_{22} = qx_{22}x_{21} & x_{11}x_{22} - x_{22}x_{11} = (q - q^{-1})x_{12}x_{21}
\end{array}
$$

(Exercise I.7.G). Moreover, we also have $x_{11}x_{22} - qx_{12}x_{21} = 1$ (Exercise I.7.H).

In view of the relations above, there is a surjective k-algebra homomorphism $\phi : \mathcal{O}_q(SL_2(k)) \to \mathcal{O}_q(G)$ sending $\bar{a}$, $\bar{b}$, $\bar{c}$, $\bar{d}$ to x_{11}, x_{12}, x_{21}, x_{22}, respectively. In addition, the equations in (I.7.10) and Exercise I.7.D imply that ϕ is a homomorphism of Hopf algebras. We shall see that ϕ is, in fact, an isomorphism, thus verifying Theorem I.7.9 in the case $n = 2$. The computations involve some automorphisms and skew derivations which arise from actions of $U_q(\mathfrak{g})$ on $\mathcal{O}_q(G)$ in the following way.

I.7.13. Recall that any bialgebra H becomes a left and right H°-module algebra (see Example I.9.21(a)). Dually, H° becomes a left and right H-module algebra, a structure which we prefer to construct directly, as follows.

First, recall the standard H-H-bimodule structure on $H^* = \mathrm{Hom}_k(H, k)$, where

$$(u.a)(x) = a(xu) \qquad\qquad (a.u)(x) = a(ux)$$

for $u \in H$, $a \in H^*$, and $x \in H$. If $a \in H^\circ$, then $\ker(a)$ contains an ideal I of H with finite codimension. For any $u \in H$, we have $Iu \subseteq I \subseteq \ker(a)$ and so $I \subseteq \ker(u.a)$, whence $u.a \in H^\circ$. Similarly, $a.u \in H^\circ$. Therefore H° is a sub-bimodule of H^*.

In terms of the Sweedler notation, the left and right H-module multiplications on H° can be written as

$$u.a = \sum a_1 a_2(u) \qquad\qquad a.u = \sum a_1(u)a_2 \qquad\qquad (1)$$

for $u \in H$ and $a \in H^\circ$. For instance, to verify the formula for $u.a$, observe that $(u.a)(x) = a(xu) = \sum a_1(x)a_2(u)$ for all $x \in H$. Since the multiplication in H° is convolution, it follows that H° is an H-module algebra on both sides, that is,

$$u.(ab) = \sum (u_1.a)(u_2.b) \qquad\qquad (ab).u = \sum (a.u_1)(b.u_2)$$

for $u \in H$ and $a, b \in H^\circ$ (Exercise I.7.I). It is immediate from (1) that any sub-bialgebra of H° is an H-H-subbimodule.

The formulas above immediately yield, in the terminology of (I.1.11),

I.7.14. Lemma. Let H be a bialgebra.

(a) If $u \in H$ is a unit such that $\Delta(u) = u \otimes u$, then the maps $\sigma : a \mapsto a.u$ and $\tau : a \mapsto u.a$ are k-algebra automorphisms of H°, with inverses given by $a \mapsto a.u^{-1}$ and $a \mapsto u^{-1}.a$, respectively.

(b) Let u, σ, and τ be as in part (a). Suppose that $e, f \in H$ are elements such that

$$\Delta(e) = u \otimes e + e \otimes 1 \qquad\qquad \Delta(f) = f \otimes u + 1 \otimes f.$$

Then the map $a \mapsto e.a$ is a left τ-derivation on H° and $a \mapsto a.e$ is a left σ-derivation, while $a \mapsto f.a$ is a right τ-derivation and $a \mapsto a.f$ is a right σ-derivation. $\square$

I.7.15. In particular, it follows from Lemma I.7.14 that the rules

$$\sigma(y) = \sum y_1(K)y_2 \qquad\qquad \tau(y) = \sum y_1 y_2(K^{-1})$$

define k-algebra automorphisms σ and τ of $\mathcal{O}_q(G)$, and that the rules

$$\delta(y) = \sum y_1(E)y_2 \qquad\qquad \partial(y) = \sum y_1 y_2(F)$$

define a left σ-derivation δ and a right τ-derivation ∂, respectively. It is easily checked that

$$\sigma(x_{1j}) = qx_{1j} \qquad\qquad \sigma(x_{2j}) = q^{-1}x_{2j}$$
$$\tau(x_{i1}) = q^{-1}x_{i1} \qquad\qquad \tau(x_{i2}) = qx_{i2}$$
$$\delta(x_{1j}) = x_{2j} \qquad\qquad \delta(x_{2j}) = 0$$
$$\partial(x_{i1}) = x_{i2} \qquad\qquad \partial(x_{i2}) = 0$$

for all j, i (Exercise I.7.J).

I.7.16. Theorem. *The map $\phi : \mathcal{O}_q(SL_2(k)) \to \mathcal{O}_q(SL_2(\mathbb{C}))_k$ defined in (I.7.12) is an isomorphism of Hopf algebras.*

Proof. Set $A = \mathcal{O}_q(SL_2(\mathbb{C}))_k$. We first check that the set

$$S = \{\overline{a}^l \overline{b}^m \overline{c}^s \mid l, m, s \geq 0\} \cup \{\overline{b}^m \overline{c}^s \overline{d}^t \mid m, s \geq 0 \text{ and } t > 0\}$$

spans $\mathcal{O}_q(SL_2(k))$ (Exercise I.7.K), and consequently that the set

$$S' = \{x_{11}^l x_{12}^m x_{21}^s \mid l, m, s \geq 0\} \cup \{x_{12}^m x_{21}^s x_{22}^t \mid m, s \geq 0 \text{ and } t > 0\}$$

spans A. (In fact, S is a basis for $\mathcal{O}_q(SL_2(k))$ – see Example I.11.8.) If we can show that S' is linearly independent, then ϕ must be an isomorphism. This is done with the help of the automorphisms σ, τ and skew derivations δ, ∂ defined in (I.7.15). We first compute several reduction formulas:

$$\delta(x_{1j}^l) = \frac{q^{2l} - 1}{q^2 - 1} x_{2j} x_{1j}^{l-1} \tag{2}$$

for $l > 0$ and $j = 1, 2$;

$$\delta(x_{11}^l x_{12}^m) = \frac{q^{2m+l+1} - q^{1-l}}{q^2 - 1} x_{11}^{l-1} x_{12}^m x_{21} + \frac{q^{2m+l} - q^l}{q^2 - 1} x_{11}^{l-1} x_{12}^{m-1} \tag{3}$$

for $l, m > 0$; and

$$\partial(x_{21}^s) = \frac{q^{-2s} - 1}{q^{-2} - 1} x_{21}^{s-1} x_{22} \tag{4}$$

for $s > 0$ (Exercise I.7.L).

Suppose that $\mathcal{S}'$ is not linearly independent. Then there is a nontrivial relation of the form

$$\sum_{l,m,s \geq 0} \alpha_{lms} x_{11}^l x_{12}^m x_{21}^s + \sum_{m,s \geq 0,\, t > 0} \beta_{mst} x_{12}^m x_{21}^s x_{22}^t = 0, \tag{5}$$

where the α_{lms} and β_{mst} are scalars, all but finitely many of which are zero. If a positive power of x_{11} and/or x_{12} appears in (5), then after applying δ and taking account of (2) and (3), we obtain a new, nontrivial relation of the same type, but with lower total degree in x_{11} and x_{12}. (The coefficients in (2) and (3) do not vanish due to our assumption that q is not a root of unity.) Continuing in this way, we eventually obtain a relation of type (5) in which x_{11} and x_{12} do not appear, that is, a nontrivial relation of the form

$$\sum_{s,t \geq 0} \gamma_{st} x_{21}^s x_{22}^t = 0. \tag{6}$$

By applying σ and τ to (6) and comparing coefficients, we see that $x_{21}^s x_{22}^t = 0$ for some s, t. Then, applying ∂^s and taking account of (4), we find that $x_{22}^{s+t} = 0$. Since the counit ε is an algebra homomorphism and $\varepsilon(x_{22}) = 1$, we have a contradiction.

Thus $\mathcal{S}'$ is linearly independent, and therefore ϕ is an isomorphism. $\square$

NOTES

Two distinct philosophies for the quantized coordinate rings $\mathcal{O}_q(G)$ have been developed. The first philosophy takes $\mathcal{O}_q(G)$ as a primary object, and seeks to construct it directly via generators and relations (usually related to some R-matrix associated with the corresponding quantized enveloping algebra $U_q(\mathfrak{g})$), while the second takes $U_q(\mathfrak{g})$ as the basic object and obtains $\mathcal{O}_q(G)$ via the representation theory of $U_q(\mathfrak{g})$. Originally, quantized coordinate rings for the classical simple Lie groups (i.e., $SL_n(\mathbb{C})$, $SO_n(\mathbb{C})$, and $Sp_n(\mathbb{C})$) were constructed according to the first philosophy, by Faddeev-Reshetikhin-Takhtadjan [54, 186] and Takeuchi [202] (cf. [91]). The former authors also showed how $\mathcal{O}_q(G)$ (for classical G), when extended by a few extra generators, becomes isomorphic to $U_q(\mathfrak{g})^\circ$. Constructions of $\mathcal{O}_q(G)$ under the second philosophy seem to have been inspired by the work of Woronowicz [211, 212], Vaksman-Soibelman [205], and Soibelman [196, 197], whose Hopf *-algebra quantized coordinate rings of compact groups were generated by certain coordinate functions. In particular, Soibelman constructed a quantized coordinate ring of $SU_n(\mathbb{C})$ from the coordinate functions of a category of finite dimensional unitarizable representations of $U_q(\mathfrak{sl}_n(\mathbb{C}))$ [**196**]. Hodges and Levasseur extended Soibelman's approach to $\mathcal{O}_q(SL_n(\mathbb{C}))$ [**94, 95**], and it was continued in Joseph's work on $\mathcal{O}_q(G)$ [**115, 116**]. This approach has now become standard, and carries the advantage that it treats the exceptional semisimple groups on the same footing with the classical ones.

EXERCISES

Exercise I.7.A. Show that $C_q(\mathfrak{g}, L)$ consists of those modules in $C_q(\mathfrak{g})$ whose weights all lie in L. Then show that $C_q(\mathfrak{g}, L)$ is closed under finite direct sums, tensor products, and duals. □

Exercise I.7.B. Let H be a Hopf algebra. Show that every functional $c \in H^\circ$ can be expressed as a coordinate function of some finite dimensional H-module. If M is a finite dimensional H-module and $\{v_i\}$ and $\{f_j\}$ are bases for M and M^*, respectively, show that the coordinate functions $c^M_{f_j,v_i}$ span $C(M)$. □

Exercise I.7.C. Prove Lemma I.7.3. □

Exercise I.7.D. Let $V = V(\varpi_1)$ with basis $\{v_1, v_2\}$ as in (I.7.10) and with dual basis $\{f_1, f_2\}$ for V^*. Show that there is a $U_q(\mathfrak{g})$-module isomorphism $V^* \to V$ such that $f_2 \mapsto v_1$ and $-qf_1 \mapsto v_2$. Set $x_{ij} = c^V_{f_i,v_j}$ and recall from Lemma I.7.3 that $S(x_{ij}) = c^{V^*}_{v_j,f_i}$. Compute that

$$S(x_{11}) = x_{22} \qquad\qquad S(x_{12}) = -q^{-1}x_{12}$$
$$S(x_{21}) = -qx_{21} \qquad\qquad S(x_{22}) = x_{11}. \quad \square$$

Exercise I.7.E. Prove Lemma I.7.11. □

Exercise I.7.F. Show that the linear transformation R defined in (I.7.12) is an isomorphism of $U_q(\mathfrak{g})$-modules. □

Exercise I.7.G. Derive the relations for the x_{ij} displayed in (I.7.12). □

Exercise I.7.H. Let $x_{ij} \in \mathcal{O}_q(G)$ as in (I.7.10). Set $d_q = x_{11}x_{22} - qx_{12}x_{21}$ and show that $\Delta(d_q) = d_q \otimes d_q$. Observe that this relation implies that d_q is a k-algebra homomorphism from $U_q(\mathfrak{g})$ to k. Now show that d_q equals the counit ε on $U_q(\mathfrak{g})$, that is, $d_q = 1$ in $\mathcal{O}_q(G)$. □

Exercise I.7.I. Let H be a bialgebra, and verify that H° becomes a left and right H-module algebra as in (I.7.13). □

Exercise I.7.J. Verify the formulas for σ, τ, δ, and ∂ displayed in (I.7.15). □

Exercise I.7.K. Show that the set

$$S = \{\bar{a}^l \bar{b}^m \bar{c}^s \mid l, m, s \geq 0\} \cup \{\bar{b}^m \bar{c}^s \bar{d}^t \mid m, s \geq 0 \text{ and } t > 0\}$$

spans $\mathcal{O}_q(SL_2(k))$. □

Exercise I.7.L. Verify the reduction formulas (2), (3), (4) in the proof of Theorem I.7.16. □

Exercise I.7.M. As noted in (I.7.7), the groups $PSL_2(\mathbb{C})$ and $SL_2(\mathbb{C})$ have the same Lie algebra, namely $\mathfrak{g} = \mathfrak{sl}_2(\mathbb{C})$, and so the algebras $A = \mathcal{O}_q(PSL_2(\mathbb{C}))_k$ and $B = \mathcal{O}_q(SL_2(\mathbb{C}))_k$ are both subalgebras of $U_q(\mathfrak{g})_k^\circ$. The corresponding lattices L are $2\mathbb{Z}\varpi_1$ and $\mathbb{Z}\varpi_1$, respectively, and so Proposition I.7.8 says that A and B are generated by the coordinate functions of $V(2\varpi_1)$ and $V(\varpi_1)$, respectively. Thus $A \subseteq B$, and we seek to describe a set of k-algebra generators for A in terms of the generators x_{ij} for B given in (I.7.10). This amounts to finding relations between the coordinate functions of $V(2\varpi_1)$ and those of $V(\varpi_1)$, which can be done as follows.

Show that $V(\varpi_1) \otimes V(\varpi_1) \cong V(2\varpi_1) \oplus V(0)$. [Hint: Look at the submodules generated by $v_1 \otimes v_1$ and $v_1 \otimes v_2 - qv_2 \otimes v_1$, respectively.] Observe that the coordinate functions of $V(0)$ are just scalar multiples of the identity in $U_q(\mathfrak{g})_k^\circ$. Hence, A is generated by the coordinate functions of $V(\varpi_1) \otimes V(\varpi_1)$. Conclude that A is generated by the 16 products $x_{ij}x_{lm}$ for $i, j, l, m = 1, 2$. □

$\mathcal{O}_q(G)$ IS A NOETHERIAN DOMAIN

Fix a connected, complex, semisimple algebraic group G along with associated data L, $\mathfrak{g}$, etc. as in the previous chapter. In particular, q remains a fixed non-root of unity in the base field k. We present a sample kit of tools that have been developed to study $\mathcal{O}_q(G)$, and use them to prove the statement in the title of the chapter.

THE ACTION OF $U_q(\mathfrak{g})$ ON $\mathcal{O}_q(G)$

I.8.1. Recall from (I.7.13) that $U_q(\mathfrak{g})^\circ$ is a $U_q(\mathfrak{g})$-$U_q(\mathfrak{g})$-bimodule, and that $\mathcal{O}_q(G)$, being a sub-bialgebra of $U_q(\mathfrak{g})^\circ$, must be a sub-bimodule. Let us see how elements of $U_q(\mathfrak{g})$ act on the left and right of coordinate functions. Thus let V be a module in $\mathcal{C}_q(\mathfrak{g})$, and let $f \in V^*$, $v \in V$. If $\{v_i\}$ and $\{f_j\}$ are dual bases for V and V^*, then $\Delta(c^V_{f,v}) = \sum_i c^V_{f,v_i} \otimes c^V_{f_i,v}$ by Lemma I.7.3. Consequently, (I.7.13) implies that

$$u.c^V_{f,v} = \sum_i f_i(uv)c^V_{f,v_i} \qquad \text{and} \qquad c^V_{f,v}.u = \sum_i f(uv_i)c^V_{f_i,v}$$

for all $u \in U_q(\mathfrak{g})$. In particular, it follows that $u.C(V) \subseteq C(V)$ and $C(V).u \subseteq C(V)$, and the following lemma is proved.

I.8.2. Lemma. (a) *For each object $V \in \mathcal{C}_q(\mathfrak{g})$, the coordinate space $C(V)$ is a $U_q(\mathfrak{g})$-$U_q(\mathfrak{g})$-subbimodule of $U_q(\mathfrak{g})^\circ$.*
    (b) *$\mathcal{O}_q(G)$ is a $U_q(\mathfrak{g})$-$U_q(\mathfrak{g})$-subbimodule of $U_q(\mathfrak{g})^\circ$.* $\square$

I.8.3. Definitions. We shall need analogues of the skew derivations defined on $\mathcal{O}_q(SL_2(\mathbb{C}))_k$ in (I.7.15). Thus, for $i = 1, \ldots, n$ and $y \in \mathcal{O}_q(G)$, define

$$\sigma_i(y) = y.K_i = \sum y_1(K_i)y_2 \qquad \tau_i(y) = K_i^{-1}.y = \sum y_1 y_2(K_i^{-1})$$
$$\delta_i(y) = y.E_i = \sum y_1(E_i)y_2 \qquad \partial_i(y) = F_i.y = \sum y_1 y_2(F_i).$$

It follows from Lemmas I.7.14 and I.8.2 that

    σ_i and τ_i are k-algebra automorphisms of $\mathcal{O}_q(G)$;
    δ_i is a left σ_i-derivation on $\mathcal{O}_q(G)$;
    ∂_i is a right τ_i-derivation on $\mathcal{O}_q(G)$.

We also need to note how the pairs σ_i, δ_i and τ_i, ∂_i interact with each other. For instance, $\delta_i \sigma_i(y) = y.(K_i E_i)$ while $\sigma_i \delta_i(y) = y.(E_i K_i)$ for $y \in \mathcal{O}_q(G)$. Since $K_i E_i = q_i^2 E_i K_i$, we find that

$$\delta_i \sigma_i = q_i^2 \sigma_i \delta_i.$$

Similarly, since $F_i K_i^{-1} = q_i^{-2} K_i^{-1} F_i$, we obtain

$$\partial_i \tau_i = q_i^{-2} \tau_i \partial_i.$$

These relations allow a convenient description of the way powers of δ_i and ∂_i act on products of elements from $\mathcal{O}_q(G)$, as follows.

I.8.4. Let (τ, δ) be a left skew derivation on a ring R, and $R[x; \tau, \delta]$ the corresponding skew polynomial ring. A (τ, δ)-*constant* in R is any element $q \in R$ such that $\tau(q) = q$ and $\delta(q) = 0$. (For instance, if $R[x; \tau, \delta]$ is a k-algebra, any scalar in k is a (τ, δ)-constant.) Suppose that q is a central (τ, δ)-constant in R. If $\delta\tau = q\tau\delta$, then the pair (τ, δ) is called a q-*skew derivation*, and $R[x; \tau, \delta]$ is called a q-*skew polynomial ring* over R. In this situation, the following q-*Leibniz Rules* hold in R and $R[x; \tau, \delta]$, where the q-binomial coefficients $\binom{n}{i}_q$ are as defined in (I.6.1):

$$\delta^n(ab) = \sum_{i=0}^{n} \binom{n}{i}_q \tau^{n-i} \delta^i(a) \delta^{n-i}(b)$$

$$x^n a = \sum_{i=0}^{n} \binom{n}{i}_q \tau^{n-i} \delta^i(a) x^{n-i}$$

for all $a, b \in R$ and $n \geq 0$ (see [**68**, Lemma 6.2]).

I.8.5. Lemma. Let $y, z \in \mathcal{O}_q(G)$. Then

$$\delta_i^m(yz) = \sum_{j=0}^{m} \binom{m}{j}_{q_i^2} \sigma_i^{m-j} \delta_i^j(y) \delta_i^{m-j}(z) \tag{1}$$

$$\partial_i^m(yz) = \sum_{j=0}^{m} \binom{m}{j}_{q_i^{-2}} \partial_i^{m-j}(y) \tau_i^{m-j} \partial_i^j(z) \tag{2}$$

for $i = 1, \ldots, n$ and $m \geq 0$.

Proof. Equation (1) is immediate from (I.8.3) and (I.8.4). Since ∂_i is a right τ_i-derivation on $\mathcal{O}_q(G)$, it is a left τ_i-derivation on the opposite ring $\mathcal{O}_q(G)^{\mathrm{op}}$. If we denote the multiplication in $\mathcal{O}_q(G)^{\mathrm{op}}$ by $*$, then a second application of the q-Leibniz Rule yields

$$\partial_i^m(z * y) = \sum_{j=0}^{m} \binom{m}{j}_{q_i^{-2}} \tau_i^{m-j} \partial_i^j(z) * \partial_i^{m-j}(y),$$

from which (2) follows. $\square$

THE $L \times L$ GRADING ON $\mathcal{O}_q(G)$

I.8.6. Definition. Since $\mathcal{O}_q(G)$ is both a left and a right $U_q(\mathfrak{g})$-module, it has weight spaces for both structures. Combining these, we obtain weight spaces indexed by pairs of weights. On the other hand, if $c_{f,v}^V$ is a coordinate function such that f and v are weight vectors, say with weights β and η, it is natural to say that $c_{f,v}^V$ has weight (β, η). In order to ensure that this matches the weight space scheme for $\mathcal{O}_q(G)$, we define weight spaces in $\mathcal{O}_q(G)$ as follows:

$$\mathcal{O}_q(G)_{\beta,\eta} = \{y \in \mathcal{O}_q(G) \mid y.K_i^{-1} = q^{(\beta,\alpha_i)}y \text{ and } K_i.y = q^{(\eta,\alpha_i)}y$$
$$\text{for all } i = 1, \ldots, n\}$$

for $\beta, \eta \in L$. We leave it as an exercise to check that under this definition, coordinate functions as above have the desired weights (Exercise I.8.A).

These weight spaces can be used to give a description of $\mathcal{O}_q(G)$ that is independent of coordinate functions and representations of $U_q(\mathfrak{g})$ (Exercise I.8.B).

I.8.7. Lemma. (a) $\mathcal{O}_q(G) = \bigoplus_{\beta,\eta \in L} \mathcal{O}_q(G)_{\beta,\eta}$.
(b) *The decomposition in* (a) *is an* $L \times L$ *grading of the algebra* $\mathcal{O}_q(G)$.
(c) *If* V *is a module in* $\mathcal{C}_q(\mathfrak{g}, L)$, *then* $C(V)$ *is a homogeneous subspace of* $\mathcal{O}_q(G)$ *with respect to the* $L \times L$ *grading.*
(d) *For* $i = 1, \ldots, n$ *and* $\beta, \eta \in L$,

$$E_i.\mathcal{O}_q(G)_{\beta,\eta} \subseteq \mathcal{O}_q(G)_{\beta,\eta+\alpha_i} \qquad \mathcal{O}_q(G)_{\beta,\eta}.E_i \subseteq \mathcal{O}_q(G)_{\beta+\alpha_i,\eta}$$
$$F_i.\mathcal{O}_q(G)_{\beta,\eta} \subseteq \mathcal{O}_q(G)_{\beta,\eta-\alpha_i} \qquad \mathcal{O}_q(G)_{\beta,\eta}.F_i \subseteq \mathcal{O}_q(G)_{\beta-\alpha_i,\eta}.$$

Proof. Exercise I.8.C. $\square$

I.8.8. The information in this subsection and the previous one will be combined as follows. Suppose that V is a module in $\mathcal{C}_q(\mathfrak{g}, L)$. Then $C(V)$ is a finite dimensional subspace of $\mathcal{O}_q(G)$, homogeneous with respect to the $L \times L$ grading by Lemma I.8.7. Thus, we can write

$$C(V) = \bigoplus_{\beta,\eta \in L} C(V)_{\beta,\eta}$$

where $C(V)_{\beta,\eta} = C(V) \cap \mathcal{O}_q(G)_{\beta,\eta}$. Since $C(V)$ is finite dimensional, only finitely many of the weight spaces $C(V)_{\beta,\eta}$ can be nonzero.

Lemma I.8.2 says that $C(V)$ is a $U_q(\mathfrak{g})$-$U_q(\mathfrak{g})$-subbimodule of $\mathcal{O}_q(G)$. In particular, it is invariant under right and left multiplication by $K_i^{\pm 1}$, E_i, F_i, and so it is invariant under the automorphisms σ_i, τ_i and the skew derivations δ_i, ∂_i defined in (I.8.3). Given the definitions of σ_i, τ_i, δ_i, ∂_i and $\mathcal{O}_q(G)_{\beta,\eta}$, it follows from Lemma I.8.7 that

$$\sigma_i\big(C(V)_{\beta,\eta}\big) = C(V)_{\beta,\eta} \qquad \delta_i\big(C(V)_{\beta,\eta}\big) \subseteq C(V)_{\beta+\alpha_i,\eta} \qquad (3)$$
$$\tau_i\big(C(V)_{\beta,\eta}\big) = C(V)_{\beta,\eta} \qquad \partial_i\big(C(V)_{\beta,\eta}\big) \subseteq C(V)_{\beta,\eta-\alpha_i}$$

for all i, β, η. Now $C(V)_{\beta+t\alpha_i,\eta}$ and $C(V)_{\beta,\eta-t\alpha_i}$ must be zero for large enough positive integers t, whence δ_i^t and ∂_i^t vanish on $C(V)_{\beta,\eta}$ for $t \gg 0$. Therefore the restrictions of δ_i and ∂_i to $C(V)$ are nilpotent.

$$\mathcal{O}_q(G) \text{ IS A DOMAIN}$$

I.8.9. Theorem. *The algebra $\mathcal{O}_q(G)$ is an integral domain.*

Proof. Suppose that there exist nonzero elements $y, z \in \mathcal{O}_q(G)$ such that $yz = 0$. According to Exercise I.8.D, there is no loss of generality in assuming that $y \in \mathcal{O}_q(G)_{\beta,\eta}$ and $z \in \mathcal{O}_q(G)_{\gamma,\nu}$ for some $\beta, \eta, \gamma, \nu \in L$. In view of Lemma I.7.6, there is a module V in $\mathcal{C}_q(\mathfrak{g}, L)$ such that $y, z \in C(V) \subseteq \mathcal{O}_q(G)$. Thus $y \in C(V)_{\beta,\eta}$ and $z \in C(V)_{\gamma,\nu}$. Since $C(V)$ has only finitely many nonzero weight spaces, we may assume that β, γ (respectively, η, ν) are maximal (respectively, minimal) among weights β', γ' (respectively, η', ν') in L such that $C(V)_{\beta',\eta'}$ and $C(V)_{\gamma',\nu'}$ contain nonzero elements y' and z' with $y'z' = 0$. (Here "maximal" and "minimal" are meant with respect to the partial ordering defined on P in (I.5.3).)

We claim that $y.E_i = z.E_i = 0$ for all i. Fix i for the moment. As we saw in (I.8.8), the restriction of δ_i to $C(V)$ is nilpotent. Hence, there exist nonnegative integers r and s maximal such that $\delta_i^r(y) \neq 0$ and $\delta_i^s(z) \neq 0$. Since $yz = 0$, equation (1) of Lemma I.8.5 shows that

$$\sum_{j=0}^{r+s} \binom{r+s}{j}_{q_i^2} \sigma_i^{r+s-j} \delta_i^j(y) \delta_i^{r+s-j}(z) = 0. \tag{4}$$

Now $\delta_i^j(y) = 0$ for $j > r$, while $\delta_i^{r+s-j}(z) = 0$ for $j < r$, and so the term corresponding to $j = r$ in (4) must also vanish. Further, $\binom{r+s}{r}_{q_i^2} \neq 0$ because q_i^2 is not a root of unity, and thus $\sigma_i^s \delta_i^r(y) \delta_i^s(z) = 0$. In view of inclusions (3), we see that

$$\sigma_i^s \delta_i^r(y) \in C(V)_{\beta+r\alpha_i,\eta} \qquad \text{and} \qquad \delta_i^s(z) \in C(V)_{\gamma+s\alpha_i,\nu}.$$

The maximality of β and γ forces $r = s = 0$, that is, $\delta_i(y) = \delta_i(z) = 0$. This establishes the claim. Similarly, we find that $F_i.y = F_i.z = 0$ for all i.

Now y and z vanish on $E_i U_q(\mathfrak{g})$ and on $U_q(\mathfrak{g})F_i$ for all i. Since $U_q(\mathfrak{g})$ is spanned by products of the form

$$E_{i_1} E_{i_2} \cdots E_{i_r} K^s F_{j_1} F_{j_2} \cdots F_{j_t}$$

with $s \in \mathbb{Z}^n$ (Exercise I.8.E), and since y vanishes on such products whenever $r > 0$ or $t > 0$, we must have $y(K^s) \neq 0$ for some $s \in \mathbb{Z}^n$. Given that $y(xK_i^{\pm 1}) = (K_i^{\pm 1}.y)(x) = q^{\pm(\eta,\alpha_i)}y(x)$ for all $x \in U_q(\mathfrak{g})$ and all i, it follows that $y(1) \neq 0$. Similarly, $z(1) \neq 0$. In other words, $\varepsilon(y) \neq 0$ and $\varepsilon(z) \neq 0$, where ε denotes the counit of $\mathcal{O}_q(G)$. But $\varepsilon : \mathcal{O}_q(G) \to k$ is a k-algebra homomorphism, and so $\varepsilon(yz) \neq 0$, contradicting the assumption that $yz = 0$.

Therefore $\mathcal{O}_q(G)$ is a domain. $\square$

R-MATRICES FOR FINITE DIMENSIONAL $U_q(\mathfrak{g})$-MODULES

If M and N are modules over an ordinary enveloping algebra, or indeed over any cocommutative Hopf algebra, the flip maps provide module isomorphisms between $M \otimes N$ and $N \otimes M$. This no longer holds over a non-cocommutative Hopf algebra such as $U_q(\mathfrak{g})$. However, for modules $M, N \in \mathcal{C}_q(\mathfrak{g})$, the tensor products $M \otimes N$ and $N \otimes M$ are nonetheless isomorphic; the isomorphisms are given by so-called R-matrices which we describe below. We have already seen a specific instance of this in (I.7.12).

Recall the definition of a Hopf pairing from (I.9.22).

I.8.10. Theorem. *There exists a unique Hopf pairing*

$$(-,-) : U_q^{\geq 0}(\mathfrak{g})^{\mathrm{op}} \times U_q^{\leq 0}(\mathfrak{g}) \to k$$

such that

$$(K_i, K_j) = q^{-(\alpha_i, \alpha_j)} \qquad\qquad (K_i, F_j) = 0$$
$$(E_i, F_j) = -\delta_{ij}(q_i - q_i^{-1})^{-1} \qquad\qquad (E_i, K_j) = 0$$

for all i, j.

This pairing is called the *Rosso-Tanisaki form*.

Proof. See, e.g., [**107**, Proposition 6.12] or [**125**, Proposition 6.34]. Relative to [**107**], for instance, we have switched the order of the factors for convenience, and have replaced $U_q^{\geq 0}(\mathfrak{g})$ by its opposite algebra so that the formulas yield a Hopf pairing. $\square$

I.8.11. Definition. In order to state some further properties of the Rosso-Tanisaki form, we need to refer to some weight spaces of $U_q(\mathfrak{g})$. When viewed as a left or right module over itself via the ring multiplication, the weight spaces of $U_q(\mathfrak{g})$ are all zero. Instead, we view $U_q(\mathfrak{g})$ as a module over itself via the adjoint action. Since $\Delta(K_i) = K_i \otimes K_i$ and $S(K_i) = K_i^{-1}$, we have $(\mathrm{ad}\, K_i)(u) = K_i u K_i^{-1}$ for $u \in U_q(\mathfrak{g})$ and $i = 1, \ldots, n$. Hence, the weight spaces for the adjoint action of $U_q(\mathfrak{g})$ on itself are given by

$$U_q(\mathfrak{g})_\lambda = \{u \in U_q(\mathfrak{g}) \mid K_i u K_i^{-1} = q^{(\lambda, \alpha_i)} u \text{ for } i = 1, \ldots, n\}$$

for $\lambda \in Q$.

Observe that $U_q(\mathfrak{g})_\lambda U_q(\mathfrak{g})_\mu \subseteq U_q(\mathfrak{g})_{\lambda+\mu}$ for all $\lambda, \mu \in Q$. The relations for $U_q(\mathfrak{g})$ immediately imply that the $K_j^{\pm 1} \in U_q(\mathfrak{g})_0$, while $E_j \in U_q(\mathfrak{g})_{\alpha_j}$ and $F_j \in U_q(\mathfrak{g})_{-\alpha_j}$ for all j. Hence, there is a decomposition

$$U_q(\mathfrak{g}) = \bigoplus_{\lambda \in Q} U_q(\mathfrak{g})_\lambda$$

which makes $U_q(\mathfrak{g})$ into a Q-graded algebra. This grading restricts to corresponding Q-gradings on the subalgebras $U_q^{\geq 0}(\mathfrak{g})$, $U_q^{\leq 0}(\mathfrak{g})$, and $U_q^{\pm}(\mathfrak{g})$. Observe that $U_q^{\geq 0}(\mathfrak{g})$ and $U_q^+(\mathfrak{g})$ are actually graded by Q^+ (that is, their components of degree λ are zero for $\lambda \notin Q^+$), while $U_q^{\leq 0}(\mathfrak{g})$ and $U_q^-(\mathfrak{g})$ are graded by $-Q^+$. Moreover, $U_q^{\pm}(\mathfrak{g})_0 = k\cdot 1$ and the components $U_q^{\pm}(\mathfrak{g})_\lambda$ are finite dimensional (Exercise I.8.F).

Finally, note that $U_q(\mathfrak{g})_\lambda M_\mu \subseteq M_{\mu+\lambda}$ for $\lambda \in Q$, $\mu \in P$, and any $U_q(\mathfrak{g})$-module M.

I.8.12. Proposition. (a) *For any distinct $\lambda, \mu \in Q^+$, the Rosso-Tanisaki form vanishes on $U_q^+(\mathfrak{g})_\lambda \times U_q^-(\mathfrak{g})_{-\mu}$.*

(b) *The restriction of the Rosso-Tanisaki form to $U_q^+(\mathfrak{g})_\lambda \times U_q^-(\mathfrak{g})_{-\lambda}$ is non-degenerate for each $\lambda \in Q^+$.*

(c) *The Rosso-Tanisaki form is a perfect pairing.*

Proof. (a) Exercise I.8.G.

(b) [**107**, Proposition 6.18 and Corollary 8.30] or [**125**, Proposition 6.37].

(c) Exercise I.8.G. $\square$

I.8.13. Definitions. Let $\lambda \in Q^+$, and choose a basis $u_{\lambda 1}, \ldots, u_{\lambda, m(\lambda)}$ for $U_q^+(\mathfrak{g})_\lambda$. Since the Rosso-Tanisaki form restricts to a nondegenerate bilinear form on the product $U_q^+(\mathfrak{g})_\lambda \times U_q^-(\mathfrak{g})_{-\lambda}$, there is a basis $v_{\lambda 1}, \ldots, v_{\lambda, m(\lambda)}$ for $U_q^-(\mathfrak{g})_{-\lambda}$ such that $(u_{\lambda i}, v_{\lambda j}) = \delta_{ij}$ for all i, j. The element

$$\theta_\lambda = \sum_i u_{\lambda i} \otimes v_{\lambda i} \in U_q^+(\mathfrak{g})_\lambda \otimes U_q^-(\mathfrak{g})_{-\lambda}$$

is called the *canonical element* for this restriction of the Rosso-Tanisaki form. It is independent of the choice of basis for $U_q^+(\mathfrak{g})_\lambda$ (Exercise I.8.H).

Note that since $U_q^+(\mathfrak{g})_0 = k\cdot 1$ (Exercise I.8.F), we must have $\theta_0 = 1 \otimes 1$.

I.8.14. Definitions. Let $M, N \in \mathcal{C}_q(\mathfrak{g})$, and note that

$$\theta_\lambda(M_\mu \otimes N_\nu) \subseteq M_{\mu+\lambda} \otimes N_{\nu-\lambda}$$

for all $\lambda \in Q^+$ and $\mu, \nu \in P$. Since M and N have only finitely many weights, all but finitely many of the elements θ_λ annihilate $M \otimes N$. Thus multiplication by the formal infinite sum $\sum_{\lambda \in Q^+} \theta_\lambda$ on elements of $M \otimes N$ is well-defined, and this multiplication gives us a linear transformation that we label

$$\theta_{M,N} : M \otimes N \to M \otimes N.$$

Now let $\tau_{M,N} : M \otimes N \to N \otimes M$ be the flip, and define a linear transformation $E_{M,N} : M \otimes N \to M \otimes N$ so that

$$E_{M,N}(a \otimes b) = q^{-(\mu,\nu)} a \otimes b \qquad \text{for } a \in M_\mu \text{ and } b \in N_\nu.$$

Finally, we define the *R-matrix for M and N* to be the linear transformation

$$R_{M,N} = \tau_{M,N}\theta_{M,N}E_{M,N} : M \otimes N \to N \otimes M.$$

Written explicitly,

$$R_{M,N}(a \otimes b) = q^{-(\mu,\nu)} \sum_{\lambda \in Q^+} \sum_i v_{\lambda i} b \otimes u_{\lambda i} a$$

$$= q^{-(\mu,\nu)} \left(b \otimes a + \sum_{\substack{\lambda \in Q^+ \\ \lambda \neq 0}} \sum_i v_{\lambda i} b \otimes u_{\lambda i} a \right) \tag{5}$$

for $a \in M_\mu$ and $b \in N_\nu$.

I.8.15. Theorem. *For any modules $M, N \in \mathcal{C}_q(\mathfrak{g})$, the R-matrix $R_{M,N}$ is a $U_q(\mathfrak{g})$-module isomorphism of $M \otimes N$ onto $N \otimes M$.*

Proof. See [**107**, Theorem 7.3]. That theorem actually proves that $\theta_{M,N}E_{M,N}\tau_{N,M}$ is a $U_q(\mathfrak{g})$-module isomorphism. Since

$$R_{M,N} = \tau_{M,N}(\theta_{M,N}E_{M,N}\tau_{N,M})\tau_{M,N},$$

the desired result follows. (See also [**125**, Proposition 8.19].)   □

COMMUTATION RELATIONS IN $\mathcal{O}_q(G)$

Recall from Lemma I.7.11 that if M and N are $U_q(\mathfrak{g})$-modules, then any $U_q(\mathfrak{g})$-module homomorphism from $M \otimes N$ to $N \otimes M$ implies some commutation relations among the coordinate functions of M and N. The isomorphisms $R_{M,N}$ of Theorem I.8.15 yield the following commutation relations. Here we view the duals of left $U_q(\mathfrak{g})$-modules as right $U_q(\mathfrak{g})$-modules in the standard way, without converting them to left $U_q(\mathfrak{g})$-modules using the antipode.

I.8.16. Theorem. *Let $\lambda, \mu \in P^+$ and $\eta, \rho, \beta, \gamma \in P$, and let*

$$u \in V(\lambda)_\eta \qquad v \in V(\mu)_\rho \qquad f \in V(\lambda)^*_\beta \qquad g \in V(\mu)^*_\gamma.$$

Let $u_{\nu i} \in U_q^+(\mathfrak{g})_\nu$ and $v_{\nu i} \in U_q^-(\mathfrak{g})_{-\nu}$ as in Definition I.8.13. Then

$$q^{-(\eta,\rho)}c_{g,v}^{V(\mu)}c_{f,u}^{V(\lambda)} + q^{-(\eta,\rho)} \sum_{\substack{\nu \in Q^+ \\ \nu \neq 0}} \sum_i c_{g,v_{\nu i}v}^{V(\mu)}c_{f,u_{\nu i}u}^{V(\lambda)}$$

$$= q^{-(\beta,\gamma)}c_{f,u}^{V(\lambda)}c_{g,v}^{V(\mu)} + \sum_{\substack{\nu \in Q^+ \\ \nu \neq 0}} q^{(\nu+\beta,\nu-\gamma)} \sum_i c_{fu_{\nu i},u}^{V(\lambda)}c_{gv_{\nu i},v}^{V(\mu)} \tag{6}$$

where all but finitely many terms in the double sums are zero.

Proof. To simplify notation, we omit superscripts $V(\lambda)$ and $V(\mu)$ on coordinate functions in this proof. Recall that we may take $u_{0i} = v_{0i} = 1$ (with only one index i in this case).

Let $R = R_{V(\lambda),V(\mu)}$, which by Theorem I.8.15 is a $U_q(\mathfrak{g})$-module isomorphism of $V(\lambda) \otimes V(\mu)$ onto $V(\mu) \otimes V(\lambda)$. Then

$$c_{g\otimes f, R(u\otimes v)} = c_{R^*(g\otimes f), u\otimes v} \tag{7}$$

by Exercise I.8.J. We shall show that the left and right sides of (7) are equal to the left and right sides of (6), respectively.

The explicit formula (5) for R yields

$$R(u \otimes v) = q^{-(\eta,\rho)} v \otimes u + q^{-(\eta,\rho)} \sum_{\substack{\nu\in Q^+ \\ \nu\neq 0}} \sum_i v_{\nu i} v \otimes u_{\nu i} u,$$

from which we see that $c_{g\otimes f, R(u\otimes v)}$ indeed equals the lefthand side of (6).

In order to work out $c_{R^*(g\otimes f), u\otimes v}$, we first expand $R^*(g \otimes f)$. For any weight vectors $x \in V(\lambda)_\sigma$ and $y \in V(\mu)_\tau$, we compute that

$$R^*(g \otimes f)(x \otimes y) = (g \otimes f)R(x \otimes y)$$
$$= q^{-(\sigma,\tau)} \sum_{\nu\in Q^+} \sum_i g(v_{\nu i}y)f(u_{\nu i}x) \tag{8}$$
$$= \sum_{\nu\in Q^+} \sum_i q^{-(\sigma,\tau)}(fu_{\nu i})(x)(gv_{\nu i})(y).$$

Observe that $fu_{\nu i}$ vanishes on all weight spaces of $V(\lambda)$ with weights different from $-\beta - \nu$, and that $gv_{\nu i}$ vanishes on weight spaces with weights different from $-\gamma + \nu$. Hence, for each term $q^{-(\sigma,\tau)}(fu_{\nu i})(x)(gv_{\nu i})(y)$ in (8), the exponent $-(\sigma,\tau)$ may be replaced by $-(-\beta - \nu, -\gamma + \nu) = (\nu + \beta, \nu - \gamma)$. Since the resulting sum is independent of the weights of x and y, we obtain

$$R^*(g \otimes f) = q^{-(\beta,\gamma)} f \otimes g + \sum_{\substack{\nu\in Q^+ \\ \nu\neq 0}} \sum_i q^{(\nu+\beta,\nu-\gamma)} fu_{\nu i} \otimes gv_{\nu i}.$$

It follows that $c_{R^*(g\otimes f), u\otimes v}$ equals the righthand side of (6), completing the proof. $\square$

We shall see that the equations (6) suffice to show that $\mathcal{O}_q(G)$ is noetherian.

$\mathcal{O}_q(G)$ IS NOETHERIAN

The following proposition is an application of the filtered-graded method discussed in Appendix I.12.

I.8.17. Proposition. *Let A be a k-algebra generated by a finite sequence of elements $u_1, \ldots, u_m$, and suppose that there exist scalars $q_{ij} \in k^\times$ and $\alpha_{ij}^{st}, \beta_{ij}^{st} \in k$ such that*

$$u_i u_j = q_{ij} u_j u_i + \sum_{s=1}^{j-1} \sum_{t=1}^{m} (\alpha_{ij}^{st} u_s u_t + \beta_{ij}^{st} u_t u_s) \qquad (9)$$

for $1 \leq j < i \leq m$. Then A is noetherian.

Proof. We claim that there is a nonnegative filtration $(A_d)_{d \geq 0}$ on A such that $\operatorname{gr} A$ is generated as a k-algebra by elements $y_1, \ldots, y_m$ satisfying $y_i y_j = q_{ij} y_j y_i$ for all $i > j$. Then $\operatorname{gr} A$ is a homomorphic image of a quantum affine space $\mathcal{O}_q(k^m)$, whence $\operatorname{gr} A$ is noetherian, and therefore A is noetherian by Theorem I.12.13.

The trick is to assign appropriate degrees to the u_i, that is, to pick positive integers $d_1, \ldots, d_m$ so that with $u_i \in A_{d_i}$ for each i, the terms in the double sum in (9) will always end up in $A_{d_i + d_j - 1}$. This means that we need to have

$$d_s + d_t < d_i + d_j \qquad \text{for } i > j > s \text{ and all } t. \qquad (10)$$

There are many possibilities; we show that the choices $d_i = 2^m - 2^{m-i}$ satisfy these conditions. First, observe that for $1 < j < m$ we have

$$2^{m-j-1} + 2^{m-j} = 3 \cdot 2^{m-j-1} < 2^{m-j+1} + 1,$$

and consequently

$$d_{j-1} + d_m = 2^{m+1} - 2^{m-j+1} - 1 < 2^{m+1} - 2^{m-j-1} - 2^{m-j} = d_{j+1} + d_j.$$

Hence, for $i > j > s$ and any t, we obtain

$$d_s + d_t \leq d_{j-1} + d_m < d_{j+1} + d_j \leq d_i + d_j,$$

and (10) is established.

Now set $A_0 = k \cdot 1$ and for $d > 0$, let A_d be the k-subspace of A spanned by the set

$$\{u_{i_1} u_{i_2} \cdots u_{i_r} \mid d_{i_1} + d_{i_2} + \cdots + d_{i_r} \leq d\}.$$

It is clear that this defines a nonnegative filtration of A. Each u_i lies in A_{d_i}, and so we can define the element

$$y_i := u_i + A_{d_i - 1} \in (\operatorname{gr} A)_{d_i}.$$

Any homogeneous component $(\operatorname{gr} A)_d$, if it is nonzero, is spanned by the cosets of those products $u_{i_1} u_{i_2} \cdots u_{i_r}$ such that $d_{i_1} + d_{i_2} + \cdots + d_{i_r} = d$, and any such coset equals $y_{i_1} y_{i_2} \cdots y_{i_r}$. Thus the y_i generate $\operatorname{gr} A$.

By hypothesis, for $i > j$ the element $u_i u_j - q_{ij} u_j u_i$ is a linear combination of products $u_s u_t$ and $u_t u_s$ with $s < j$. Condition (10) then implies that $y_i y_j = q_{ij} y_j y_i$. This verifies the claim and therefore concludes the proof. $\square$

I.8.18. Theorem. *The algebra $\mathcal{O}_q(G)$ is noetherian.*

Proof. As we saw in Proposition I.7.8, L^+ is a finitely generated semigroup, and if $\lambda_1, \ldots, \lambda_r$ generate L^+, then the coordinate functions of the $V(\lambda_i)$ generate $\mathcal{O}_q(G)$. For each $i = 1, \ldots, r$, let B_i and B_i^* be bases for $V(\lambda_i)$ and $V(\lambda_i)^*$, respectively, consisting of weight vectors. Then the finite set

$$X = \{ c_{f,u}^{V(\lambda_i)} \mid i = 1, \ldots, r, \ f \in B_i^*, \text{ and } u \in B_i \}$$

generates $\mathcal{O}_q(G)$. For $\beta, \eta \in P$, let

$$X(\beta, \eta) = \{ c_{f,u}^{V(\lambda_i)} \in X \mid f \in V(\lambda_i)_\beta^* \text{ and } u \in V(\lambda_i)_\eta \}.$$

Since the bases B_i and B_i^* consist of weight vectors, X is the union of the $X(\beta, \eta)$.

Let $\mathcal{B}$ denote the union of the weight sets of the $V(\lambda_i)^*$. Since $\mathcal{B}$ is finite, its elements can be listed in a sequence $\beta_1, \ldots, \beta_b$ in such a way that each β_l is maximal in $\mathcal{B} \setminus \{ \beta_1, \ldots, \beta_{l-1} \}$. Hence,

$$\beta_l < \beta_{l'} \implies l > l'$$

for any $l, l' \in \{ 1, \ldots, b \}$. Similarly, the union of the weight sets of the $V(\lambda_i)$ can be listed in a sequence $\eta_1, \ldots, \eta_c$ such that

$$\eta_h < \eta_{h'} \implies h > h'$$

for any $h, h' \in \{ 1, \ldots, c \}$.

Now X is the disjoint union of the subsets $X(\beta_l, \eta_h)$. List these subsets in lexicographic order with respect to the index pairs (l, h), that is,

$$X(\beta_1, \eta_1), \ X(\beta_1, \eta_2), \ \ldots, \ X(\beta_1, \eta_c), \ X(\beta_2, \eta_1), \ \ldots, \ X(\beta_2, \eta_c), \ \ldots,$$
$$X(\beta_b, \eta_1), \ \ldots, \ X(\beta_b, \eta_c).$$

Then list the elements of each $X(\beta_l, \eta_h)$ arbitrarily, and concatenate these lists according to the order above. We thus have listed the elements of X in a sequence $x_1, \ldots, x_m$ with the following property:

(∗) Suppose that $x_j \in X(\beta_l, \eta_h)$ and $x_s \in X(\beta_{l'}, \eta_{h'})$. If either $\beta_l < \beta_{l'}$, or $\beta_l = \beta_{l'}$ and $\eta_h < \eta_{h'}$, then $s < j$.

Because of (∗), Theorem I.8.16 shows that there exist $q_{ij} \in k^\times$ and $\alpha_{ij}^{st}, \beta_{ij}^{st} \in k$ such that

$$x_i x_j = q_{ij} x_j x_i + \sum_{s=1}^{j-1} \sum_{t=1}^{m} (\alpha_{ij}^{st} x_s x_t + \beta_{ij}^{st} x_t x_s)$$

for $1 \leq j < i \leq m$. Therefore Proposition I.8.17 implies that $\mathcal{O}_q(G)$ is noetherian. $\square$

NOTES

That $\mathcal{O}_q(G)$ is a noetherian domain was proved by Joseph assuming either q is transcendental over $\mathbb{Q}$ or $q \in \mathbb{C}^\times$ is not a root of unity [115, 116]; we have followed his method in proving Theorem I.8.9. The existence of a Hopf pairing similar to the Rosso-Tanisaki form was first proved by Drinfel'd in the case of $U_\hbar(\mathfrak{g})$ [51]; the version adapted to $U_q(\mathfrak{g})$ is due to Rosso [189] and Tanisaki [203]. It has long been known that R-matrix isomorphisms between tensor products of $U_q(\mathfrak{g})$-modules imply commutation relations for coordinate functions in $\mathcal{O}_q(G)$. Key relations of this form were computed modulo certain ideals by Soibelman [197] and Levendorskii-Soibelman [140]; complete relations, as in Theorem I.8.16, were given by Hodges-Levasseur-Toro [96]. Apparently new here is the use of these relations to prove that $\mathcal{O}_q(G)$ is noetherian; in doing so, we have adapted some of the refiltering techniques of Bueso, Gómez-Torrecillas and Lobillo [26] to prove Proposition I.8.17.

EXERCISES

Exercise I.8.A. Suppose that V is a module in $\mathcal{C}_q(\mathfrak{g}, L)$, and let $f \in V_\beta^*$ and $v \in V_\eta$ for some $\beta, \eta \in L$. Show that $c_{f,v}^V \in \mathcal{O}_q(G)_{\beta,\eta}$. □

Exercise I.8.B. Define weight spaces for $U_q(\mathfrak{g})^\circ$ just as in (I.8.6):

$$U_q(\mathfrak{g})^\circ_{\beta,\eta} = \{y \in U_q(\mathfrak{g})^\circ \mid y.K_i^{-1} = q^{(\beta,\alpha_i)}y \text{ and } K_i.y = q^{(\eta,\alpha_i)}y$$
$$\text{for all } i = 1, \ldots, n\}$$

for $\beta, \eta \in P$. Show that $\mathcal{O}_q(G) = \bigoplus_{\beta,\eta \in L} U_q(\mathfrak{g})^\circ_{\beta,\eta}$. [Hints: Given $y \in U_q(\mathfrak{g})^\circ_{\beta,\eta}$ with $\beta, \eta \in L$, let I be the largest left ideal of $U_q(\mathfrak{g})$ contained in $\ker(y)$, and note that $V = U_q(\mathfrak{g})/I$ is a finite dimensional $U_q(\mathfrak{g})$-module. Show that the coset $v = 1 + I$ has weight η, and infer that $V \in \mathcal{C}_q(\mathfrak{g}, L)$. Finally, express y as a coordinate function of V.] □

Exercise I.8.C. Prove Lemma I.8.7. (Recall from (I.12.7) the definition of a grading of an algebra by a group.) □

Exercise I.8.D. Let A be an algebra graded by a free abelian group G of finite rank. Pick a total ordering on G which is compatible with the group operation (e.g., the lexicographic ordering with respect to some choice of basis), and define the *highest term* of any nonzero element $y \in A$ to be the homogeneous component y_g where g is the greatest element of G (with respect to the chosen ordering) for which $y_g \neq 0$. Show that if y and z are nonzero elements of A with highest terms y_g and z_h, then $(yz)_{gh} = y_g z_h$. Conclude that if A is not a domain, then it must contain two nonzero homogeneous elements whose product is zero. □

Exercise I.8.E. Show that $U_q(\mathfrak{g})$ is spanned by products of the form

$$E_{i_1} E_{i_2} \cdots E_{i_r} K_{l_1}^{\pm 1} K_{l_2}^{\pm 1} \cdots K_{l_s}^{\pm 1} F_{j_1} F_{j_2} \cdots F_{j_t}$$

(where r, s, t are allowed to be zero). $\quad\square$

Exercise I.8.F. Show that $U_q^{\pm}(\mathfrak{g})_0 = k \cdot 1$, that each $U_q^{\pm}(\mathfrak{g})_\lambda$ is finite dimensional, and that $U_q^{\pm}(\mathfrak{g})_\lambda = 0$ for $\lambda \notin \pm Q^+$. Show also that $U_q^{\geq 0}(\mathfrak{g})_\lambda = U_q^{\leq 0}(\mathfrak{g})_{-\lambda} = 0$ for $\lambda \notin Q^+$, while $U_q^{\geq 0}(\mathfrak{g})_0 = U_q^{\leq 0}(\mathfrak{g})_0 = k[K_1^{\pm 1}, \ldots, K_n^{\pm 1}]$. $\quad\square$

Exercise I.8.G. Prove parts (a) and (c) of Proposition I.8.12. [Hint: For (c), use the last part of the previous exercise.] $\quad\square$

Exercise I.8.H. Let $(-,-) : U \times V \to k$ be a nondegenerate bilinear form where U and V are finite dimensional vector spaces over k. Show that there is an isomorphism $\phi : V \to U^*$ such that $\phi(v)(u) = (u,v)$ for $v \in V$ and $u \in U$. Consequently, there is a vector space isomorphism $\widetilde{\phi} : U \otimes V \to \operatorname{End}_k(U)$ such that $\widetilde{\phi}(u \otimes v)(u') = (u',v)u$ for $u, u' \in U$ and $v \in V$. If $u_1, \ldots, u_m$ is a basis for U and $v_1, \ldots, v_m$ is the corresponding dual basis for V with respect to $(-,-)$, show that $\widetilde{\phi}(\sum_{i=1}^m u_i \otimes v_i)$ equals the identity map in $\operatorname{End}_k(U)$. Conclude that the *canonical element* $\sum_{i=1}^m u_i \otimes v_i$ is independent of the choice of the basis $\{u_i\}$. $\quad\square$

Exercise I.8.I. Show that the functionals $f u_{\nu i}$ and $g v_{\nu i}$ occurring in Theorem I.8.16 lie in $U_q^{\geq 0}(\mathfrak{g})f$ and $U_q^{\leq 0}(\mathfrak{g})g$, respectively. [Hint: Observe that $S(U_q^{\geq 0}(\mathfrak{g})) = U_q^{\geq 0}(\mathfrak{g})$ and $S(U_q^{\leq 0}(\mathfrak{g})) = U_q^{\leq 0}(\mathfrak{g})$.] $\quad\square$

Exercise I.8.J. Let M and N be modules over a Hopf algebra H. Let $u \in M$ and $f \in N^*$, let $\phi : M \to N$ be an H-module homomorphism, and let $\phi^* : N^* \to M^*$ be the transpose map. Show that $c_{f, \phi(u)}^N = c_{\phi^*(f), u}^M$. $\quad\square$

APPENDIX I.9

BIALGEBRAS AND HOPF ALGEBRAS

As usual, our base field k will remain fixed throughout, and unadorned tensor products $\otimes$ will mean $\otimes_k$. Given vector spaces V and W, the *flip map* $\tau : V \otimes W \to W \otimes V$ is the linear transformation sending $v \otimes w \mapsto w \otimes v$ for $v \in V$ and $w \in W$.

COALGEBRAS

The definition of a k-coalgebra is obtained by dualising the familiar definition of a k-algebra. To achieve this, the latter should be thought of as a k-vector space A equipped with k-linear maps *multiplication* $m : A \otimes A \to A$ (sending $a \otimes b \mapsto ab$) and *unit* $u : k \to A$ (sending $\alpha \mapsto \alpha \cdot 1$) satisfying various axioms which can be expressed in the form of commutative diagrams.

I.9.1. Definitions. A *k-coalgebra* (with *counit*) is a k-vector space C together with k-linear maps, *comultiplication* or *coproduct* $\Delta : C \to C \otimes C$ and *counit* $\varepsilon : C \to k$, such that the following diagrams are commutative.

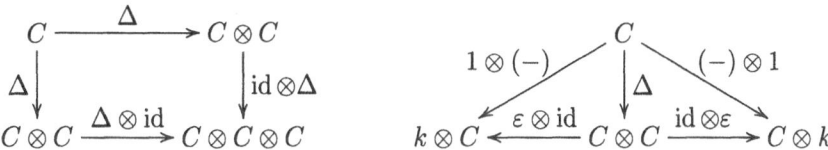

We'll write Δ_C and ε_C to denote the coproduct and counit on the coalgebra C, but will often omit the suffices when no confusion is likely. A *morphism* f from a coalgebra C to a coalgebra D is a linear map such that $\Delta_D \circ f = (f \otimes f) \circ \Delta_C$ and $\varepsilon_D \circ f = \varepsilon_C$. A subspace I of C is a *coideal* if $\Delta(I) \subseteq C \otimes I + I \otimes C$ and $\varepsilon(I) = 0$. In this case, C/I is a coalgebra with the induced coproduct and counit. A *cocommutative* coalgebra is one for which $\tau \circ \Delta = \Delta$.

I.9.2. Notation. Following the *Sweedler summation notation*, we shall denote the coproduct $\Delta(c)$ of an element c of a coalgebra C by $\sum c_1 \otimes c_2$. Here c_1 and c_2 refer to variable elements of C, not uniquely determined. The subscripts 1 and 2 indicate the positions of these elements in the tensor product expression for $\Delta(c)$. Several variants of this notation are also in common use: for example one can put brackets around the suffices, write $\sum_{(c)}$ in place of $\sum$, or adopt a more minimalist approach by omitting the summation symbol altogether.

As an example of the use of the Sweedler notation, observe that the counit axiom can be expressed in the form $\sum \varepsilon(c_1)c_2 = \sum c_1\varepsilon(c_2) = c$ for $c \in C$.

I.9.3. It's already apparent from their definitions that there is a duality between algebras and coalgebras. To make this clear, we write V^* for the vector space dual $\mathrm{Hom}_k(V, k)$ of a vector space V. If $\alpha : V \to W$ is linear then its *transpose* $\alpha^* : W^* \to V^*$ is the linear map given by $\alpha^*(f) = f \circ \alpha$ for $f \in W^*$.

I.9.4. Lemma. *If C is a coalgebra then C^* is an algebra with multiplication Δ^* (that is, $\Delta^*|_{C^* \otimes C^*}$) and unit ε^*. If C is cocommutative then C^* is commutative.* □

The multiplication Δ^* in C^* is called *convolution* and is sometimes written in the form $f * f'$; explicitly, $(f * f')(c) = \sum f(c_1)f'(c_2)$ for $c \in C$. However, we will mostly write ff' rather than $f * f'$.

I.9.5. Definition. The dual of Lemma I.9.4 fails since, if V is an infinite dimensional vector space, $V^* \otimes V^*$ is a proper subspace of $(V \otimes V)^*$, so that the dual of the multiplication map on an infinite dimensional algebra A need not take all values in $A^* \otimes A^*$. This is remedied by defining the *finite dual* or *Hopf dual* of the k-algebra A to be

$$A^\circ = \{f \in A^* : f(I) = 0 \text{ for some ideal } I \text{ of } A \text{ with } \dim_k(A/I) < \infty\}.$$

Then Lemma I.9.4 still applies with C° replacing C^*, and moreover we have

I.9.6. Lemma. *If A is an algebra with multiplication m and unit u than A° is a coalgebra with $\Delta = m^*$ and $\varepsilon = u^*$. If A is commutative then A° is cocommutative.* □

BIALGEBRAS

I.9.7. Definitions. A k-*bialgebra* is a k-vector space B equipped with linear maps m, u, Δ, ε such that (B, m, u) is an algebra, (B, Δ, ε) is a coalgebra, and either

(1) Δ and ε are algebra morphisms, or
(2) m and u are coalgebra morphisms.

It is easy to see that (1) and (2) are equivalent conditions. A map $f : B \to D$ of bialgebras is a *bialgebra morphism* if it is both an algebra and a coalgebra morphism. The definition and construction of *biideals* and *quotient bialgebras* is clear.

A bialgebra B has a canonical *trivial module*, namely the field k equipped with the B-module structure given by the counit ε; thus $b.\alpha = \varepsilon(b)\alpha$ for $b \in B$ and $\alpha \in k$.

I.9.8. The comultiplication in a bialgebra B allows tensor products of B-modules to be made into B-modules. Suppose that V and W are left B-modules, and view the module multiplication operations as algebra homomorphisms $\mu_V : B \to \mathrm{End}_k(V)$ and $\mu_W : B \to \mathrm{End}_k(W)$. Then there is an algebra homomorphism

$$B \xrightarrow{\Delta} B \otimes B \xrightarrow{\mu_V \otimes \mu_W} \mathrm{End}_k(V) \otimes \mathrm{End}_k(W) \xrightarrow{\subseteq} \mathrm{End}_k(V \otimes W),$$

which turns $V \otimes W$ into a left B-module. The formula for the module multiplication is $b(v \otimes w) = \sum b_1 v \otimes b_2 w$ for $b \in B$, $v \in V$, and $w \in W$.

HOPF ALGEBRAS

I.9.9. Definitions. A bialgebra $(H, m, u, \Delta, \varepsilon)$ is a *Hopf algebra* if there exists a k-linear map $S : H \to H$ such that

$$m \circ (S \otimes \mathrm{id}) \circ \Delta = m \circ (\mathrm{id} \otimes S) \circ \Delta = u \otimes \varepsilon,$$

that is,

$$\sum (Sh_1)h_2 = \varepsilon(h)1_H = \sum h_1(Sh_2)$$

for all $h \in H$. The map S is called the *antipode* of H. Note that the defining condition on the antipode is equivalent to requiring that S is an inverse to id_H under the *convolution product* $*$, defined on $f, g \in \mathrm{Hom}_k(H, H)$ (as in H^*) by

$$(f * g)(h) = \sum f(h_1)g(h_2)$$

for $h \in H$. Hence, the antipode in a Hopf algebra is unique. A map of Hopf algebras $f : H \to G$ is a *morphism of Hopf algebras* if it is a bialgebra morphism such that $f \circ S_H = S_G \circ f$. *Hopf ideals* and *quotient Hopf algebras* have the obvious definitions.

The antipode of a Hopf algebra is an algebra antihomomorphism and a coalgebra antihomomorphism (e.g., [**122**, Theorem III.3.4]). It need not be bijective, though all the examples of Hopf algebras with which we will be concerned have bijective antipodes. If H is a Hopf algebra, then so is the Hopf dual H°, with antipode S^*.

I.9.10. Just as the comultiplication in a bialgebra allows tensor products of modules to become modules, the antipode in a Hopf algebra H allows (vector space) duals of H-modules to become H-modules on the same side. In particular, if V is a left H-module, then V^* becomes a left H-module, with $(hf)(v) = f((Sh)v)$ for $h \in H$, $f \in V^*$, and $v \in V$.

I.9.11. Examples. Note that the axioms ensure that, for a Hopf algebra H, the maps Δ, ϵ and S are determined by their values on a set of algebra generators for H.

(a) Let $\mathfrak{g}$ be a k-Lie algebra. Then the *enveloping algebra* $U(\mathfrak{g})$ is a Hopf algebra in which

$$\Delta(x) = x \otimes 1 + 1 \otimes x \qquad\qquad Sx = -x \qquad\qquad \varepsilon(x) = 0$$

for $x \in \mathfrak{g}$. For example, let $\mathfrak{g} = \mathfrak{sl}_2(k)$, with k-basis

$$e = \begin{pmatrix} 0 & 1 \\ 0 & 0 \end{pmatrix} \qquad\qquad f = \begin{pmatrix} 0 & 0 \\ 1 & 0 \end{pmatrix} \qquad\qquad h = \begin{pmatrix} 1 & 0 \\ 0 & -1 \end{pmatrix}.$$

Then $U(\mathfrak{g})$ is the k-algebra with generators e, f and h, and relations

$$he - eh = 2e \qquad\qquad hf - fh = -2f \qquad\qquad ef - fe = h.$$

(b) Let G be a group. The *group algebra* kG is a Hopf algebra with

$$\Delta(g) = g \otimes g \qquad\qquad Sg = g^{-1} \qquad\qquad \varepsilon(g) = 1$$

for $g \in G$. For instance, the simplest infinite noncommutative example is the group algebra over k of the *infinite dihedral group*

$$G = \langle a, x : xax = a^{-1},\ x^2 = 1 \rangle.$$

In this case kG is the k-algebra generated by a, a^{-1}, x, subject to the above relations and, of course, $aa^{-1} = 1 = a^{-1}a$.

(c) The algebra $\mathcal{O}(M_n(k)) = k[X_{ij} : 1 \leq i, j \leq n]$ of *polynomial functions on the space of $n \times n$ matrices over k* is a bialgebra. As an algebra, $\mathcal{O}(M_n(k))$ is the polynomial k-algebra on the n^2 indeterminates X_{ij}, where, of course, X_{ij} denotes the $(i, j)^{th}$ coordinate function on $M_n(k)$. The matrix operations in $M_n(k)$ induce a bialgebra structure on $\mathcal{O}(M_n(k))$. So the coproduct is the dual of matrix multiplication – that is, $\Delta(X_{ij}) = \sum_{t=1}^{n} X_{it} \otimes X_{tj}$, and $\varepsilon(X_{ij}) = \delta_{ij}$.

One can check that the determinant D of the $n \times n$ matrix (X_{ij}) satisfies $\varepsilon(D) = 1$ and $\Delta(D) = D \otimes D$. (We say that D is a *grouplike* element of $\mathcal{O}(M_n(k))$ – see (I.9.24).) Thus $\langle D - 1 \rangle$ is a biideal of $\mathcal{O}(M_n(k))$, and the factor bialgebra

$$\mathcal{O}(SL_n(k)) = \mathcal{O}(M_n(k))/\langle D - 1 \rangle$$

is a Hopf algebra, with SX_{ij} the $(ij)^{th}$ entry of $(X_{ij})^{-1}$ (modulo $D - 1$). In particular, $\mathcal{O}(SL_2(k)) = k[a, b, c, d : ad - bc = 1]$, where we have written a for X_{11}, b for X_{12} and so on. In this notation,

$$\begin{pmatrix} \Delta(a) & \Delta(b) \\ \Delta(c) & \Delta(d) \end{pmatrix} = \begin{pmatrix} a \otimes a + b \otimes c & a \otimes b + b \otimes d \\ d \otimes c + c \otimes a & d \otimes d + c \otimes b \end{pmatrix}$$

$$\begin{pmatrix} \varepsilon(a) & \varepsilon(b) \\ \varepsilon(c) & \varepsilon(d) \end{pmatrix} = \begin{pmatrix} 1 & 0 \\ 0 & 1 \end{pmatrix} \qquad\qquad \begin{pmatrix} Sa & Sb \\ Sc & Sd \end{pmatrix} = \begin{pmatrix} d & -b \\ -c & a \end{pmatrix}.$$

Similarly, one can extend Δ and ε from $\mathcal{O}(M_n(k))$ to the localization

$$\mathcal{O}(M_n(k))[D^{-1}] = \mathcal{O}(GL_n(k)),$$

and this bialgebra too becomes a Hopf algebra with a similar description of the antipode. These Hopf algebras, respectively the coordinate rings of $SL_n(k)$ and $GL_n(k)$, are special cases of the next class of examples.

(d) Let G be an *affine algebraic group over k* – that is, a group which is also a k-affine algebraic variety such that the group operations of multiplication and taking inverses are morphisms of algebraic varieties. Then the *coordinate ring* $\mathcal{O}(G)$ *of G* (that is, the algebra of polynomial functions on G), is a Hopf algebra with

$$\Delta(f)(x,y) = f(xy) \qquad \varepsilon(f) = f(1_G) \qquad (Sf)(x) = f(x^{-1})$$

for $f \in \mathcal{O}(G)$ and $x, y \in G$. (Note that in defining Δ we are identifying the algebra $\mathcal{O}(G) \otimes \mathcal{O}(G)$ with $\mathcal{O}(G \times G)$.)

Of the above examples, (a) and (b) are cocommutative, but are commutative only when the Lie algebra or group involved is Abelian; and (c) and (d) are commutative but not cocommutative unless $n = 1$ in (c) or the group G is Abelian in (d).

I.9.12. Definitions. Let H be a Hopf algebra. The *(left) adjoint action of H on itself* is given by

$$\mathrm{ad}(h)(k) = \mathrm{ad}^\ell(h)(k) = \sum h_1 k S(h_2),$$

for all $h, k \in H$. The *right adjoint action* is given by

$$\mathrm{ad}^r(h)(k) = \sum S(h_1)kh_2,$$

for all $h, k \in H$. By the fact that S is an antihomomorphism it is clear that ad (respectively, ad^r) is a k-algebra homomorphism (respectively, anti-homomorphism) from H to the ring of linear transformations $\mathrm{End}_k(H)$, where multiplication in the latter is given by composition of maps. A subset of H which is closed under one or the other of these actions is called *left* or *right normal*, as appropriate, and *normal* if it is closed under both actions.

I.9.13. Examples. (a) When $H = kG$ is a group algebra, a subset A of H is normal if and only if $g^{-1}Ag \subseteq A$ for all $g \in G$.

(b) When $H = U(\mathfrak{g})$ is an enveloping algebra, a subset A of H is normal if and only if $xa - ax \in A$ for all $a \in A$, $x \in \mathfrak{g}$.

I.9.14. Definitions. It is sometimes useful to twist the adjoint actions by an automorphism τ of H. This yields the *left and right twisted adjoint actions* ad_τ and ad_τ^r, given by the rules

$$\mathrm{ad}_\tau(h)(k) = \sum \tau(h_1)kS(h_2) \qquad \text{and} \qquad \mathrm{ad}_\tau^r(h)(k) = \sum S(h_1)k\tau(h_2).$$

Comodules

I.9.15. Definitions. The definition of a comodule is got by dualising the definition of a module. Thus, for a coalgebra C, a *right C-comodule* is a k-vector space V with a linear map $\rho : V \to V \otimes C$ such that the following diagrams commute:

There is an analogous definition of a *left comodule*. The Sweedler notation for a right comodule takes the form $\rho(v) = \sum v_0 \otimes v_1 \in V \otimes C$. Suffices -1 and 0 are used for left comodules, adhering to the convention that non-zero indices indicate elements of C. If V and W are right C-comodules with structure maps ρ_V and ρ_W respectively, then a linear map $f : V \to W$ is a *comodule morphism* if $\rho_W \circ f = (f \otimes \mathrm{id}) \circ \rho_V$.

The most basic example of a comodule for a coalgebra C is C itself, which is both a right and a left C-comodule thanks to the coproduct Δ. Any subspace M of C such that $\Delta(M) \subseteq M \otimes C$ is also a comodule; it's a *subcomodule* of C in an obvious sense, and is called a *right coideal* of C. Similar remarks apply on the left.

The connection between modules and comodules is clarified by

I.9.16. Lemma. *Let A be an algebra and C a coalgebra.*

(a) If (V, ρ) is a right C-comodule, then V becomes a left C^*-module with $f.v = \sum f(v_1)v_0$ for $f \in C^*$ and $v \in V$.

(b) Let V be a left A-module. Then V can be made into a right $A°$-comodule whose associated left A-module (as above) is V if and only if $\dim_k(Av) < \infty$ for all $v \in V$ (that is, V is a locally finite dimensional A-module).

Proof. (a) This is a straightforward calculation.

(b) If V is a right $A°$-module whose associated left A-module is V, local finite dimensionality follows from the formula in (a) – for $v \in V$, the submodule Av is contained in the k-span of the v_0s. Conversely, assume that V is a locally finite-dimensional A-module, and let $v \in V$, with $v_1, \ldots, v_m$ a k-basis for Av. (Here the subscripts $1, \ldots, m$ are ordinary indices, not Sweedler notation.) Define elements f_i of A^* by $av = \sum_{i=1}^m f_i(a)v_i$, for $a \in A$. Then $f_i \in A°$, and we define $\rho(v) = \sum v_i \otimes f_i \in V \otimes A°$. One checks that $\rho(v)$ is independent of the choice of basis for Av, and that ρ defines the desired right $A°$-comodule structure on V. $\square$

I.9.17. Dually to the situation in (I.9.8), tensor products of comodules over a bialgebra B become comodules. Thus, if (V, ρ_V) and (W, ρ_W) are right B-comodules, then the map

$$V \otimes W \xrightarrow{\rho_V \otimes \rho_W} V \otimes B \otimes W \otimes B \xrightarrow{\mathrm{id} \otimes \tau \otimes \mathrm{id}} V \otimes W \otimes B \otimes B \xrightarrow{\mathrm{id} \otimes \mathrm{id} \otimes m_B} V \otimes W \otimes B$$

gives a right B-comodule structure on $V \otimes W$. Note that $v \otimes w \mapsto \sum v_0 \otimes w_0 \otimes v_1 w_1$ for $v \in V$, $w \in W$.

MODULE AND COMODULE ALGEBRAS

I.9.18. Definitions. Let H be a bialgebra and A an algebra. A *left H-module algebra structure on A* is a linear map $\lambda : H \otimes A \to A$ such that

(1) A is a left H-module via $h.a = \lambda(h \otimes a)$ for $h \in H$ and $a \in A$;
(2) $h.(ab) = \sum (h_1.a)(h_2.b)$ for $h \in H$ and $a, b \in A$; and
(3) $h.1_A = \varepsilon(h)1_A$.

Note that (2) and (3) simply say that the multiplication and unit maps m_A and u_A on A are H-module maps (where $A \otimes A$ is made into a left H-module as in (I.9.8), and k is viewed as the trivial H-module). So it is clear how to dualise the definition: A *right H-comodule algebra structure* on an algebra A is a map $\rho : A \to A \otimes H$ such that

(4) (A, ρ) is a right H-comodule, and
(5) m_A and u_A are right H-comodule maps.

It's not hard to check [**122**, Proposition III.7.2] that, given condition (4), A is a right H-comodule algebra if and only if

(5') ρ is an algebra homomorphism.

The correspondences of Lemma I.9.16 extend to the present context. Thus one can easily prove

I.9.19. Lemma. *Let H be a bialgebra and A an algebra.*

(a) If A has a right H-comodule algebra structure, then it has a left H°-module algebra structure.

(b) There exists a left H-module algebra structure on A with $\dim_k H.a < \infty$ for all $a \in A$ if and only if there exists a right H°-comodule algebra structure on A. $\square$

In checking particular examples it's often useful to have the following easy results available.

I.9.20. Lemma. *Let H be a bialgebra generated (as an algebra) by a subset X, and A an algebra generated by a subset Y.*

(a) Suppose that A is a left H-module with $h.1_A = \varepsilon(h)1_A$ for all $h \in H$. Then A is an H-module algebra if and only if $x.(ab) = \sum (x_1.a)(x_2.b)$ for all $x \in X$ and $a, b \in A$.

(b) Let $\rho_A : A \to A \otimes H$ be an algebra homomorphism. Then A is a right H-comodule algebra if and only if ρ_A satisfies the comodule conditions with the elements of Y in its argument. $\square$

I.9.21. Examples. **(a)** Every bialgebra H is itself a right (or left) H-comodule algebra, with $\rho = \Delta$. And by Lemma I.9.19, this means that H is a left and right H°-module algebra – explicitly, the actions are

$$f.h = \sum h_1 f(h_2) \qquad \text{and} \qquad h.f = \sum f(h_1) h_2$$

for $h \in H$ and $f \in H^\circ$. These actions are sometimes called the left and right *hit* actions, denoted $f \rightharpoonup h$ and $h \leftharpoonup f$.

(b) Generalising the first part of (a), if $\pi : H \to F$ is a morphism of bialgebras, then H is a right F-comodule algebra via $(\text{id}_H \otimes \pi) \circ \Delta_H$. A similar definition applies on the left.

(c) Let H be kG, a group algebra. From Lemma I.9.20(a), since $\Delta(g) = g \otimes g$ for $g \in G$, the H-module algebras are simply the algebras A for which there is a homomorphism of groups from G to $\text{Aut}_{k\text{-alg}}(A)$. And the (right) kG-comodule algebras are precisely the G-graded algebras A (see Definitions I.12.7), with graded components $A_g = \{a \in A : \rho_A(a) = a \otimes g\}$.

(d) Let U be the enveloping algebra of a k-Lie algebra $\mathfrak{g}$. In this case Lemma I.9.20(a) shows that the U-module algebras are simply those k-algebras A admitting a $\mathfrak{g}$-module structure with the elements of $\mathfrak{g}$ acting as derivations, i.e., admitting a Lie algebra homomorphism $\mathfrak{g} \to \text{Der}_k(A, A)$.

(e) Let $\mathfrak{g} = \mathfrak{sl}_2(k)$ with notation as in Example I.9.11(a), let $U = U(\mathfrak{g})$, and let $A = k[x, y] = S(V)$, the symmetric algebra of the two-dimensional irreducible U-module $V = kx \oplus ky$, where

$$e.x = 0 \qquad e.y = x \qquad f.x = y \qquad f.y = 0 \qquad h.x = x \qquad h.y = -y.$$

Then A becomes a U-module algebra by (d), since the action of $\mathfrak{g}$ on V extends to a Lie algebra homomorphism $\mathfrak{g} \to \text{Der}_k(A, A)$ such that

$$e \mapsto x\frac{\partial}{\partial y} \qquad\qquad f \mapsto y\frac{\partial}{\partial x} \qquad\qquad h \mapsto x\frac{\partial}{\partial x} - y\frac{\partial}{\partial y}.$$

(f) Lemma I.9.20(b) similarly makes it easy to check that $A = k[x, y]$ is a right $\mathcal{O}(SL_2(k))$-comodule algebra, with

$$\rho_A(x) = x \otimes a + y \otimes c \qquad \text{and} \qquad \rho_A(y) = x \otimes b + y \otimes d.$$

(g) Any Hopf algebra H is itself a left and a right H-module algebra under the left and right adjoint actions defined in (I.9.12). This is a straightforward exercise using the Hopf algebra axioms and the Sweedler notation.

Dual pairings

I.9.22. Definitions. Comparing Examples (e) and (f) of (I.9.21) and noting Lemma I.9.19, one might guess that there is a duality between the two Hopf algebras $U(\mathfrak{sl}_2(k))$ and $\mathcal{O}(SL_2(k))$ involved here. This is indeed the case, and the duality applies to all simply connected semisimple groups and their Lie algebras over algebraically closed fields of characteristic 0. To formulate the statement, we define a *dual pairing* or *Hopf pairing* of two Hopf k-algebras U and H to be a bilinear form $(-,-) : H \times U \to k$ such that, for all $u, v \in U$ and $f, h \in H$,

(1) $(h, uv) = \sum (h_1, u)(h_2, v)$;
(2) $(fh, u) = \sum (f, u_1)(h, u_2)$;
(3) $(1, u) = \varepsilon(u)$ and $(h, 1) = \varepsilon(h)$;
(4) $(h, Su) = (Sh, u)$.

In fact, condition (4) follows from the other three [**125**, Section 1.2.5, Proposition 9].

Skew versions of a dual pairing can be defined in several ways, e.g., as a dual pairing between H^{op} and U, or – in case the antipode S of U is bijective – a dual pairing between H and the *co-opposite* Hopf algebra

$$U^{\mathrm{cop}} = (U, m, u, \tau \circ \Delta, \varepsilon, S^{-1}).$$

Given a Hopf pairing as above, the induced maps $U \to H^*$ and $H \to U^*$ actually map $U \to H^\circ$ and $H \to U^\circ$, and the latter maps are Hopf algebra homomorphisms. The pairing is *perfect* (or *nondegenerate*) if the above maps are both injections.

I.9.23. Theorem. *Suppose that k is algebraically closed of characteristic 0. Let $U = U(\mathfrak{sl}_2(k))$ and $H = \mathcal{O}(SL_2(k))$, and let a, b, c, d be k-algebra generators for H as in Example I.9.11(c). Let $V = kx \oplus ky$ be the two-dimensional irreducible U-module as in Example I.9.21(e), and $\phi : U \to M_2(k)$ the corresponding algebra homomorphism. There exists a bilinear form $(-,-) : (ka \oplus kb \oplus kc \oplus kd) \times U \to k$ such that*

$$\phi(u) = \begin{pmatrix} (a, u) & (b, u) \\ (c, u) & (d, u) \end{pmatrix}$$

for $u \in U$, and $(-,-)$ extends uniquely to a perfect Hopf pairing of U and H. □

Notes on the proof. (a) To prove the theorem, observe first that the definition given realises a, b, c, d as elements of U°, so there is a unique extension to an algebra homomorphism ψ from $k[a, b, c, d]$ to the (commutative) k-algebra U°. One confirms that $ad - bc - 1$ is in the kernel of ψ. This means that the induced algebra homomorphism $H \to U^\circ$ defines a bilinear form on $H \times U$ such that (2) and (3) of the definition of a Hopf pairing hold; one then checks (1) and (4) on generators, which is easily seen to be enough. Details can be found in [**122**, Section V.7].

(b) It is in fact the case that the image of ψ is U°. Moreover, this conclusion, that there is a perfect Hopf pairing between $U(\mathfrak{g})$ and $\mathcal{O}(G)$, with

$$\mathcal{O}(G) \cong U(\mathfrak{g})^\circ,$$

is valid (when k is algebraically closed of characteristic 0) for any semisimple, connected, simply connected group G and $\mathfrak{g} = \mathrm{Lie}(G)$, [**93**, Theorem 3.1]. However, it's not true – even in characteristic 0 – that $\mathcal{O}(G)^\circ \cong U(\mathfrak{g})$; in fact, for *any* affine algebraic k-group G with Lie algebra $\mathfrak{g}$,

$$U(\mathfrak{g}) \cong \{f \in \mathcal{O}(G)^* : f(\mathfrak{m}_1^n) = 0 \text{ for some } n > 0\},$$

where $\mathfrak{m}_1 = \{\phi \in \mathcal{O}(G) : \phi(1) = 0\}$, the augmentation ideal of $\mathcal{O}(G)$ [**1**, p. 198].

GROUPLIKE ELEMENTS AND WINDING AUTOMORPHISMS

I.9.24. Definitions. A *grouplike element* in a coalgebra is any element g such that $\Delta(g) = g \otimes g$ and $\varepsilon(g) = 1$. If B is a bialgebra, then the set $G(B)$ of grouplike elements of B forms a monoid under the multiplication in B. In the case of a Hopf algebra H, the monoid $G(H)$ is actually a group; the inverse of $g \in G(H)$ is Sg (cf. [**122**, Proposition III.3.7]).

Now let H be a Hopf algebra. The group $G(H^\circ)$ of grouplike elements of H° is just the *character group* $X(H)$ of H; that is, it is the group of k-algebra homomorphisms from H to k. In this group, the inverse of a character λ is $S^*\lambda = \lambda \circ S$. Set $I = \bigcap\{\ker\lambda : \lambda \in X(H)\}$, and observe that I is a Hopf ideal of H (because if V and W are 1-dimensional H-modules then so are $V \otimes W$ and V^*). Thus H/I is a semiprime commutative Hopf algebra, so that, when H is affine, $G(H^\circ) = X(H)$ is an affine algebraic group over k.

It's not hard to see that the elements of $G(H^\circ)$ are linearly independent in H° – apply any k-linear combination of them to suitably chosen elements of H – so that the subalgebra of H° generated by $G(H^\circ)$ is just the group algebra $kG(H^\circ)$ of $G(H^\circ)$, and is a sub-Hopf algebra of H°.

I.9.25. Definitions. Again let H be a Hopf algebra. We've already noted in Example I.9.21(a) how H is a left and a right H°-module algebra. For $\lambda \in X(H)$, the H°-module multiplications of λ with elements of H define *left and right winding automorphisms* τ_λ^ℓ and τ_λ^r on H, where

$$\tau_\lambda^\ell(h) = h.\lambda = \sum \lambda(h_1)h_2 \qquad \text{and} \qquad \tau_\lambda^r(h) = \lambda.h = \sum h_1\lambda(h_2)$$

for $h \in H$. We have followed the convention that the adjectives 'left' and 'right' here refer to the components of $\Delta(h)$ on which λ acts. Some authors use the opposite convention.

The map $\tau^\ell : X(H) \to \mathrm{Aut}_{k\text{-alg}}(H)$ is a group antihomomorphism, while $\tau^r : X(H) \to \mathrm{Aut}_{k\text{-alg}}(H)$ is a group homomorphism. Since $\varepsilon \circ \tau^\ell_\lambda = \varepsilon \circ \tau^r_\lambda = \lambda$ for $\lambda \in X(H)$, the maps τ^ℓ and τ^r are monomorphisms. A group homomorphism $X(H) \to \mathrm{Aut}_{k\text{-alg}}(H)$ can be obtained from left winding automorphisms by using the map $\lambda \mapsto \tau^\ell_{S*\lambda}$.

NOTES

This appendix is a summary of standard material, adopting current points of view and notation. For more detail, see, e.g., [122] or [163], in addition to the classics [1] and [201].

APPENDIX I.10

R-MATRICES

Throughout this appendix, V will denote a finite dimensional vector space over k and $R : V \otimes V \to V \otimes V$ will be an invertible linear transformation. Recall that the *flip* $\tau : V \otimes V \to V \otimes V$ is the linear transformation sending $v \otimes w \mapsto w \otimes v$ for all $v, w \in V$.

I.10.1. Given a basis $e_1, \ldots, e_n$ for V, the natural basis to use for $V \otimes V$ consists of the pure tensors $e_i \otimes e_j$. In this case, we write $\left(R_{ij}^{lm} \right)$ for the matrix of R with respect to this basis, using the following convention: $R(e_i \otimes e_j) = \sum_{l,m} R_{ij}^{lm} e_l \otimes e_m$ for all i, j. We think of $\left(R_{ij}^{lm} \right)$ as an $n^2 \times n^2$ matrix with rows indexed by the pairs (l, m) and columns indexed by the pairs (i, j). These pairs are assumed to be listed in lexicographic order – for instance, in the order $(1, 1)$, $(1, 2)$, $(2, 1)$, $(2, 2)$ in case $n = 2$.

The action of R on $V \otimes V$ can be extended in three natural ways to actions on $V \otimes V \otimes V$. For $s, t \in \{1, 2, 3\}$, let R_{st} denote the (invertible) linear transformation on $V \otimes V \otimes V$ which acts like R on the s-th and t-th components while acting like the identity on the remaining component. Thus $R_{12} = R \otimes \mathrm{id}$ and $R_{23} = \mathrm{id} \otimes R$, while $R_{13} = (\mathrm{id} \otimes \tau)(R \otimes \mathrm{id})(\mathrm{id} \otimes \tau)$. More explicitly, if $e_1, \ldots, e_n$ is a basis for V, then

$$R_{12}(e_i \otimes e_j \otimes v) = \sum_{l,m} R_{ij}^{lm} e_l \otimes e_m \otimes v$$

$$R_{23}(v \otimes e_i \otimes e_j) = \sum_{l,m} R_{ij}^{lm} v \otimes e_l \otimes e_m$$

$$R_{13}(e_i \otimes v \otimes e_j) = \sum_{l,m} R_{ij}^{lm} e_l \otimes v \otimes e_m$$

for all i, j and all $v \in V$.

I.10.2. Definition. We distinguish two types of *Quantum Yang-Baxter Equation* that the transformation R might satisfy:

$$R_{12} R_{13} R_{23} = R_{23} R_{13} R_{12} \qquad \text{and} \qquad R_{12} R_{23} R_{12} = R_{23} R_{12} R_{23}.$$

The first equation is the *original QYBE*, while the second is called the *braid form* of the QYBE or just the *braid equation*. The transformation R satisfies the original QYBE if and only if τR (or $R \tau$) satisfies the braid form. Solutions to either QYBE are often called *R-matrices*.

I.10.3. Examples. The identity transformation and the flip are easily seen to satisfy both forms of the QYBE. An example satisfying the original QYBE, which recurs in the study of quantized algebras, can be written as follows, taking the case that V is 2-dimensional:

$$
(R_{ij}^{lm}) = \begin{pmatrix} q & 0 & 0 & 0 \\ 0 & 1 & 0 & 0 \\ 0 & q - q^{-1} & 1 & 0 \\ 0 & 0 & 0 & q \end{pmatrix},
$$

where q is a nonzero scalar in k. By our conventions, this means the transformation R satisfies

$$
\begin{aligned}
R(e_1 \otimes e_1) &= q e_1 \otimes e_1 & R(e_1 \otimes e_2) &= e_1 \otimes e_2 + (q - q^{-1}) e_2 \otimes e_1 \\
R(e_2 \otimes e_1) &= e_2 \otimes e_1 & R(e_2 \otimes e_2) &= q e_2 \otimes e_2.
\end{aligned}
$$

Corresponding to this, the matrix

$$
((R\tau)_{ij}^{lm}) = \begin{pmatrix} q & 0 & 0 & 0 \\ 0 & 0 & 1 & 0 \\ 0 & 1 & q - q^{-1} & 0 \\ 0 & 0 & 0 & q \end{pmatrix}
$$

satisfies the braid form of the QYBE.

I.10.4. Definitions. Recall that the *tensor algebra on V* is the k-algebra

$$
T(V) = \bigoplus_{n=0}^{\infty} V^{\otimes n},
$$

where $V^{\otimes 0}$ is identified with k and the multiplication is induced from the rule

$$
(v_1 \otimes v_2 \otimes \cdots \otimes v_m)(w_1 \otimes w_2 \otimes \cdots \otimes w_n) = v_1 \otimes v_2 \otimes \cdots \otimes v_m \otimes w_1 \otimes w_2 \otimes \cdots \otimes w_n.
$$

If $e_1, \ldots, e_n$ is a basis for V, then $T(V)$ is a free k-algebra on $e_1, \ldots, e_n$. This algebra is naturally $\mathbb{Z}_{\geq 0}$-graded (see Definitions I.12.7 and Example I.12.8(d)) where $T(V)_n = V^{\otimes n}$ for $n \geq 0$.

A *quadratic algebra*, or *algebra with quadratic relations*, is any algebra obtained from a tensor algebra $T(V)$ by factoring out an ideal generated by a subset of $T(V)_2 = V \otimes V$. Such a factor inherits the grading from $T(V)$. A classical example is the *symmetric algebra on V*, namely the algebra

$$
S(V) := T(V)/\langle v \otimes w - w \otimes v \mid v, w \in V \rangle.
$$

This is a polynomial algebra over V in $\dim V$ indeterminates.

I.10.5. Definition. Given the transformation R on $V \otimes V$, the R-*symmetric algebra of* V is the quadratic algebra

$$S_R(V) := T(V)/\langle z - R(z) \mid z \in V \otimes V \rangle.$$

For instance, if R is the flip then $S_R(V) = S(V)$, while if R is the identity then $S_R(V) = T(V)$. Any quantum affine space $\mathcal{O}_q(k^n)$ (see (I.2.1)) can be written as an R-symmetric algebra where $V = k^n$ with basis $e_1, \ldots, e_n$ and $R(e_i \otimes e_j) = q_{ij} e_j \otimes e_i$ for all i, j.

I.10.6. Many quantized coordinate rings were first presented on the basis of the following *Fadeev-Reshetikhin-Takhtadjan construction* relative to an R-matrix. In fact, an R-matrix is not necessary, as the construction requires only a linear transformation on a tensor square. We give the 'braid version' of the construction, appropriate for R-matrices satisfying the braid version of the QYBE; to obtain the original version requires switching certain indices. Keep V and R as above.

Since V is finite dimensional, we may identify $(V \otimes V)^*$ with $V^* \otimes V^*$, so that the transpose R^* of R becomes a linear transformation on $V^* \otimes V^*$. Set $E = V \otimes V^*$ (which can be identified with $\mathrm{End}_k(V)$, if desired). We also define transformations R_{13} and R_{24}^* on $E \otimes E$ as follows: R_{13} acts as R on the first and third factors of $E \otimes E = V \otimes V^* \otimes V \otimes V^*$ and as the identity on the other factors, while R_{24}^* acts as R^* on the second and fourth factors and as the identity on the others. Alternatively, $R_{13} = \tau_{23}(R \otimes \mathrm{id} \otimes \mathrm{id})\tau_{23}$ and $R_{24}^* = \tau_{23}(\mathrm{id} \otimes \mathrm{id} \otimes R^*)\tau_{23}$, where $\tau_{23} = \mathrm{id} \otimes \tau \otimes \mathrm{id}$ interchanges terms in the second and third tensor factors.

The *coordinate ring of quantum matrices associated with* R or the *FRT construction applied to* R is the quadratic algebra

$$A_R(V) := T(E)/\langle R_{13}(z) - R_{24}^*(z) \mid z \in E \otimes E \rangle.$$

This algebra is sometimes denoted $\mathcal{O}_R(\mathrm{End}_k(V))$, or $\mathcal{O}_R(M_n(k))$ if $\dim V = n$. Since R is invertible, the image of $R_{13} - R_{24}^*$ coincides with that of $(\mathrm{id} - R_{24}^* R_{13}^{-1})$, and hence $A_R(V) = S_T(E)$ where $T = R_{24}^* R_{13}^{-1}$.

Given a basis $e_1, \ldots, e_n$ for V, let $f_1, \ldots, f_n$ be the corresponding dual basis for V^*. We then use the tensors $x_{ij} = e_i \otimes f_j$ as a basis for E, and we write X_{ij} for the image of x_{ij} in $A_R(V)$. If (R_{ij}^{lm}) is the matrix for R as above, then $R^*(f_i \otimes f_j) = \sum_{l,m} R_{lm}^{ij} f_l \otimes f_m$ for all i, j. Hence,

$$(R_{13} - R_{24}^*)(x_{ij} \otimes x_{st}) = \sum_{l,m} R_{is}^{lm} x_{lj} \otimes x_{mt} - \sum_{l,m} R_{lm}^{jt} x_{il} \otimes x_{sm}$$

for all i, j, s, t, and so

$$\sum_{l,m} R_{is}^{lm} X_{lj} X_{mt} = \sum_{l,m} R_{lm}^{jt} X_{il} X_{sm}$$

in $A_R(V)$ for all i, j, s, t.

The algebra $A_R(V)$ also has a bialgebra structure, such that

$$\Delta(X_{ij}) = \sum_{l=1}^{n} X_{il} \otimes X_{lj} \qquad\qquad \varepsilon(X_{ij}) = \delta_{ij}$$

for all i, j. Having taken the braid version of the FRT construction, we gain the advantage that $S_R(V)$ is a left comodule algebra for $A_R(V)$, and $S_{R^*}(V^*)$ is a right comodule algebra for $A_R(V)$, with comodule structure maps such that

$$E_i \mapsto \sum_{j=1}^{n} X_{ij} \otimes E_j \qquad\qquad F_j \mapsto \sum_{i=1}^{n} F_i \otimes X_{ij}$$

for all i, j, where E_i and F_j denote the images of e_i and f_j in $S_R(V)$ and $S_{R^*}(V^*)$.

For example, if $\dim_k(V) = 2$ and R is the transformation described in (I.10.3), then $A_{R_\tau}(V)$ is isomorphic to the quantum matrix algebra $\mathcal{O}_q(M_2(k))$ (see (I.1.7)).

NOTES

See [**34**, §7.5,], [**112**, §8.5], [**130**, Introduction], [**146**, §4.4], for example, for discussions of the origins of and other details concerning the QYBE. The FRT construction was presented by Fadeev, Reshetikhin, and Takhtadjan in [**54**, **186**].

APPENDIX I.11

THE DIAMOND LEMMA

Suppose that A is a k-algebra presented by generators and relations. Then A is spanned by monomials in the generators, and so some set of monomials will form a basis for A (as a vector space over k). The problem is to find such a set of monomials; the Diamond Lemma provides a method.

I.11.1. Any presentation of A by generators and relations can be given as an identification

$$F/\langle w_\sigma - f_\sigma \mid \sigma \in \Sigma \rangle = A,$$

where $F = k\langle X \rangle$ is the free algebra on a set X, the $f_\sigma \in F$, and the w_σ are *words*, that is, products of elements from X. Thus $w_\sigma \in W$, where W is the free monoid on X. Since W is a basis for F, the cosets $\overline{w}$ for $w \in W$ span A. The set

$$S = \{(w_\sigma, f_\sigma) \mid \sigma \in \Sigma\} \subseteq W \times F$$

is called a *reduction system*, since it records a selection of substitutions $(aw_\sigma b \mapsto af_\sigma b)$ which can be applied to expressions in which w_σs occur.

To make the substitution process effective, we need

(1) Some way to say that expressions 'improve' after suitable substitutions, and that after sufficiently many such improvements, we reach expressions that cannot be improved.

(2) Some means to ensure that expressions can be unambiguously reduced to 'fully improved' expressions.

Item (1) will be managed in terms of a partial ordering on W satisfying the DCC; a word $w' \in W$ will be an 'improvement' on a word w if $w' < w$ with respect to the ordering. For item (2), we identify a minimal set of ambiguities to be resolved.

I.11.2. Given $\sigma \in \Sigma$ and $a, b \in W$, let $r_{a,\sigma,b} : F \to F$ be the k-linear map sending $aw_\sigma b \mapsto af_\sigma b$ and fixing all other words. This map is called an *elementary reduction*. In general, a *reduction* is any composition of finitely many elementary reductions.

An element $f \in F$ is *irreducible* (with respect to our given system of reductions) provided that $r(f) = f$ for all (elementary) reductions r.

I.11.3. A *semigroup ordering* on W is a partial order $\leq$ such that

$$b < b' \quad \Longrightarrow \quad abc < ab'c$$

for all $a, b, b', c \in W$. We will require a semigroup ordering which satisfies the DCC. The most typical example in practice is the length-lexicographic ordering, described as follows.

Suppose (for convenience) that X is finite, say $X = \{x_1, x_2, \ldots, x_n\}$. The *lexicographic ordering* on W (relative to the given order in which the x_i are listed) is just the lexicographic order on sequences of positive integers transferred to monomials in W via indices. That is,

$$a = x_{i(1)} x_{i(2)} \cdots x_{i(s)} \quad \leq_{\text{lex}} \quad b = x_{j(1)} x_{j(2)} \cdots x_{j(t)}$$

if and only if either

- $s \leq t$ and $i(l) = j(l)$ for all $l \leq s$, or
- there is some $u \leq \min\{s, t\}$ such that $i(l) = j(l)$ for all $l < u$ while $i(u) < j(u)$.

This is a semigroup ordering, but it does not satisfy the DCC. The *length-lexicographic ordering* on W is the following modification of the lexicographic ordering: $a \leq b$ if and only if either

- $\text{length}(a) < \text{length}(b)$, or
- $\text{length}(a) = \text{length}(b)$ and $a \leq_{\text{lex}} b$.

This is a semigroup ordering on W and it not only satisfies the DCC, but is a well-ordering.

A semigroup ordering $\leq$ on W is *compatible with* S provided that for all $\sigma \in \Sigma$, the element f_σ is a linear combination of words $w < w_\sigma$.

I.11.4. An *overlap ambiguity* is a 5-tuple (a, b, c, σ, τ) in $W^3 \times \Sigma^2$ such that $ab = w_\sigma$ and $bc = w_\tau$. The ambiguity lies in the fact that abc has two immediate reductions:

$$r_{1,\sigma,c}(abc) = f_\sigma c \qquad \text{and} \qquad r_{a,\tau,1}(abc) = a f_\tau.$$

An *inclusion ambiguity* is a 5-tuple (a, b, c, σ, τ) in $W^3 \times \Sigma^2$ such that $abc = w_\sigma$ and $b = w_\tau$. In this case,

$$r_{1,\sigma,1}(abc) = f_\sigma \qquad \text{and} \qquad r_{a,\tau,c}(abc) = a f_\tau c.$$

When σ and τ are clear, we refer to abc itself as the ambiguity.

Note that if all the w_σ have the same length, then the only possible inclusion ambiguities would have the form $(1, b, 1, \sigma, \tau)$. Thus, if the w_σ are distinct words of the same length, there are no inclusion ambiguities (apart from the trivial case $(1, b, 1, \sigma, \sigma)$, where there is nothing to resolve).

I.11.5. An overlap (respectively, inclusion) ambiguity (a, b, c, σ, τ) is *resolvable* (with respect to our given reduction system) if and only if there exist reductions r, r' such that $r(f_\sigma c) = r'(a f_\tau)$ (respectively, $r(f_\sigma) = r'(a f_\tau c)$). The resolutions of these ambiguities can be displayed in the following diamond patterns:

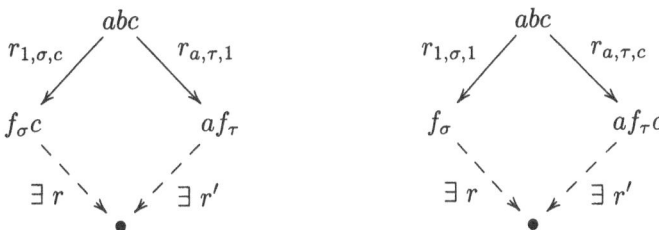

I.11.6. Diamond Lemma. *Let $F = k\langle X \rangle$ be a free algebra on a set X and W the free monoid on X. Let $S = \{(w_\sigma, f_\sigma) \mid \sigma \in \Sigma\}$ be a reduction system for F, and $\leq$ a semigroup ordering on W which is compatible with S and satisfies the DCC. Assume that all overlap and inclusion ambiguities are resolvable. Then the cosets $\overline{w}$, for irreducible words $w \in W$, form a basis for the factor algebra $F/\langle w_\sigma - f_\sigma \mid \sigma \in \Sigma \rangle$.* $\square$

Proof. [**13**, Theorem 1.2] $\square$

I.11.7. Example. A straightforward application of the Diamond Lemma is to compute a basis for the quantum matrix algebra $\mathcal{O}_q(M_2(k))$ (see (I.1.7)), as follows. Let $F = k\langle x_1, x_2, x_3, x_4 \rangle$, which we map onto $\mathcal{O}_q(M_2(k))$ by sending x_1, x_2, x_3, x_4 to a, b, c, d (in that order). The kernel of this map is the ideal generated by the relations coming from the following reduction system S:

$$\big\{ (x_2 x_1, q^{-1} x_1 x_2),\ (x_3 x_1, q^{-1} x_1 x_3),\ (x_4 x_2, q^{-1} x_2 x_4),$$
$$(x_4 x_3, q^{-1} x_3 x_4),\ (x_3 x_2, x_2 x_3),\ (x_4 x_1, x_1 x_4 - \hat{q} x_2 x_3) \big\},$$

where $\hat{q} = q - q^{-1}$. Note that S has been arranged to be compatible with the length-lexicographic ordering on words in x_1, x_2, x_3, x_4.

There are no inclusion ambiguities, and only four overlap ambiguities, which are resolvable as shown in Diagram I.11.7 below.

The irreducible monomials (with respect to the current reduction system) are those of the form $x_1^i x_2^j x_3^l x_4^m$. Therefore the Diamond Lemma implies that the set

$$\{ a^i b^j c^l d^m \mid i, j, l, m \in \mathbb{Z}^+ \}$$

is a basis for $\mathcal{O}_q(M_2(k))$. This is one method of proceeding to show that $\mathcal{O}_q(M_2(k))$ is an iterated skew polynomial ring in the variables a, b, c, d (cf. Example I.1.16).

$$x_3x_2x_1 : \begin{cases} (x_2x_3)x_1 \to x_2(q^{-1}x_1x_3) \to q^{-1}(q^{-1}x_1x_2)x_3 \\ x_3(q^{-1}x_1x_2) \to q^{-1}(q^{-1}x_1x_3)x_2 \to q^{-2}x_1(x_2x_3) \end{cases}$$

$$x_4x_2x_1 : \begin{cases} (q^{-1}x_2x_4)x_1 \to q^{-1}x_2(x_1x_4 - \hat{q}x_2x_3) \\ \qquad\qquad \to q^{-1}(q^{-1}x_1x_2)x_4 - q^{-1}\hat{q}x_2^2x_3 \\ x_4(q^{-1}x_1x_2) \to q^{-1}(x_1x_4 - \hat{q}x_2x_3)x_2 \\ \qquad\qquad \to q^{-1}x_1(q^{-1}x_2x_4) - q^{-1}\hat{q}x_2(x_2x_3) \end{cases}$$

$$x_4x_3x_1 : \begin{cases} (q^{-1}x_3x_4)x_1 \to q^{-1}x_3(x_1x_4 - \hat{q}x_2x_3) \\ \qquad\qquad \to q^{-1}(q^{-1}x_1x_3)x_4 - q^{-1}\hat{q}(x_2x_3)x_3 \\ x_4(q^{-1}x_1x_3) \to q^{-1}(x_1x_4 - \hat{q}x_2x_3)x_3 \\ \qquad\qquad \to q^{-1}x_1(q^{-1}x_3x_4) - q^{-1}\hat{q}x_2x_3^2 \end{cases}$$

$$x_4x_3x_2 : \begin{cases} (q^{-1}x_3x_4)x_2 \to q^{-1}x_3(q^{-1}x_2x_4) \to q^{-2}(x_2x_3)x_4 \\ x_4(x_2x_3) \to (q^{-1}x_2x_4)x_3 \to q^{-1}x_2(q^{-1}x_3x_4). \end{cases}$$

DIAGRAM I.11.7

I.11.8. Example. In dealing with the algebra $\mathcal{O}_q(SL_2(k))$ (see (I.1.9)), a bit of care is required with the order of the generators and with the relations. We have two relations involving $\bar{a}\bar{d}$, namely $\bar{a}\bar{d} - \bar{d}\bar{a} = \hat{q}\bar{b}\bar{c}$ and $\bar{a}\bar{d} - q\bar{b}\bar{c} = 1$. We could choose an ordering in which $\bar{b}\bar{c} < \bar{a}\bar{d}$ and $\bar{d}\bar{a} < \bar{a}\bar{d}$, but then the equation $\hat{q}\bar{b}\bar{c} + \bar{d}\bar{a} = \bar{a}\bar{d} = q\bar{b}\bar{c} + 1$ would represent an inclusion ambiguity that is not resolvable. To get around this problem, replace the two relations above by the relations $\bar{a}\bar{d} - q\bar{b}\bar{c} = \bar{d}\bar{a} - q^{-1}\bar{b}\bar{c} = 1$. Then we should choose an ordering such that $\bar{b}\bar{c} < \bar{a}\bar{d}$ and $\bar{b}\bar{c} < \bar{d}\bar{a}$, for instance, the length-lexicographic ordering with $\bar{b} < \bar{a} < \bar{d} < \bar{c}$. This works out fine, as follows.

Thus let $F = k\langle x_1, x_2, x_3, x_4 \rangle$, mapped onto $\mathcal{O}_q(SL_2(k))$ by sending the letters x_1, x_2, x_3, x_4 to $\bar{b}$, $\bar{a}$, $\bar{d}$, $\bar{c}$, and let S be the reduction system

$$\{(x_2x_1, qx_1x_2),\ (x_3x_1, q^{-1}x_1x_3),\ (x_4x_2, q^{-1}x_2x_4),\ (x_4x_3, qx_3x_4),$$
$$(x_4x_1, x_1x_4),\ (x_2x_3, qx_1x_4 + 1),\ (x_3x_2, q^{-1}x_1x_4 + 1)\}.$$

This time there are 8 overlap ambiguities:

$$x_4x_2x_1,\ x_4x_2x_3,\ x_4x_3x_1,\ x_4x_3x_2,\ x_2x_3x_1,\ x_2x_3x_2,\ x_3x_2x_1,\ x_3x_2x_3.$$

All of these are resolvable; for instance,

$$x_4x_2x_1 : \begin{cases} (q^{-1}x_2x_4)x_1 \to q^{-1}x_2(x_1x_4) \to q^{-1}(qx_1x_2)x_4 \\ x_4(qx_1x_2) \to q(x_1x_4)x_2 \to qx_1(q^{-1}x_2x_4) \end{cases}$$

$$x_4x_2x_3 : \begin{cases} (q^{-1}x_2x_4)x_3 \to q^{-1}x_2(qx_3x_4) \to (qx_1x_4 + 1)x_4 \\ x_4(qx_1x_4 + 1) \to q(x_1x_4)x_4 + x_4. \end{cases}$$

The irreducible monomials are those of the forms $x_1^i x_2^j x_4^m$ and $x_1^i x_3^l x_4^m$. Therefore the set

$$\{\bar{b}^i \bar{a}^j \bar{c}^m \mid i, j, m \in \mathbb{Z}_{\geq 0}\} \cup \{\bar{b}^i \bar{d}^l \bar{c}^m \mid i, l, m \in \mathbb{Z}_{\geq 0} \text{ and } l > 0\}$$

is a basis for $\mathcal{O}_q(SL_2(k))$.

NOTES

The original Diamond Lemma was proved by Newman [172] and used to provide a method for showing the existence of normal forms in a wide range of mathematical theories. Bergman presented this result in a form tailored to rings given by generators and relations [13], and our account of the Diamond Lemma is excerpted from his paper.

FILTERED AND GRADED RINGS

Filtrations and gradings on rings and modules are both ways of assigning degrees with behaviour analogous to degrees of polynomials. We shall need to consider not only filtrations and gradings indexed by $\mathbb{Z}$, but also those indexed by more general (semi-) groups, in particular filtrations and degrees indexed by $\mathbb{Z}_{\geq 0}^n$ or $\mathbb{Z}^n$.

$\mathbb{Z}$-FILTERED RINGS AND MODULES

I.12.1. Definitions. A $\mathbb{Z}$-*filtration* (often just called a *filtration*) on a ring R is an indexed family $(R_i)_{i \in \mathbb{Z}}$ of additive subgroups of R such that

(1) $R_i \subseteq R_j$ for all $i \leq j$;
(2) $1 \in R_0$;
(3) $R_i R_j \subseteq R_{i+j}$ for all i, j;
(4) $\bigcup_{i \in \mathbb{Z}} R_i = R$.

Sometimes filtrations are defined using only conditions (1)–(3); in that case, a filtration satisfying (4) is called *exhaustive*. Another requirement that is often imposed is the condition $\bigcap_{i \in \mathbb{Z}} R_i = 0$. An alternative notation is to name the filtration, say $\mathcal{F}$, and to label the corresponding subgroups R_i in the form $\mathcal{F}_i R$.

We adopt the convention that if R is a k-algebra, any filtration on R is assumed to consist of k-subspaces.

A filtration on R is called *discrete* if there is an integer ν such that $R_i = 0$ for all $i < \nu$. In case $R_i = 0$ for all $i < 0$, the filtration is said to be *nonnegative*, or *positive*. One also refers to a $\mathbb{Z}_{\geq 0}$-*filtration* in this case.

A *filtered ring* consists of a ring together with a particular filtration. If R is a filtered ring and $\bigcap_{i \in \mathbb{Z}} R_i = 0$, we can define a degree function on R by setting

$$\deg(r) = \min\{i \in \mathbb{Z} \mid r \in R_i\}$$

for nonzero elements $r \in R$.

I.12.2. Examples. (a) Any ring R has a *trivial filtration*, in which $R_i = 0$ for all $i < 0$ while $R_i = R$ for all $i \geq 0$.

(b) Any skew polynomial ring $T = R[x; \tau, \delta]$ can be given a filtration with $T_n = 0$ for $n < 0$ and $T_n = R + Rx + \cdots + Rx^n$ for $n \geq 0$.

(c) Suppose that X is a set of k-algebra generators for a k-algebra A. Put $A_0 = k$ (that is, $k \cdot 1$), and for $i = 1, 2, \ldots$ let A_i be the k-linear span of all products of i or fewer elements from X. Then $A_0 \subseteq A_1 \subseteq \cdots$ is a nonnegative filtration on A. For instance, if $A = k[x_1, \ldots, x_n]$ is a polynomial ring over k and $X = \{x_1, \ldots, x_n\}$, then A_i (for $i \geq 0$) consists of zero together with all polynomials with total degree at most i.

(d) Consider the skew polynomial algebra $A = k[x][y; \delta]$ where $k[x]$ is a polynomial ring over k and δ is the derivation $x^3 d/dx$ on $k[x]$. Since A is generated by x and y, it has a nonnegative filtration of the form described in the previous example. However, A_2 then contains the element $yx - xy = x^3$, and so we cannot say that A_2 consists only of expressions with total degree at most 2.

This problem can be avoided by working with a suitable *weighted* total degree on standard monomials $x^s y^t$; the simplest choice for this example is to assign x degree 1 and y degree 2. In other words, we now define A_i (for $i \geq 0$) to be the k-linear span of those monomials $x^s y^t$ for which $s + 2t \leq i$. It is an easy exercise to check that this defines a filtration on A.

(e) Let J be a proper ideal in a ring R. The *J-adic filtration* on R is given by setting $R_i = J^{-i}$ for $i \leq 0$ while $R_i = R$ for $i > 0$.

I.12.3. Definitions. Let R be a $\mathbb{Z}$-filtered ring and M a left R-module. A *filtration* on M (compatible with the given filtration on R) is an indexed family $(M_i)_{i \in \mathbb{Z}}$ of additive subgroups of M such that

(1) $M_i \subseteq M_j$ for all $i \leq j$;
(2) $R_i M_j \subseteq M_{i+j}$ for all i, j;
(3) $\bigcup_{i \in \mathbb{Z}} M_i = M$.

If R is a k-algebra, we assume that the M_i are k-subspaces of M. The terms *discrete* and *nonnegative* apply to module filtrations in the same way as to ring filtrations.

A *filtered R-module* consists of an R-module together with a particular filtration.

If M is a finitely generated left R-module, say with generators $x_1, \ldots, x_n$, then M has a *standard filtration* with respect to this choice of generators, where $M_i = R_i x_1 + \cdots + R_i x_n$ for all i. Note that if the filtration on R is discrete (nonnegative), then the filtration on M just described is discrete (nonnegative).

FILTRATIONS OVER ORDERED MONOIDS

Observe that the definition of a $\mathbb{Z}$-filtration does not require all the available structure of $\mathbb{Z}$, only the order relation, the zero, and the addition. Hence, filtrations can equally well be indexed by any monoid (i.e., a semigroup with identity) equipped with a suitable ordering, as follows.

I.12.4. Definitions. Let S be a monoid, not necessarily abelian. Unless the operation in S is already given as addition, we will write S multiplicatively and denote its identity by 1. As in (I.11.3), a *semigroup ordering* on S is a partial ordering $\leq$

such that

$$x < y \implies axb < ayb$$

for all $x, y, a, b \in S$. If S has cancellation ($axb = ayb$ implies $x = y$), e.g., if S is a submonoid of a group, then the ordering condition above is equivalent to the condition ($x \leq y \implies axb \leq ayb$). For many applications, one requires an *admissible ordering* on S; this means a semigroup ordering which is a total ordering and satisfies $1 \leq s$ for all $s \in S$.

A *partially ordered monoid* is a pair consisting of a monoid S and a specified semigroup ordering $\leq$ on S. Unless there is a potential ambiguity about the choice of the ordering, we refer to S itself as the partially ordered monoid. In case $\leq$ is a total ordering, or an admissible ordering, we say that S is a *totally ordered monoid*, or an *admissibly ordered monoid*, respectively.

For example, $\mathbb{Z}$ and $\mathbb{Z}_{\geq 0}$, equipped with the canonical order relation, are both totally ordered monoids. With respect to the product (componentwise) orderings, $\mathbb{Z}^n$ and $\mathbb{Z}_{\geq 0}^n$ become partially ordered monoids. The most common total ordering used on $\mathbb{Z}^n$ and $\mathbb{Z}_{\geq 0}^n$ is the lexicographic ordering, as in (I.11.3). The canonical ordering on $\mathbb{Z}_{\geq 0}$ and the lexicographic ordering on $\mathbb{Z}_{\geq 0}^n$ are both admissible.

I.12.5. Definitions. Let S be a partially ordered monoid. An *S-filtration* on a ring R is an indexed family $\left(R_s\right)_{s \in S}$ of additive subgroups of R such that

(1) $R_s \subseteq R_t$ for all $s \leq t$;
(2) $1 \in R_1$;
(3) $R_s R_t \subseteq R_{st}$ for all s, t;
(4) $\bigcup_{s \in S} R_s = R$.

Of course if the operation on S is addition, conditions (2) and (3) read $1 \in R_0$ and $R_s R_t \subseteq R_{s+t}$, respectively. In case R is a k-algebra, we assume that the R_s are k-subspaces of R.

As in (I.12.1), an S-filtration on R can be used to define a degree function, in case S is well-ordered and $\bigcap_{s \in S} R_s = 0$. Namely,

$$\deg(r) = \min\{s \in S \mid r \in R_s\}$$

for nonzero $r \in R$.

If R is equipped with a given S-filtration, then (compatible) *S-filtrations* on R-modules are defined in the obvious way, analogous to (I.12.3). As above, when dealing with modules over a filtered ring, it is often advantageous to allow the module filtrations a larger indexing monoid than the ring filtration. In particular, if R has a $\mathbb{Z}_{\geq 0}^n$-filtration (with respect to some ordering), one often has to work with $\mathbb{Z}^n$-filtered R-modules.

I.12.6. Examples. Here are some standard examples of $\mathbb{Z}_{\geq 0}^n$-filtrations, where $\mathbb{Z}_{\geq 0}^n$ is assumed to be equipped with some semigroup ordering. Let $\epsilon_1, \ldots, \epsilon_n$ denote the standard basis elements of $\mathbb{Z}^n_{\geq 0}$, that is, $(1,0,0,\ldots,0)$, $(0,1,0,\ldots,0)$, $\ldots$, $(0,0,\ldots,0,1)$.

(a) If $R = T[x_1,\ldots,x_n]$ is a polynomial ring, there is a $\mathbb{Z}^n_{\geq 0}$-filtration on R where each R_s is the T-submodule generated by the set

$$\{x_1^{m_1} x_2^{m_2} \cdots x_n^{m_n} \mid (m_1,\ldots,m_n) \leq s\}.$$

(b) More generally, if A is a k-algebra generated by n elements $a_1,\ldots,a_n$, we can give A a $\mathbb{Z}^n_{\geq 0}$-filtration where $A_0 = k \cdot 1$ and A_s, for $s \neq 0$, is the k-subspace spanned by the set

$$\{a_{i_1} a_{i_2} \cdots a_{i_l} \mid \epsilon_{i_1} + \epsilon_{i_2} + \cdots + \epsilon_{i_l} \leq s\}.$$

(c) For instance, if $A_1(k) = k[x][y; d/dx]$ is the first Weyl algebra over k, and $\mathbb{Z}^2_{\geq 0}$ is equipped with the componentwise ordering, then with respect to the filtration described above, each

$$A_1(k)_{(d,e)} = \sum_{i=0}^{d} \sum_{j=0}^{e} kx^i y^j.$$

On the other hand, if $\mathbb{Z}^2_{\geq 0}$ is given the lexicographic ordering, then

$$A_1(k)_{(d,e)} = \sum_{i=0}^{d-1} x^i k[y] + \sum_{j=0}^{e} kx^d y^j.$$

GRADED RINGS AND MODULES

I.12.7. Definitions. Let S be a monoid (written multiplicatively unless otherwise specified). An S-*grading* on a ring R is an abelian group decomposition $R = \bigoplus_{s \in S} R_s$ such that $1 \in R_1$ and $R_s R_t \subseteq R_{st}$ for all $s, t \in S$. The ring R together with a specified S-grading is called an S-*graded ring*. We say that R is *strongly S-graded* provided the grading satisfies $R_s R_t = R_{st}$ for all $s, t \in S$.

As with filtrations, in case R is a k-algebra, we assume that the subgroups making up a grading of R are k-subspaces.

For a $\mathbb{Z}^n$-grading, it is natural to use additive notation: $R = \bigoplus_{s \in \mathbb{Z}^n} R_s$ with $1 \in R_0$ and $R_s R_t \subseteq R_{s+t}$ for all s, t. A $\mathbb{Z}$-graded ring is often called just a *graded ring*, and a $\mathbb{Z}_{\geq 0}$-graded ring is called a *positively graded ring*. A *connected graded k-algebra* is a $\mathbb{Z}_{\geq 0}$-graded k-algebra A such that $A_0 = k$.

Now assume that $R = \bigoplus_{s \in S} R_s$ is an S-graded ring. The subgroup R_s of R is called the *homogeneous component of degree s* (with respect to the given grading). Any element $x \in R$ can be uniquely expressed in the form $x = \sum_{s \in S} x_s$ where each $x_s \in R_s$ and all but finitely many of the x_s are zero. The element x_s is called the *homogeneous component of x of degree s*. The element x itself is *homogeneous of degree s* provided $x \in R_s$, that is, provided $x_t = 0$ for all $t \neq s$.

An ideal I of R (right, left, or two-sided) is *homogeneous* (or a *graded ideal*) provided $I = \bigoplus_{s \in S}(I \cap R_s)$. Note that I is homogeneous if and only if I can be generated by a set of homogeneous elements. When I is a homogeneous ideal of R, we can write $I = \bigoplus_{s \in S} I_s$ with $I_s = I \cap R_s$, and there is an *induced S-grading* $R/I = \bigoplus_{s \in S}(R/I)_s$ where $(R/I)_s = (R_s + I)/I$.

There are graded versions of many standard ring-theoretic concepts, often distinguished from the originals by use of the prefix 'gr-'. For example, a *gr-prime ideal* of R is any proper homogeneous ideal P such that for all homogeneous ideals I and J, we have $IJ \subseteq P$ only if $I \subseteq P$ or $J \subseteq P$. A *gr-prime ring* is a (nonzero) graded ring in which 0 is a gr-prime ideal, i.e., all products of nonzero homogeneous ideals are nonzero. Some other gr-conditions are discussed in (II.3.1).

I.12.8. Examples. (a) The *trivial grading* on a ring R, with respect to any monoid S, is given by setting $R_1 = R$ and $R_s = 0$ for all $s \neq 1$.

(b) There are several standard $\mathbb{Z}_{\geq 0}$-gradings for a polynomial ring $R = T[x_1, \ldots, x_n]$. First, there is the grading given by setting R_i (for $i \geq 0$) equal to the T-submodule generated by all monomials of total degree i. Note that the degree function with respect to this grading coincides with the total degree. It can also be convenient to use a weighted grading, as in Example I.12.2(d) above. Thus, let $w_1, \ldots, w_n$ be any integers, and for $i \in \mathbb{Z}$ let R_i be the T-submodule generated by those monomials $x_1^{a_1} x_2^{a_2} \cdots x_n^{a_n}$ for which $w_1 a_1 + w_2 a_2 + \cdots + w_n a_n = i$.

(c) A polynomial ring $R = T[x_1, \ldots, x_n]$ also has a natural $\mathbb{Z}_{\geq 0}^n$-grading, where $T_{(m_1, \ldots, m_n)} = T x_1^{m_1} x_2^{m_2} \cdots x_n^{m_n}$. In the same manner, a Laurent polynomial ring $T[x_1^{\pm 1}, \ldots, x_n^{\pm 1}]$ has a natural $\mathbb{Z}^n$-grading.

(d) A free algebra $F = k\langle x_1, \ldots, x_n \rangle$ has a natural $\mathbb{Z}_{\geq 0}$-grading where F_i is the k-linear span of all words in $\{x_1, \ldots, x_n\}$ of length i. If I is a homogeneous ideal of F, then F/I inherits an induced grading as in (I.12.7). For example, since the defining relations of many quantized coordinate rings are homogeneous (of degree 2), these algebras inherit natural gradings. E.g., $\mathcal{O}_q(M_2(k))$ has the nonnegative grading where $\mathcal{O}_q(M_2(k))_i$ (for $i \geq 0$) is the k-linear span of those monomials $a^r b^s c^t d^u$ such that $r + s + t + u = i$. Since the quantum determinant in this algebra is homogeneous (of degree 2), the $\mathbb{Z}_{\geq 0}$-grading on $\mathcal{O}_q(M_2(k))$ extends uniquely to a $\mathbb{Z}$-grading on the localization $\mathcal{O}_q(GL_2(k))$.

I.12.9. Definitions. Let S be a monoid, $R = \bigoplus_{s \in S} R_s$ an S-graded ring, and M a left R-module. An *S-grading* on M (compatible with the given grading on R) is, of course, an abelian group decomposition $M = \bigoplus_{s \in S} M_s$ such that $R_s M_t \subseteq M_{s+t}$ for all $s, t \in S$. A *graded R-module* is an R-module equipped with a particular grading.

ASSOCIATED GRADED RINGS AND MODULES

I.12.10. Let S be a partially ordered monoid and R an S-filtered ring. For $s \in S$, set

$$R_s^- := \sum_{\substack{t \in S \\ t < s}} R_t$$

(with the convention that $R_s^- = 0$ if s is a minimal element of S). Note that if S is totally ordered, then R_s^- is actually the *union* of the subgroups R_t for $t < s$. Now form the additive quotient groups $\mathrm{gr}(R)_s := R_s/R_s^-$, and set

$$\mathrm{gr}(R) := \bigoplus_{s \in S} \mathrm{gr}(R)_s = \bigoplus_{s \in S} R_s/R_s^-.$$

Next, observe that $R_s R_t^- + R_s^- R_t \subseteq R_{st}^-$ for all $s, t \in S$. Hence, there is a well-defined multiplication operation

$$\mathrm{gr}(R)_s \times \mathrm{gr}(R)_t \to \mathrm{gr}(R)_{st}$$

such that $[x + R_s^-][y + R_t^-] = xy + R_{st}^-$ for $x \in R_s$ and $y \in R_t$. These operations induce a multiplication operation on $\mathrm{gr}(R)$ in the obvious manner, and $\mathrm{gr}(R)$ becomes an S-graded ring, with unit $1 + R_1^-$. It is called the *associated graded ring* of R (with respect to the given filtration).

Similarly, if M is an S-filtered left R-module, there is an *associated graded module*

$$\mathrm{gr}(M) := \bigoplus_{s \in S} \mathrm{gr}(M)_s = \bigoplus_{s \in S} M_s/M_s^-$$

which is a left $\mathrm{gr}(R)$-module with $[x + R_s^-][m + M_t^-] = xm + M_{st}^-$ for $x \in R_s$ and $m \in M_t$.

I.12.11. Examples. In the examples below, given an element $x \in A_i \setminus A_{i-1}$, we will write $\bar{x}$ for the coset $x + A_{i-1}$ in $\mathrm{gr}(A)_i$.

(a) If R is a ring equipped with the trivial filtration, then $\mathrm{gr}(R)$ is just R equipped with the trivial grading.

(b) If R is an S-graded ring, for some partially ordered monoid S, then there is an S-filtration $\mathcal{F}$ on R such that $\mathcal{F}_t R = \bigoplus_{s \leq t} R_s$ for all $t \in S$. The associated graded ring $\mathrm{gr}(R)$ with respect to this filtration is just R with its original grading.

(c) Let $T = R[x; \tau, \delta]$ be a skew polynomial ring, equipped with the $\mathbb{Z}$-filtration discussed in Example I.12.2(b). Since $x \in T_1$ while $xr - \tau(r)x = \delta(r) \in T_0$ for $r \in R$, we have $\bar{x} \cdot \bar{r} = \overline{\tau(r)} \cdot \bar{x}$ in $\mathrm{gr}(T)_1$. In fact, $\mathrm{gr}(T) \cong R[\bar{x}; \tau]$.

(d) Let $A = U(\mathfrak{g})$ where $\mathfrak{g}$ is a finite dimensional Lie algebra over k. Choose a basis $x_1, \ldots, x_n$ for $\mathfrak{g}$, and note that the x_j generate A as a k-algebra. Equip A with the $\mathbb{Z}$-filtration described in Example I.12.2(c): thus A_i is the k-linear span of all products of i or fewer of the x_j. It follows from the Poincaré-Birkhoff-Witt Theorem that $\mathrm{gr}(A)$ in this case is a commutative polynomial ring $k[\overline{x}_1, \ldots, \overline{x}_n]$ (see [**158**, Corollary 1.7.5]).

(e) Let $A = k[x][y; x^3 d/dx]$ as in Example I.12.2(d). We consider three different $\mathbb{Z}$-filtrations on A.

(i) First, view A as the k-algebra generated by x and y, and equip it with the filtration described in Example I.12.2(c); in other words, A_i is the k-linear span of all products (in any order) of i or fewer xs and ys. Since $x^3 = yx - xy \in A_2$, while $x \in A_1 \setminus A_0$ and $x^2 \in A_2 \setminus A_1$, we see that $\overline{x} \cdot \overline{x^2} = 0$ in $\mathrm{gr}(A)$, that is, $\overline{x}^3 = 0$. On the other hand, $x^3 \in A_2 \setminus A_1$, and so $\overline{x^3}$ is a nonzero element of $\mathrm{gr}(A)_2$.

(ii) Next, give A the second filtration discussed in Example I.12.2(d), so that A_i is the k-linear span of those monomials $x^s y^t$ for which $s + 2t \le i$. This is the filtration associated (as in example (b) above) with a grading for which the i-th component is the k-linear span of those monomials $x^s y^t$ with $s + 2t = i$. Thus in this case, $\mathrm{gr}(A) \cong A$.

(iii) Finally, consider the filtration in which x has degree 1 while y has degree 3; thus A_i is the k-linear span of those monomials $x^s y^t$ for which $s + 3t \le i$. In this case, $x \in A_1 \setminus A_0$ and $y \in A_3 \setminus A_2$ while $yx - xy = x^3 \in A_3$, so we obtain $\overline{y} \cdot \overline{x} = \overline{x} \cdot \overline{y}$ in $\mathrm{gr}(A)$. In fact, $\mathrm{gr}(A) \cong k[\overline{x}, \overline{y}]$, a commutative polynomial ring.

(f) Let $A = k[[x]]$ be the algebra of power series in one variable over k, and equip A with the $\langle x \rangle$-adic $\mathbb{Z}$-filtration, so that $A_i = \langle x^{-i} \rangle$ for $i \le 0$ and $A_i = A$ for $i > 0$. Then $\mathrm{gr}(A) \cong k[x]$.

TRANSFER FROM $\mathrm{gr}(R)$ TO R

As Examples (c), (d), and (e)(iii) of (I.12.11) show, the associated graded ring of a filtered ring R can have a simpler structure that of R. In such cases, one would like to transfer properties from $\mathrm{gr}(R)$ (where they are typically easier to verify) back to R. Here are some sample results of this type.

I.12.12. Lemma. *Let S be a well-ordered monoid and R an S-filtered ring such that $\bigcap_{s \in S} R_s = 0$. If $\mathrm{gr}(R)$ is an integral domain (respectively, a prime ring) then R is an integral domain (respectively, a prime ring).*

Proof. First assume that $\mathrm{gr}(R)$ is a domain, and let x, y be nonzero elements of R. Let $s, t \in S$ be minimal such that $x \in R_s$ and $y \in R_y$. Then $x \notin R_s^-$ and $y \notin R_t^-$, whence $\overline{x}$ is a nonzero element of $\mathrm{gr}(R)_s$ and $\overline{y}$ is a nonzero element of $\mathrm{gr}(R)_t$. Since $\mathrm{gr}(R)$ is a domain, $\overline{x} \cdot \overline{y} \ne 0$. By definition, $\overline{x} \cdot \overline{y}$ is the coset $xy + R_{st}^-$ in R_{st}/R_{st}^-, so from $\overline{x} \cdot \overline{y} \ne 0$ we conclude that $xy \notin R_{st}^-$. In particular, $xy \ne 0$. Thus, R is a domain.

The proof for prime rings is similar. $\square$

I.12.13. Theorem. *Let n be a positive integer, equip $\mathbb{Z}_{\ge 0}^n$ with an admissible ordering, and let R be a $\mathbb{Z}_{\ge 0}^n$-filtered ring. If $\mathrm{gr}(R)$ is right noetherian (respectively,* *right artinian), then R is right noetherian (respectively, right artinian). In fact, it suffices to assume that* $\mathrm{gr}(R)$ *has ACC (respectively, DCC) on homogeneous right ideals.*

Proof. For the case when $n = 1$ see, e.g., [**158**, Proposition 1.6.7, Theorem 1.6.9]. The general case of the noetherian result is proved in [**65**, Theorem 1.5], and the proof of the artinian case is essentially the same. □

An immediate corollary of Lemma I.12.12 and Theorem I.12.13 is that for any finite dimensional Lie algebra $\mathfrak{g}$ over k, the enveloping algebra $U(\mathfrak{g})$ is a noetherian domain [**158**, Corollaries 1.7.4, 1.7.5].

In the theorem below, r.gl.dim R denotes the right global (homological) dimension of the ring R (I.15.1).

I.12.14. Theorem. *Let n be a positive integer, equip $\mathbb{Z}_{\geq 0}^n$ with an admissible ordering, and let R be a $\mathbb{Z}_{\geq 0}^n$-filtered ring. Then* r.gl.dim $R \leq$ r.gl.dim $\mathrm{gr}(R)$.

Proof. See [**158**, Corollary 7.6.18] or [**191**, Corollary 5.1.40] for the case when $n = 1$, and [**66**, Corollary 2.8] for the general case. □

COCYCLE TWISTS

For reference, we outline how the multiplication in a graded k-algebra can be twisted by suitable cocycles to obtain new k-algebras.

I.12.15. Definitions. Let S be a semigroup. A *2-cocycle on S (with values in $k^\times$)* is a map $c : S \times S \to k^\times$ such that

$$c(x, yz)c(y, z) = c(xy, z)c(x, y)$$

for all $x, y, z \in S$. In case S is an abelian group, an *alternating bicharacter on S (with values in $k^\times$)* is a map $c : S \times S \to k^\times$ such that

$$c(xy, z) = c(x, z)c(y, z) \qquad \text{and} \qquad c(y, x) = c(x, y)^{-1}$$

for all $x, y, z \in S$. Any alternating bicharacter is a 2-cocycle.

I.12.16. Let S be a monoid and A an S-graded k-algebra. Fix a 2-cocycle $c : S \times S \to k^\times$ such that $c(1, 1) = 1$. Let A^* be a copy of A, viewed as an S-graded vector space over k, and use $a \mapsto a^*$ to denote the natural S-graded vector space isomorphism of A onto A^*. Define a product on A^* so that $a^* b^* = c(x, y)(ab)^*$ for homogeneous elements $a \in A_x$ and $b \in A_y$, and extend by linearity. Then A^* is an S-graded k-algebra, called the *twist of A by c*; the map $a \mapsto a^*$ is then called the *twist map.*

For example, let $A = k[x_1, \ldots, x_n]$ be a commutative polynomial ring over k, and let $\boldsymbol{q} = (q_{ij}) \in k^\times$ be a multiplicatively antisymmetric $n \times n$ matrix. There is a natural $\mathbb{Z}_{\geq 0}^n$-grading on A, where $A_{(a_1,\ldots,a_n)} = kx_1^{a_1} x_2^{a_2} \cdots x_n^{a_n}$ for $(a_1,\ldots,a_n) \in \mathbb{Z}_{\geq 0}^n$. Define $c : \mathbb{Z}_{\geq 0}^n \times \mathbb{Z}_{\geq 0}^n \to k^\times$ by the rule

$$c\big((a_1,\ldots,a_n),(b_1,\ldots,b_n)\big) = \prod_{i>j} q_{ij}^{a_i b_j},$$

and check that c is a 2-cocycle on $\mathbb{Z}_{\geq 0}^n$. Then the twist of A by c is the quantum affine space $\mathcal{O}_q(k^n)$.

NOTES

General references for nonnegatively filtered and graded rings and modules include the books by Năstăsescu and Van Oystaeyen [170, 171], as well as [158, §7.6]. For multifiltered rings, see, e.g., [26] and its references. The cocycle twists described in (I.12.16) were introduced by Artin, Schelter, and Tate [8].

APPENDIX I.13

POLYNOMIAL IDENTITY ALGEBRAS

I.13.1. Definitions. A ring R is a *polynomial identity ring* (or *PI ring* for short) if R satisfies a monic polynomial $f \in \mathbb{Z}\langle X \rangle$. Here, $\mathbb{Z}\langle X \rangle$ is the free $\mathbb{Z}$-algebra on a finite set $X = \{x_1, \ldots, x_m\}$, and to say that R satisfies $f = f(x_1, \ldots, x_m)$ means $f(r_1, \ldots, r_m) = 0$ for all $r_1, \ldots, r_m \in R$. That f is *monic* means that at least one of the monomials of highest degree in f has coefficient 1; here degree refers to total degree. The *minimal degree* of a PI ring R is the least degree of a monic polynomial identity for R.

The polynomial f is *multilinear* if f is nonzero and has the form

$$f = \sum_{\sigma \in S_m} a_\sigma x_{\sigma(1)} \cdots x_{\sigma(m)}$$

with each $a_\sigma \in \mathbb{Z}$. In the special case where each coefficient a_σ equals the sign of σ (namely, 1 or -1 depending on whether σ is even or odd), we obtain the *standard identity of degree m*, denoted s_m.

A fundamental observation is

Proposition. *If the ring R satisfies a monic polynomial identity of degree d, then R satisfies a monic multilinear identity of degree at most d.*

Proof. [**158**, Proposition 13.1.9] □

I.13.2. Examples of PI rings. 1. Commutative rings are of course PI rings of minimal degree 2, satisfying the identity $xy - yx = 0$.

2. The **Amitsur-Levitzki Theorem** states that if d is a positive integer and A is a non-zero commutative ring, then the ring $R = M_d(A)$ satisfies the standard identity s_{2d} ([**158**, Theorem 13.3.3], [**181**, Theorem 5.1.9]). In fact, the minimal degree of R is $2d$, because of [**158**, Proposition 13.3.2].

3. If R is a ring which is a finitely generated module over a commutative subring, then R is a PI ring [**158**, Corollary 13.1.13].

4. If R is a PI ring with minimal degree d then the subrings and factor rings of R are PI rings, of minimal degree at most d.

5. Direct products and direct sums of semiprime PI rings of minimal degree at most d are PI rings of minimal degree at most d. To see this, one needs the fact that any semiprime PI ring of minimal degree at most d satisfies the standard identity s_d (Corollary I.13.3(3)).

6. [**191**, Proposition 1.7.8] Let R be a PI ring and let S be a nonempty multiplicatively closed set of central elements of R. Then the localization RS^{-1} is also a PI ring. Suppose that S consists of non zero divisors in R. Then

$$\text{minimal degree of } R \;=\; \text{minimal degree of } RS^{-1}.$$

These claims follow easily from Proposition I.13.1.

7. Not every PI ring is a finitely generated module over a commutative ring [**191**, Example 5.1.18]; nor is every PI ring a subring of a matrix ring over a commutative ring ([**181**, Theorem 6.4.4], [**191**, Example 3.2.48]).

8. Lie algebras in positive characteristic. Let k be a field of positive characteristic p, and let $\mathfrak{g}$ be a finite dimensional Lie algebra over k. We say that $\mathfrak{g}$ is *restricted* if $\mathfrak{g}$ is a subalgebra of $\mathfrak{gl}_n(k) = M_n(k)$ for some n, and for each $x \in \mathfrak{g}$, $x^{[p]} \in \mathfrak{g}$ also, (where here $x^{[p]}$ is used to denote the pth power of the matrix x). One can define an abstract notion of a restricted Lie algebra which does not depend on any assumed embedding in $M_n(k)$, and in fact it can be shown that the abstract definition is equivalent to the one given above.

Theorem. *Let $\mathfrak{g}$ be a finite dimensional restricted Lie algebra over k, with basis $\{x_1, \ldots, x_m\}$. Let $Z_0 = k\langle x^p - x^{[p]} : x \in \mathfrak{g}\rangle \subseteq U(\mathfrak{g})$.*

1. $Z_0 = k[x_1^p - x_1^{[p]}, \ldots, x_m^p - x_m^{[p]}]$ is a central sub-Hopf algebra of $U(\mathfrak{g})$, isomorphic to the polynomial algebra over k in m indeterminates.

2. $U(\mathfrak{g})$ is a free Z_0-module of rank p^m.

Proof. (Sketch) Let λ_y and ρ_y denote left and right multiplication by y in the associative k-algebra A. Then $\text{ad}(y)(x) = yx - xy = (\lambda_y - \rho_y)(x)$ for $x \in A$, so that $\text{ad}(y)^p(x) = (\lambda_y - \rho_y)^p(x) = y^p x - x y^p$ by the Binomial Theorem in characteristic p. Thus, taking $A = M_n(k)$, we obtain $\text{ad}(y)^p = \text{ad}(y^{[p]})$ for all $y \in M_n(k)$. Embedding $\mathfrak{g}$ in $M_n(k)$ we deduce that $\text{ad}(x)^p - \text{ad}(x^{[p]}) = 0$ on $M_n(k)$. But the same argument as above, applied this time in $A = U(\mathfrak{g})$, gives $\text{ad}(x)^p = \text{ad}(x^p)$ on $U(\mathfrak{g})$. So it follows that $\text{ad}(x^p - x^{[p]}) = 0$ on $\mathfrak{g}$ and hence on $U(\mathfrak{g})$; that is, $x^p - x^{[p]}$ lies in the centre of $U(\mathfrak{g})$.

That $\Delta(Z_0) \subseteq Z_0 \otimes Z_0$ follows also from the Binomial Theorem in characteristic p, and the remaining requirements for being a Hopf subalgebra are clear. To see that Z_0 is generated by $\{x_i^p - x_i^{[p]} : 1 \leq i \leq m\}$, use the fact that the map $\psi : \mathfrak{g} \to Z_0 : x \mapsto x^p - x^{[p]}$ is semilinear, meaning that for $x, y \in \mathfrak{g}$ and $\alpha \in k$,

$$\psi(x + y) = \psi(x) + \psi(y) \qquad \text{and} \qquad \psi(\alpha x) = \alpha^p \psi(x).$$

This is proved in [**105**, Chapter V, Formula (63), p. 187].

Finally, the remaining claims follow from the Poincaré-Birkhoff-Witt Theorem for $U(\mathfrak{g})$. $\square$

9. $\mathfrak{sl}_2(k)$ in positive characteristic. When $\mathfrak{g} = \mathfrak{sl}_2(k)$ with basis e, f, h as in (I.9.11)(a), $e^{[p]} = f^{[p]} = 0$ and $h^{[p]} = h$, so that the central sub-Hopf algebra $Z_0 \subseteq U(\mathfrak{sl}_2(k))$ as above is given by

$$Z_0 = k[e^p, f^p, h^p - h].$$

I.13.3. Primitive and prime PI rings. A *central simple algebra* over a field F is a simple F-algebra R whose centre is just F (that is, $F \cdot 1$) and which is finite dimensional over F. To say that R is a central simple algebra without mention of a particular field just means that R is a central simple algebra over its centre Z. Clearly then R is a complete matrix ring over a central simple division ring by the Artin-Wedderburn Theorem, and by a classical result [**191**, Corollary 2.3.25],

$$\dim_Z(R) = n^2$$

for some positive integer n. In this case n is called the *PI degree* of R.

Fundamental to the theory of PI rings is

Kaplansky's Theorem. *Let R be a primitive PI ring of minimal degree d. Then d is even and R is a central simple algebra of dimension $(d/2)^2$ over its centre.*

Proof. [**158**, Theorem 13.3.8] □

Notice in particular that this implies that the primitive ideals of a PI ring are maximal. The generalisation of Kaplansky's theorem to prime rings is

Posner's Theorem. *Let R be a prime PI ring with centre Z and with minimal degree d. Let $S = Z \setminus \{0\}$, let $Q = RS^{-1}$, and let $F = ZS^{-1}$ denote the quotient field of Z. Then Q is a central simple algebra with centre F, and $\dim_F(Q) = (d/2)^2$.*

Proof. [**158**, Theorem 13.6.5], [**191**, Theorem 6.1.30] □

Observe that Posner's Theorem tells us in particular that every prime PI ring R is a prime (right and left) Goldie ring, whose complete ring of fractions Fract(R) can be got by inverting the non-zero central elements of R. In the light of the above theorems the definition at the beginning of this subsection is extended by saying that the *PI degree* of a prime PI ring R, denoted PI-deg(R), is the square root of the vector space dimension of its Goldie quotient ring over the field of fractions of its centre. By Posner's Theorem,

$$\text{PI-deg}(R) = \tfrac{1}{2}(\text{minimal degree of } R).$$

In particular, it follows that PI-deg(R/P) $\leq$ PI-deg(R) for all prime ideals P of R. We do not give the definition of PI degree for non-prime PI rings since that would require further technical developments (see [**191**, Vol. II, p. 98]).

Two immediate but important consequences of Posner's Theorem are given in the first two statements of the

Corollary. 1. *Let R be a prime PI ring with centre Z. If I is a nonzero ideal of R then $I \cap Z \neq 0$.*

2. *Let R be a prime PI ring. If S is a subring of $\mathrm{Fract}(R)$ with $R \subseteq S$, then S is a prime PI ring and*

$$\mathrm{PI\text{-}deg}(S) = \mathrm{PI\text{-}deg}(R).$$

3. *A semiprime PI ring R satisfies a polynomial identity of degree at most d if and only if R satisfies s_d.*

Proof. As noted above, 1 and 2 are immediate from Posner's Theorem.

3. Let R be a semiprime PI ring satisfying a polynomial identity of degree at most d. Since R is semiprime, the intersection of its prime ideals is 0. Hence, to see that R satisfies s_d, it will be enough to show that all its prime factors satisfy s_d. Thus we may assume, without loss of generality, that R is a prime ring. By Posner's Theorem, R is contained in a simple algebra Q with centre F such that $\dim_F(Q) \leq (d/2)^2$. Let L be a maximal subfield of Q, and form $\overline{Q} := Q \otimes_F L$. It follows that $\overline{Q} \cong M_n(L)$ for some $n \leq d/2$ [**158**, Corollary 13.3.5]. By the Amitsur-Levitzki Theorem (see Example I.13.2(2)), $\overline{Q}$ satisfies s_{2n}. Since $R \subseteq Q \subseteq \overline{Q}$, the ring R satisfies s_{2n} and therefore s_d. $\square$

Part 3 of this corollary can also be proved using Kaplansky's Theorem in place of Posner's, as in [**181**, Theorem 6.4.1] or [**191**, Theorem 6.1.26].

I.13.4. The Artin-Tate Lemma. As we shall see, many of the affine algebras arising in quantum groups and in Lie theory are finite modules over their centres (and hence PI rings) when either the quantizing parameter is a non-trivial root of unity or the ground field has positive characteristic. In such cases we want to know that the centre of the algebra is also affine, so as to be able to make use of a geometric perspective and geometric techniques. That this is so is an immediate consequence of the following famous result:

Artin-Tate Lemma. *Let $K \subseteq B \subseteq A$ be rings. Suppose that K is Noetherian, that B is central in A and that A is an affine K-algebra and a finitely generated B-module. Then B is an affine K-algebra.*

Proof. [**158**, Lemma 13.9.10] $\square$

I.13.5. Affine PI algebras over algebraically closed fields. Kaplansky's Theorem (I.13.3) has striking consequences when A is a prime affine PI algebra over an algebraically closed field k. Thus, let $\overline{A}$ be any primitive factor of such an algebra A, and let $\overline{Z}$ be the centre of $\overline{A}$. By Kaplansky's Theorem, $\overline{Z}$ is a field and $\dim_{\overline{Z}}(\overline{A}) = (\mathrm{PI\text{-}deg}(\overline{A}))^2 \leq (\mathrm{PI\text{-}deg}(A))^2$. Then by the Artin-Tate Lemma I.13.4, $\overline{Z}$ is an affine k-algebra, and hence $\dim_k(\overline{Z}) < \infty$. Since k is algebraically closed, $\overline{Z} = k$ and thus $\dim_k(\overline{A}) \leq (\mathrm{PI\text{-}deg}(A))^2$. By the Artin-Wedderburn Theorem we conclude that $\overline{A} = M_d(k)$ for some $d \leq \mathrm{PI\text{-}deg}(A)$. We have therefore proved

Theorem. *Let A be a prime affine PI algebra over the algebraically closed field k, with $\mathrm{PI\text{-}deg}(A) = n$.*

1. *If S is a primitive factor of A then*

$$S \cong M_t(k)$$

for some integer t bounded above by n.

2. *Let V be an irreducible A-module. Then V is a vector space over k of dimension t, where $t \leq n$, and $A/\operatorname{ann}_A(V) \cong M_t(k)$.* $\square$

In fact the upper bounds in both parts of the theorem are attained – see Lemma III.1.2(2).

NOTES

A comprehensive reference for the theory of PI rings is [**191**], but most of the above can also be found in [**158**]. One point perhaps worthy of special note is that Posner's theorem as stated in (I.13.3) is a considerably stronger and more precise result than that actually proved by Posner. In the form stated here it's due to Formanek, but the labelling we've used follows the current conventions in the literature (and the advice of Formanek).

SKEW POLYNOMIAL RINGS
SATISFYING A POLYNOMIAL IDENTITY

In this short appendix, we prove a result of Jøndrup on the PI degree of skew polynomial algebras in characteristic 0, and illustrate its relevance in the settings of interest in these notes. Exceptionally, we make use in this section of a few lemmas from Part II.

I.14.1. Jøndrup's Theorem. Recall the notation $R[x; \tau, \delta]$ for a skew polynomial ring with coefficient ring R introduced in (I.1.11). Clearly $R[x; \tau, \delta]$ is a PI ring only if R is also, by Example I.13.2(4).

Theorem. *Let R be a prime PI algebra over a field of characteristic 0, and let $T = R[x; \tau, \delta]$ be a skew polynomial ring.*

1. Suppose that τ is the identity on the centre Z of R. Then there exists a unit $u \in \mathrm{Fract}(R)$ such that τ is given by conjugation by u, and T is a PI ring if and only if $u\delta$ is an inner derivation on $\mathrm{Fract}(R)$. In this case, $R[x'; \tau]$ is a PI ring and

$$\mathrm{PI\text{-}deg}(T) = \mathrm{PI\text{-}deg}(R) = \mathrm{PI\text{-}deg}(R[x'; \tau]).$$

2. Suppose that τ is not the identity on Z. Then T is a PI ring if and only if $\tau|_Z$ has finite order. In this case, $R[x'; \tau]$ is a PI ring and

$$\mathrm{PI\text{-}deg}(T) = \mathrm{PI\text{-}deg}(R[x'; \tau]).$$

3. If T satisfies a polynomial identity, then so does $R[x'; \tau]$, and

$$\mathrm{PI\text{-}deg}(T) = \mathrm{PI\text{-}deg}(R[x'; \tau]).$$

Proof. By Lemma I.1.12(b), all the rings featuring in the theorem are prime. Thus, in view of Lemma II.5.6 together with Posner's Theorem (I.13.3) and Corollary I.13.3(2), there is no harm in replacing R by $\mathrm{Fract}(R)$ throughout – that is, we shall assume without loss of generality that R is a central simple algebra.

1. Suppose that τ is the identity map on Z. Then τ is given by conjugation by some unit $u \in R$ by the Noether-Skolem theorem [**191**, Theorem 3.1.2]. Thus, by Lemma II.5.5(b),

$$T = R[ux; u\delta],$$

where $u\delta$ is an ordinary derivation on R.

Suppose first that $u\delta$ is an inner derivation. By Lemma II.5.5(c),

$$T = R[ux - a] \cong R[z],$$

for some $a \in R$, where $R[z]$ is an ordinary polynomial ring. Hence in this case, T is a PI ring and PI-deg(T) = PI-deg(R). Moreover, Lemma II.5.5(b) shows that $R[x';\tau] = R[ux'] \cong R[z]$, whence $R[x';\tau]$ is a PI ring with the same PI degree as R.

Now suppose that $u\delta$ is not inner. In this case, since R is simple it follows that T is simple ([**83**, Proposition 1.14], [**158**, Theorem 1.8.4]). But the chain of left ideals $Tx \supset Tx^2 \supset \cdots \supset Tx^i \supset \cdots$ shows that T is not Artinian. Therefore by Kaplansky's Theorem T cannot be a PI ring.

2. Suppose that τ is not trivial on Z; thus there exists $c \in Z$ such that $c - \tau(c)$ is a nonzero element of Z. By [**68**, Lemma 2.4], δ is an inner τ-derivation, and so

$$T = R[x';\tau] \tag{$*$}$$

by Lemma II.5.5(c). Now T is a finitely generated free module over $T' := Z[x';\tau|_Z]$ and so T is a subring of a complete matrix ring over T'. Thus T is a PI ring if and only if T' is a PI ring. Being a principal ideal domain ([**83**, Theorem 1.11], [**158**, Theorem 1.2.9(ii)]), T' is a PI ring if and only if it is a finite module over its centre [**158**, Theorem 13.9.16]. The latter is easily seen to happen if and only if $\tau|_Z$ has finite order. In view of $(*)$ this completes the proof of 2.

3. This is immediate from 1 and 2. $\square$

The theorem above fails comprehensively in positive characteristic. For example, if char$(k) = p > 0$, it is known that the Weyl algebra $A_1(k) = k[x][y; d/dx]$ has PI-degree p, whereas $k[x][y']$ is commutative.

The following slight improvement of a result of De Concini and Procesi [**44**, Theorem 6.4] is easily deduced from the theorem above.

Corollary. *Let k be a field of characteristic 0, let $n \geq 1$, and let T be a PI algebra over k which is an iterated skew polynomial ring of the form*

$$T = k[x_1][x_2; \tau_2, \delta_2] \cdots [x_n; \tau_n, \delta_n].$$

Assume that there exist elements $q_{ij} \in k^\times$ such that

$$\tau_i(x_j) = q_{ij}x_j$$

for all i, j with $1 \leq j < i \leq n$. Write $q_{ji} = q_{ij}^{-1}$ for $j < i$ and $q_{ii} = 1$ for all i, and set $q = (q_{ij})$. Then $\mathcal{O}_q(k^n)$ is a PI algebra and

$$\text{PI-deg}(T) = \text{PI-deg}(\mathcal{O}_q(k^n)).$$

Proof. Induct on the maximum index $m \leq n$ such that $\delta_i = 0$ for $i = 1, \ldots, m$. If $m = n$, then $T = \mathcal{O}_q(k^n)$ and we are done. Hence, we may assume that $m < n$. Set $S = k[x_1][x_2; \tau_2, \delta_2] \cdots [x_{n-1}; \tau_{n-1}, \delta_{n-1}]$.

Since $S[x_n; \tau_n, \delta_n] = T$ is a PI ring, the theorem shows that $S[x_n'; \tau_n]$ is a PI ring, with the same PI degree as T. Using the relations $\tau_i(x_j) = q_{ij} x_j$, we can rewrite $S[x_n'; \tau_n]$ as an iterated skew polynomial ring of the form

$$S[x_n'; \tau_n] = k[x_n'][x_1; \tau_1', \delta_1'][x_2; \tau_2', \delta_2'] \cdots [x_{n-1}; \tau_{n-1}', \delta_{n-1}'],$$

where the restrictions of τ_i' and δ_i' to $k\langle x_1, \ldots, x_{i-1}\rangle$ agree with τ_i and δ_i, while $\tau_i'(x_n') = q_{ni}^{-1} x_n'$ and $\delta_i'(x_n') = 0$. Our inductive hypothesis now applies to $S[x_n'; \tau_n]$. Using that result and reordering the variables yields the desired conclusion. $\square$

I.14.2. Quantum affine spaces satisfying a PI.

In view of Corollary I.14.1, it is of interest to know when a quantum affine space $\mathcal{O}_q(k^n)$ is a PI ring, and what its PI degree is. This can be answered as follows.

Proposition. *Let $q = (q_{ij})$ be a multiplicatively antisymmetric $n \times n$ matrix over k.*

1. The algebra $\mathcal{O}_q(k^n)$ is a PI ring if and only if all the q_{ij} are roots of unity. In this case, there exist a primitive root of unity $q \in k^\times$ and integers a_{ij} such that $q_{ij} = q^{a_{ij}}$ for all i, j.

2. Suppose that $q_{ij} = q^{a_{ij}}$ for all i, j, where $q \in k^\times$ is a primitive ℓ-th root of unity and the $a_{ij} \in \mathbb{Z}$. Let h be the cardinality of the image of the homomorphism

$$\mathbb{Z}^n \xrightarrow{(a_{ij})} \mathbb{Z}^n \xrightarrow{\pi} (\mathbb{Z}/\ell\,\mathbb{Z})^n,$$

where π denotes the canonical epimorphism. Then $\mathrm{PI\text{-}deg}(\mathcal{O}_q(k^n)) = \sqrt{h}$.

Proof. 1. If all the q_{ij} are roots of unity, there exists a positive integer ℓ such that $q_{ij}^\ell = 1$ for all i, j. It follows that x_i^ℓ lies in the centre Z of $\mathcal{O}_q(k^n)$ for all i. Consequently, $\mathcal{O}_q(k^n)$ is a finitely generated module over Z, and thus it is a PI ring. Note that the q_{ij} generate a finite subgroup Q of the multiplicative group $k^\times$. Then Q must be cyclic, say $Q = \langle q \rangle$, the generator q must be a primitive root of unity, and each $q_{ij} = q^{a_{ij}}$ for some $a_{ij} \in \mathbb{Z}$.

Now suppose that some q_{ij} is not a root of unity. Then $i \neq j$, and after renumbering the generators we may assume that q_{12} is not a root of unity. In this case, one easily checks that $\mathcal{O}_{q_{12}}((k^\times)^2)$ is a simple ring with centre k. By Kaplansky's Theorem (I.13.3), it cannot be a PI ring, and so by Corollary I.13.3(2), $\mathcal{O}_{q_{12}}(k^2)$ cannot be a PI ring. Since $\mathcal{O}_{q_{12}}(k^2)$ is a subalgebra of $\mathcal{O}_q(k^n)$, we conclude that $\mathcal{O}_q(k^n)$ is not a PI ring.

2. [**44**, Proposition 7.1]. $\square$

I.14.3. Examples. **1.** The single parameter quantum affine space $\mathcal{O}_q(k^n)$ can be written in the form $\mathcal{O}_q(k^n)$ where $q = (q^{a_{ij}})$ and (a_{ij}) is the antisymmetric integer matrix with $a_{ij} = 1$ for all $i < j$. Now suppose that q is a primitive ℓ-th root of unity, and let's apply the algorithm of Proposition I.14.2(2). We first perform the following row reductions on the matrix (a_{ij}):

- For $i = 1, \ldots, n-1$, subtract the $(i+1)$-st row from the i-th row.
- If n is even, add the (new) first, third, $\ldots$, $(n-1)$-st rows to the last row.
- If n is odd, add the (new) first, third, $\ldots$, $(n-2)$-nd rows to the last row.

This yields the matrix

$$(\tilde{a}_{ij}) = \begin{pmatrix} 1 & 1 & 0 & 0 & \cdots & 0 & 0 & 0 \\ 0 & 1 & 1 & 0 & \cdots & 0 & 0 & 0 \\ 0 & 0 & 1 & 1 & \cdots & 0 & 0 & 0 \\ & \vdots & & & \ddots & & \vdots & \\ 0 & 0 & 0 & 0 & \cdots & 1 & 1 & 0 \\ 0 & 0 & 0 & 0 & \cdots & 0 & 1 & 1 \\ 0 & 0 & 0 & 0 & \cdots & 0 & 0 & z \end{pmatrix}$$

where $z = 1$ if n is even, while $z = 0$ if n is odd. Since the images of the homomorphisms $\mathbb{Z}^n \to \mathbb{Z}^n$ given by (a_{ij}) and $(\tilde{a}_{ij})$ coincide, the number h is either l^n (if n is even) or l^{n-1} (if n is odd). Therefore the proposition shows that

$$\text{PI-deg}(\mathcal{O}_q(k^n)) = \ell^{\lfloor n/2 \rfloor},$$

where $\lfloor n/2 \rfloor$ is the integer part of $n/2$.

2. Let $A = \mathcal{O}_q(M_2(k))$, with generators a, b, c, d as in (I.1.7), and assume that q is a primitive ℓ-th root of unity, where $\ell > 1$. It follows immediately from the relations that b^ℓ and c^ℓ are central in A, that a^ℓ commutes with a, b, c, and that d^ℓ commutes with b, c, d.

Write $A = k[a][b; \tau_2][c; \tau_3][d; \tau_4, \delta_4]$; one easily verifies that $\delta_4\tau_4 = q^2\tau_4\delta_4$. Since $(q^2)^\ell = 1$, we have $\binom{\ell}{i}_{q^2} = 0$ for $0 < i < \ell$, and so it follows from the q-Leibniz Rule (I.8.4) that

$$d^\ell a = \tau_4^\ell(a)d^\ell + \delta_4^\ell(a) = ad^\ell + (q^{-1} - q)\delta_4^{\ell-1}(bc) = ad^\ell.$$

Thus, d^ℓ is central in A. (This can also be checked directly.) Now A can also be written in the form $k[d][c; \tau_2'][b; \tau_3'][a; \tau_4', \delta_4']$, with $\delta_4'\tau_4' = q^{-2}\tau_4'\delta_4'$. Repeating the argument above, we find that a^ℓ is central in A. Therefore A is a finitely generated module over its centre, whence A is a PI ring.

By Corollary I.14.1, the PI degree of A is the same as that of $\mathcal{O}_q(k^4)$ where

$$q = \begin{pmatrix} 1 & q & q & 1 \\ q^{-1} & 1 & 1 & q \\ q^{-1} & 1 & 1 & q \\ 1 & q^{-1} & q^{-1} & 1 \end{pmatrix}.$$

It then follows via Proposition I.14.2 that $\text{PI-deg}(A) = \ell$.

3. If $q \in k^\times$ is a primitive ℓ-th root of unity where ℓ is odd, then

$$\text{PI-deg}(\mathcal{O}_q(M_n(k))) = \ell^{n(n-1)/2}$$

[**106**, Theorem 5.7].

NOTES

The theorem and corollary in (I.14.2) were proved by Jøndrup in [111], which built on earlier work of De Concini and Procesi [44, Chapters 6 and 7], which was directed more specifically towards applications to quantized enveloping algebras. Some other techniques for the analysis of skew polynomial rings can be found in [68, 77].

APPENDIX I.15

HOMOLOGICAL CONDITIONS

Throughout this appendix, let R be a noetherian ring.

I.15.1. Projective, injective, and global dimensions. The projective and injective dimensions of a module M will be denoted by $\mathrm{p.dim}(M)$ and $\mathrm{inj.dim}(M)$, respectively. The *right global dimension* of a ring R, denoted $\mathrm{r.gl.dim}(R)$, is the supremum of the projective dimensions of the right R-modules. The *left global dimension* is defined analogously. Recall that the right and left global dimensions of a noetherian ring coincide [**190**, Corollary 9.23]. Hence, we just write $\mathrm{gl.dim}(R)$ for this dimension. It is also known that if $\mathrm{inj.dim}(_RR)$ and $\mathrm{inj.dim}(R_R)$ are both finite, then these numbers coincide [**216**, Lemma A]. In this case, we just write $\mathrm{inj.dim}(R)$ for the common value.

I.15.2. Grade. The *grade* (or *j-number*) of a finitely generated R-module M is defined to be

$$j(M) := \inf\{j \geq 0 \mid \mathrm{Ext}_R^j(M, R) \neq 0\}.$$

We note that finite projective dimension implies finite grade, since if $M \neq 0$ and $\mathrm{p.dim}(M) = n < \infty$, then $\mathrm{Ext}_R^n(M, R) \neq 0$ (cf. [**190**, Exercise 9.6]). In particular, if $\mathrm{gl.dim}(R) < \infty$, then every nonzero finitely generated R-module has finite grade. The latter conclusion is also known to hold when R has finite injective dimension [**137**, Remark 2.2(1)].

I.15.3. Auslander-Gorenstein and Auslander-regular rings. The ring R satisfies the *Auslander condition* if $\mathrm{Ext}_R^i(N, R) = 0$ for all R-submodules N of $\mathrm{Ext}_R^j(M, R)$ whenever $0 \leq i < j$ and M is a finitely generated (right or left) R-module. (It follows from our noetherian assumption that any such N is finitely generated.) This condition implies that $j(N) \geq j(M)$ whenever M is a finitely generated (right or left) R-module and N is an R-submodule of any $\mathrm{Ext}_R^l(M, R)$.

The ring R is *Auslander-regular* provided $\mathrm{gl.dim}(R) < \infty$ and R satisfies the Auslander condition. Similarly, R is *Auslander-Gorenstein* provided $\mathrm{inj.dim}(R) < \infty$ (on both sides) and R satisfies the Auslander condition.

Theorem. *If R is an Auslander-regular (respectively, Auslander-Gorenstein) noetherian ring, then so is any skew polynomial ring $R[x; \tau, \delta]$.*

Proof. [**53**, Theorem 4.2] $\quad\square$

For example, since the field k is clearly Auslander-regular, it follows that the algebras $\mathcal{O}_q(k^n)$, $\mathcal{O}_{\lambda,p}(M_n(k))$, and $A_n^{Q,\Gamma}(k)$ are all Auslander-regular.

I.15.4. Macaulay and Cohen-Macaulay rings. These concepts involve the *Krull dimension* and *Gel'fand-Kirillov dimension*, denoted K.dim and GK.dim, respectively. We refer the reader to [**83**], [**127**], and [**158**] for definitions and properties of these dimensions.

The ring R is said to be *Macaulay* provided $j(M)+\mathrm{K.dim}(M) = \mathrm{K.dim}(R)$ for every nonzero finitely generated R-module M. Similarly, if R is an algebra then it is *Cohen-Macaulay* (or *GK-Cohen-Macaulay*, or just *CM*) if $j(M)+\mathrm{GK.dim}(M) = \mathrm{GK.dim}(R)$ for every nonzero finitely generated R-module M. It is known that if R is an affine noetherian PI algebra, then $\mathrm{GK.dim}(M) = \mathrm{K.dim}(M)$ for all finitely generated R-modules M. Thus, in this case, the Macaulay and Cohen-Macaulay conditions coincide.

It is a classical result that every commutative affine algebra of finite injective dimension is both Auslander-Gorenstein and Cohen-Macaulay. For typical noncommutative noetherian rings, the latter conditions are much stronger than finite injective dimension, but there are some connections in special cases, as follows. (See [**158**, 5.1.1] for the concept of a maximal order. Connected graded k-algebras and gr-prime ideals are defined in (I.12.7). A *normal element* in a ring R is any element x satisfying $xR = Rx$.)

Theorem 1. *Let A be a connected graded noetherian PI algebra with finite injective dimension. Then A is Auslander-Gorenstein and Cohen-Macaulay, and*

$$\mathrm{GK.dim}(A) = \mathrm{K.dim}(A) = \mathrm{cl.K.dim}(A) = \mathrm{inj.dim}(A).$$

If, moreover, A has finite global dimension, then A is a domain and a maximal order in its quotient division ring.

Proof. [**198**, Theorem 1.1, Corollary 1.2] $\square$

Theorem 2. *Let A be a connected graded noetherian algebra with finite injective dimension. Assume that for every non-maximal gr-prime ideal P of A, the factor algebra A/P contains a nonzero homogeneous normal element of positive degree. Then A is Auslander-Gorenstein and Cohen-Macaulay. If, moreover, A has finite global dimension, then A is a domain and a maximal order in its quotient division ring.*

Proof. [**217**, Theorem 0.2] $\square$

Finally, here is a useful lemma for checking the Auslander-regular and Cohen-Macaulay conditions:

Lemma. *Let $S = R[x;\tau,\delta]$ where R is an Auslander-regular, Cohen-Macaulay, noetherian ring.*

(a) *If R is a connected graded algebra and $\tau(R_i) = R_i$ for all graded components R_i, then S is Cohen-Macaulay.*

(b) *If z is a central regular element of R, then R/zR is Auslander-Gorenstein and Cohen-Macaulay.*

Proof. [**138**, Lemma, p. 184] □

For instance, we can view $\mathcal{O}_q(k^n)$ as a connected graded k-algebra where all the generators x_i have degree 1. By induction, part (a) of the lemma together with Theorem I.15.3 then shows that $\mathcal{O}_q(k^n)$ is Auslander-regular and Cohen-Macaulay. Similarly, $\mathcal{O}_{\lambda,\boldsymbol{p}}(M_n(k))$ is Auslander-regular and Cohen-Macaulay. It then follows from part (b) of the lemma that $\mathcal{O}_{\lambda,\boldsymbol{p}}(SL_n(k))$ is Auslander-Gorenstein and Cohen-Macaulay [**138**, Corollary]. In fact, since there is an isomorphism $\mathcal{O}_{\lambda,\boldsymbol{p}}(GL_n(k)) \cong \mathcal{O}_{\lambda,\boldsymbol{p}}(SL_n(k))[z^{\pm 1}]$ (Lemma I.2.9 and Exercise I.2.K), one has

$$\mathrm{gl.dim}\,\mathcal{O}_{\lambda,\boldsymbol{p}}(SL_n(k)) \leq \mathrm{gl.dim}\,\mathcal{O}_{\lambda,\boldsymbol{p}}(GL_n(k)) \leq \mathrm{gl.dim}\,\mathcal{O}_{\lambda,\boldsymbol{p}}(M_n(k)) \leq n^2,$$

and thus $\mathcal{O}_{\lambda,\boldsymbol{p}}(SL_n(k))$ is actually Auslander-regular [**138**, Corollary]. We shall discuss these examples in more detail in Chapter II.9.

NOTES

A comprehensive reference for homological algebra is [**190**]; a number of items that we need are also found in [**158**]. For a more detailed discussion of the Auslander conditions, see [**15**]. A short but clear survey of the key basic properties of the Auslander and Cohen-Macaulay conditions can be found in [**137**]. In the past few years it has proved fruitful to introduce these concepts in a somewhat more sophisticated setting, namely in the language of derived categories, and for rings admitting dualizing complexes rather than those with finite injective dimension. A starting point from which to explore this range of ideas is [**215**].

APPENDIX I.16

LINKS AND BLOCKS

I.16.1. Artinian rings. Let Λ be an Artinian ring. Recall that Λ is said to be *indecomposable* if Λ cannot be split as the direct sum of two non-zero rings. In general we can decompose Λ as

$$\Lambda \;=\; \Lambda_1 \oplus \cdots \oplus \Lambda_t, \tag{1}$$

a finite direct sum of indecomposable (and necessarily Artinian) rings. Equivalently,

$$1_\Lambda \;=\; e_1 + \cdots + e_t \tag{2}$$

where the e_i are *pairwise orthogonal primitive central idempotents*. (Central idempotents are *orthogonal* if their product is zero; a central idempotent is *primitive* if it is non-zero and it cannot be written as the sum of two orthogonal non-zero central idempotents.) It's clear how to get (2) from (1), and to get (1) from (2), take $\Lambda_i = \Lambda e_i$. These decompositions are *unique* (up to permutation of the indices) – indeed if Ω is *any* ring summand of Λ then there is a nonempty subset $\mathcal{J}_\Omega$ of $\{1, \dots, t\}$ such that

$$\Omega \;=\; \bigoplus_{j \in \mathcal{J}_\Omega} \Lambda_j.$$

In fact, $\mathcal{J}_\Omega = \{i \in \{1, \dots, t\} : e_i \Omega \neq 0\}$. We call $\Lambda_1, \dots, \Lambda_t$ the *blocks* of Λ.

If V is an indecomposable (left) Λ-module there exists a unique $i \in \{1, \dots, t\}$ such that $e_i V \neq 0$. Thus $e_i V = V$ and we say that V *belongs to the block* Λ_i. Now let V and W be arbitrary Λ-modules, and consider the (unique) block decompositions of V and W, namely $V = \bigoplus_i e_i V$ and $W = \bigoplus_i e_i W$. Then $\mathrm{Hom}_\Lambda(W, V) = \bigoplus_{i=1}^t \mathrm{Hom}_\Lambda(e_i W, e_i V)$. Thus the representation theory of Λ can be studied one block at a time.

The same terminology can be applied also to maxspec Λ – thus we shall say that a maximal ideal M of Λ *belongs to the block* Λ_i if and only if the irreducible module it annihilates belongs to Λ_i. Hence, M belongs to Λ_i if and only if $e_i \notin M$, in which case $e_j \in M$ for all $j \neq i$ and so $M = M e_i \oplus \Lambda(1 - e_i)$, where $M e_i$ is a maximal ideal of Λ_i. Thus, $M, N \in$ maxspec Λ are in the same block if and only if, for $1 \leq i \leq t$, we have $e_i \in M$ if and only if $e_i \in N$.

For comparison with the results in (III.9.2) consider the case where Λ is an *Artin algebra* – that is, Λ is a finitely generated module over an Artinian subring of its centre $Z(\Lambda)$. It's an easy exercise to see that then $Z(\Lambda)$ is Artinian, the centre of each indecomposable component Λ_i is a local ring, and each maximal ideal of Λ contracts to a maximal ideal of $Z(\Lambda)$. Thus, we can restate the description of the blocks in maxspec Λ in this case as follows: When Λ is an Artin algebra,

$M, N \in$ maxspec Λ are in the same block

$\Longleftrightarrow \quad M \cap Z(\Lambda) \quad = \quad N \cap Z(\Lambda)$

$\Longleftrightarrow \quad M$ and N contain the same primitive central idempotents.

I.16.2. Noetherian rings. The concept of a block can fruitfully be extended to arbitrary prime ideals of a Noetherian ring – one is led to the ideas of a "link" between prime ideals and a "clique" of prime ideals, which have been intensively studied over the past 20 years. However in order to avoid technical complications we'll restrict our attention here to the special case of finite dimensional modules over a Noetherian k-algebra A, (k being a field as usual). First, let V and W be two irreducible (left) A-modules of finite k-dimension, with $\operatorname{ann}_A(V) = M$ and $\operatorname{ann}_A(W) = N$. We say that V *is linked to* W, and write $V \rightsquigarrow W$, if and only if $\operatorname{Ext}^1_A(W, V) \neq 0$. And we say that M *is linked to* N, and write $M \rightsquigarrow N$, if and only if $MN \neq M \cap N$.

Lemma. *With the above hypotheses and notation, $V \rightsquigarrow W$ if and only if $M \rightsquigarrow N$.*

Proof. Note that A/MN is a finite dimensional k-algebra.

($\Longrightarrow$): Suppose that $\operatorname{Ext}^1_A(W, V) \neq 0$, and let U be a non-split extension of V by W. Then $(M \cap N)U \neq 0$ (because $A/(M \cap N)$ is a semisimple ring), whereas $MNU = 0$, so that $M \rightsquigarrow N$.

($\Longleftarrow$): Suppose that MN is strictly contained in $M \cap N$. Let $U = A/MN$ and $U' = N/MN$, viewed as left A-modules; $U' \neq 0$ due to our assumptions. Since U' and U/U' are annihilated by M and N respectively, $U' \cong \bigoplus_{i=1}^m V$ and $U/U' \cong \bigoplus_{j=1}^n W$ for some $m, n > 0$. On the other hand, $(M \cap N)U \neq 0$, so U' cannot be a direct summand of U. This implies that $\operatorname{Ext}^1_A(U/U', U') \neq 0$, and therefore $\operatorname{Ext}^1_A(W, V) \neq 0$, as required. $\square$

In general, the linkage relation defined above is of course neither reflexive, symmetric, nor transitive. However we can consider the relation on finite dimensional irreducible A-modules (or, equivalently, on maximal ideals of finite vector space codimension) obtained as the reflexive, symmetric and transitive closure of the linkage relation. The equivalence classes of the relation so constructed are then the *blocks* of A. To make this definition precise it's convenient to think pictorially. Thus we form a directed graph whose vertices are labelled by the isomorphism classes of finite dimensional irreducible (left) A-modules, (or, equivalently, by the annihilators of these modules). There is then a directed edge (i.e., an arrow) from the point V to the point W if and only if $\operatorname{Ext}^1_A(W, V) \neq 0$. Let

$\mathcal{M} = \{M \in \text{maxspec } A : \dim_k(A/M) < \infty\}$. Then the *blocks* of A (or of $\mathcal{M}$, or of the set of isomorphism classes of finite dimensional irreducible A-modules) are the labels of the vertices in the connected components of the above graph. Thus we have a partition $\mathcal{M} = \bigsqcup\{\mathcal{M}_i : i \in \mathcal{I}\}$ of $\mathcal{M}$ into blocks.

An A-module V is *locally finite dimensional* if every finitely generated submodule of V is finite dimensional. In that case, every finitely generated submodule of V has a composition series; the collected composition factors of the finitely generated submodules of V are called *the composition factors of V*. We shall say that V *belongs to the block* $\mathcal{M}_i$ if each of its composition factors is annihilated by an ideal in $\mathcal{M}_i$. As the last part of the following result shows, this terminology generalises the more familiar Artinian set-up recalled in (I.16.1).

Theorem and Definition. *Let A be a Noetherian k-algebra, let*

$$\mathcal{M} = \{M \in \text{maxspec } A : \dim_k(A/M) < \infty\},$$

and let $\mathcal{M} = \bigsqcup_{i \in \mathcal{I}} \mathcal{M}_i$ be the partition of $\mathcal{M}$ into blocks.

1. Let V be a locally finite dimensional (left) A-module. Then there exists a unique decomposition

$$V = \bigoplus_{i \in \mathcal{I}} V_i \tag{3}$$

where each V_i belongs to $\mathcal{M}_i$. We call (3) the block decomposition of V.

2. The partition of $\mathcal{M}$ into blocks $\mathcal{M}_i$ is the unique finest partition for which 1 holds.

3. Let V and W be locally finite dimensional A-modules, with block decompositions $V = \bigoplus_{i \in \mathcal{I}} V_i$ and $W = \bigoplus_{i \in \mathcal{I}} W_i$. Then

$$\text{Hom}_A(W, V) = \prod_{i \in \mathcal{I}} \text{Hom}_A(W_i, V_i).$$

4. Suppose that $\dim_k(A) < \infty$. The blocks of $\text{maxspec } A$ as defined in (I.16.1) and (I.16.2) coincide.

Proof. 1. For $i \in \mathcal{I}$, let V_i be the sum of all submodules of V belonging to $\mathcal{M}_i$. It is clear that V_i itself belongs to $\mathcal{M}_i$, and that this is the only possibility for V_i in any decomposition of the form (3). All composition factors of V_i belong to $\mathcal{M}_i$, whereas those of $\sum_{j \neq i} V_j$ belong to $\bigsqcup_{j \neq i} \mathcal{M}_j$. It follows that $V_i \cap (\sum_{j \neq i} V_j) = 0$ for all i, and so we obtain a submodule $V' = \bigoplus_{i \in \mathcal{I}} V_i$ of V.

To prove that $V' = V$, we just need $W = \bigoplus_{i \in \mathcal{I}} (W \cap V_i)$ for all finitely generated submodules W of V, and so we may assume for this part of the proof that V is finitely generated.

If $V' \neq V$, choose a submodule $V'' \supset V'$ in V such that V''/V' is irreducible. Then V''/V' belongs to some $\mathcal{M}_j$. Set $W = \bigoplus_{i \neq j} V_i$, and note that V''/W belongs to $\mathcal{M}_j$. Now any composition factor S of V''/W belongs to $\mathcal{M}_j$, whereas any composition factor T of W belongs to $\bigsqcup_{i \neq j} \mathcal{M}_i$. Then T cannot be linked to S, that is, $\mathrm{Ext}^1_A(S, T) = 0$. It follows that $\mathrm{Ext}^1_A(V''/W, W) = 0$, and so $V'' = W \oplus X$ for some submodule X. But then X belongs to $\mathcal{M}_j$ and so $X \subseteq V_j$, whence $V'' = V'$, contradicting our choice of V''.

Therefore $V' = V$, as desired.

2. Suppose that $\mathcal{M} = \bigsqcup_{j \in \mathcal{J}} \mathcal{N}_j$ such that every locally finite dimensional A-module V has a decomposition $V = \bigoplus_{j \in \mathcal{J}} V_j$ where each V_j belongs to $\mathcal{N}_j$. It follows from the lemma above that for distinct $j, l \in \mathcal{J}$, no maximal ideal in $\mathcal{N}_j$ can be linked to a maximal ideal in $\mathcal{N}_l$. Consequently, each $\mathcal{M}_i$ is contained in some $\mathcal{N}_{j(i)}$, and therefore each $\mathcal{N}_j$ is a union of some of the $\mathcal{M}_i$.

3. For $i \neq j$, the modules V_j and W_i have no composition factors in common, and so $\mathrm{Hom}_A(W_i, V_j) = 0$. Hence,

$$\mathrm{Hom}_A(W, V) = \prod_{i \in \mathcal{I}} \mathrm{Hom}_A(W_i, V) = \prod_{i \in \mathcal{I}} \mathrm{Hom}_A(W_i, V_i).$$

4. If A is finite dimensional, we can apply part 1 to the left module ${}_A A$; thus $A = \bigoplus_{i \in \mathcal{I}} A_i$ where each A_i is the largest left ideal belonging to $\mathcal{M}_i$. For any $x \in A$, the left ideal $A_i x$ also belongs to $\mathcal{M}_i$, and so $A_i x \subseteq A_i$. Thus, these A_i are ideals of A, and hence they are generated by central idempotents. It follows that if $M \in \mathcal{M}_i$ and $N \in \mathcal{M}_j$ with $i \neq j$, then M and N contain different central idempotents. Since every central idempotent in A is a sum of primitive ones, M and N cannot belong to the same block in the sense of (I.16.1).

Conversely, let M and N be maximal ideals of A that belong to different blocks in the sense of (I.16.1). We claim that M and N cannot be linked. Write $1_A = e_1 + \cdots + e_t$ where the e_i are orthogonal primitive central idempotents; then there are indices $m \neq n$ such that $M = M e_m + A(1 - e_m)$ and $N = N e_n + A(1 - e_n)$. It follows that

$$MN = NM = M e_m + N e_n + A(1 - e_m - e_n) = M \cap N,$$

whence $M \nrightarrow N$ and $N \nrightarrow M$, as claimed. In other words, linked maximal ideals of A belong to the same block in the sense of (I.16.1). This implies that maximal ideals in the same $\mathcal{M}_i$ must belong to the same block in the sense of (I.16.1). $\square$

I.16.3. Examples. 1. If A is a commutative Noetherian k-algebra, each block of A consists of a single maximal ideal. For, let M and N be in the set $\mathcal{M}$. If $(M \cap N)/MN \neq 0$, the annihilator of this module must be proper – so $M + N \neq R$ and hence $M = N$. So there are no links between distinct maximal ideals of A.

2. Let $\mathcal{U}$ be the enveloping algebra of the two-dimensional complex solvable non-abelian Lie algebra $\mathfrak{g} = \{\mathbb{C}x \oplus \mathbb{C}y : [xy] = y\}$. Then

$$\mathcal{M} = \{M_\lambda := \langle y, \ x - \lambda \rangle : \lambda \in \mathbb{C}\}$$

and, using the fact that $xy = y(x + 1)$ in $\mathcal{U}$, one calculates that

$$M_\lambda \rightsquigarrow M_\mu \iff \mu = \lambda \text{ or } \mu = \lambda - 1.$$

Thus each block of $\mathcal{U}$ has the form

$$\{M_{\lambda+j} : j \in \mathbb{Z}\}$$

for a fixed $\lambda \in \mathbb{C}$.

3. Let $\mathcal{U}$ be the enveloping algebra of a finite dimensional complex semisimple Lie algebra $\mathfrak{g}$. Since all finite dimensional $\mathcal{U}$-modules are completely reducible by Weyl's Theorem (e.g., [**98**, Theorem 6.3]), it follows from Lemma I.16.2 that the blocks of $\mathcal{U}$ are all singletons, and indeed that there are no links between maximal ideals of $\mathcal{U}$ of finite codimension.

4 For a noetherian ring which is a finite module over its centre, we'll discuss the nature of the links between maximal ideals and the blocks of maximal ideals in (III.9.1)–(III.9.3).

NOTES

The fundamental properties of idempotents and blocks in Artinian rings and algebras are explained in detail in many books. A particularly clear and careful treatment is given in [**183**, Chapter 6]. The concept of a link between prime ideals of a noetherian ring, their cliques – that is – blocks, the associated localization theory and representation theory were intensively studied by many people in the 1970s and 1980s. An introduction to the theory and references for further reading can be found in [**83**, Chapters 11, 12].

EXERCISES

Exercise I.16.A. Confirm the statements made in Example I.16.3(2).

PART II. GENERIC QUANTIZED COORDINATE RINGS

CHAPTER II.1

THE PRIME SPECTRUM

Our focus in Part II will be on quantized coordinate rings, and most of the discussion will be concentrated on *generic* cases – those in which suitable parameters are not roots of unity. The non-generic situation, in which different phenomena occur and which require different methods of investigation, will be addressed in Part III.

The overall theme of Part II is the study of prime and primitive ideals in the algebras of interest, a study which has structural, representation-theoretic, and geometric grounds. Concerning the first point, note that perhaps the most important aspect of the internal structure of a noetherian algebra A is its ideal theory. The prime ideals of A form key components of the ideal theory, while being more accessible (both theoretically and computationally) than ideals in general. Moreover, the tools developed to analyze prime ideals also shed light on other aspects of the structure of A. From a representation-theoretic perspective, the primary objects of interest are the irreducible representations of A, which are reflected in the structure of A by their annihilators, the primitive ideals. Following a programme laid out by Dixmier for enveloping algebras and later generalized to other algebras, a basic first step in the representation theory of A is the identification of its primitive ideals. Primitive ideals are prime, and it generally proves easier to identify prime ideals than primitive ones. Hence, the task becomes one of finding criteria that say which prime ideals are primitive. As we shall see, the well known answer for enveloping algebras – the *Dixmier-Moeglin equivalence* – is also valid for generic quantized coordinate rings.

Finally, geometric motivation for studying prime ideals in quantized coordinate rings comes precisely from the fact that these algebras are viewed as "quantum analogues" of the classical coordinate rings of geometric objects. Recall that in the coordinate ring of an affine algebraic variety V over an algebraically closed field, the maximal ideals correspond to points of V and the prime ideals to irreducible subvarieties. Thus, if A is a quantized coordinate ring of V, the collections of primitive and prime ideals of A can be viewed as "quantum remnants" of V and its scheme of irreducible subvarieties, respectively. While we shall not pursue the analogy extensively here, we will see that the prime ideals in quantized coordinate rings do support interresting geometric structure.

To begin Part II, we review the basic structure of the collection of prime ideals of a ring, and begin discussing what this structure looks like in the case of quantized coordinate rings.

PRIME SPECTRA AND ZARISKI TOPOLOGIES

II.1.1. Definitions. The *prime spectrum* of a ring R, denoted $\operatorname{spec} R$, is the set of all prime ideals of R. For all ideals I of R, define

$$V(I) = \{P \in \operatorname{spec} R \mid P \supseteq I\} \qquad \text{and} \qquad W(I) = \{P \in \operatorname{spec} R \mid P \not\supseteq I\}.$$

Observe that finite unions and arbitrary intersections of $V(I)$s are again sets of the form $V(I)$. Moreover, $\varnothing = V(R)$ and $\operatorname{spec} R = V(0)$. Thus, the sets $V(I)$ are the closed sets for a topology on $\operatorname{spec} R$, called the *Zariski topology*. (This topology can also be found under the names *Stone topology*, *Jacobson topology*, and *hull-kernel topology*.) The Zariski-open subsets of $\operatorname{spec} R$ are, of course, the sets $W(I)$. When R is noetherian, $\operatorname{spec} R$ is compact, but usually not Hausdorff (Exercise II.1.A).

Two subspaces of $\operatorname{spec} R$ are of particular importance. The *maximal spectrum* (or *maximal ideal space*) of R, denoted $\operatorname{maxspec} R$ (or just $\max R$), is the collection of all maximal ideals of R, while the *primitive spectrum* of R, denoted $\operatorname{prim} R$, is the collection of all (left) primitive ideals of R. These sets are given the relative Zariski topologies inherited from $\operatorname{spec} R$; for instance, the Zariski-closed subsets of $\operatorname{prim} R$ are the sets

$$V(I) \cap \operatorname{prim} R = \{P \in \operatorname{prim} R \mid P \supseteq I\}$$

for ideals I of R.

II.1.2. Example. Let $A = \mathcal{O}_q(k^2)$ where k is algebraically closed and q is not a root of unity. Since $A/\langle y \rangle \cong k[x]$, the prime ideals of A containing y consist of $\langle y \rangle$ together with the maximal ideals $\langle x - \alpha, y \rangle$ for $\alpha \in k$. Similarly, the primes containing x consist of $\langle x \rangle$ together with the maximal ideals $\langle x, y - \beta \rangle$ for $\beta \in k$.

To deal with the remaining primes of A requires a localization. Recall from Exercise I.2.A that x and y generate a denominator set in A, and that the localization $A[x^{-1}, y^{-1}] = \mathcal{O}_q((k^\times)^2)$. Since q is not a root of unity, $\mathcal{O}_q((k^\times)^2)$ is a simple ring, from which it follows that any nonzero prime ideal of A must contain x or y (Exercise II.1.B). Therefore $\operatorname{spec} A$ can be pictured as in Diagram II.1.2 below, where the solid lines represent inclusions.

It is easy to describe a number of Zariski-closed subsets of $\operatorname{spec} A$, such as the following:

$$\left\{\text{finitely many of the } \langle x, y - \beta \rangle\right\} \cup \left\{\text{finitely many of the } \langle x - \alpha, y \rangle\right\}$$
$$\left\{\langle x \rangle\right\} \cup \left\{\langle x, y - \beta \rangle \mid \beta \in k\right\} \cup \left\{\text{finitely many of the } \langle x - \alpha, y \rangle\right\}$$
$$\left\{\langle y \rangle\right\} \cup \left\{\langle x - \alpha, y \rangle \mid \alpha \in k\right\} \cup \left\{\text{finitely many of the } \langle x, y - \beta \rangle\right\} \tag{1}$$
$$\left\{\langle x \rangle, \langle y \rangle\right\} \cup \left\{\langle x, y - \beta \rangle \mid \beta \in k\right\} \cup \left\{\langle x - \alpha, y \rangle \mid \alpha \in k\right\}.$$

In fact, the sets listed in (1), together with $\varnothing$ and spec A, are the only closed subsets of spec A (Exercise II.1.C).

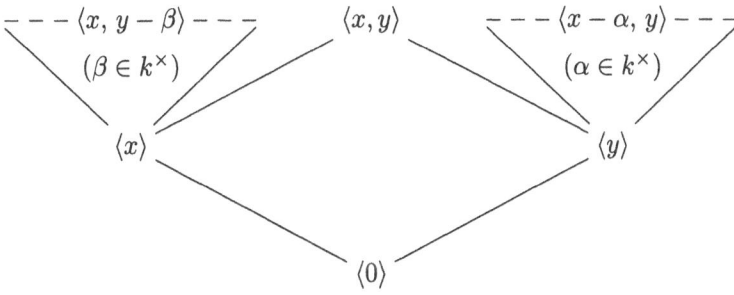

DIAGRAM II.1.2.   spec $\mathcal{O}_q(k^2)$

II.1.3. Example. Let $A = \mathcal{O}_q(SL_2(k))$ where k is algebraically closed and q is not a root of unity. A computation of spec A in this case is somewhat more work than in the previous example, but it can be done with similar techniques. For an outline, see Exercise II.1.D; with the help of some tools that we develop later, the work can be simplified (see Example II.2.14(b)). The complete picture of spec A is shown in Diagram II.1.3 below.

II.1.4. Given an algebra A, one would like to be able to 'compute' spec A and prim A insofar as possible. The first part of this goal is just to describe all the prime and primitive ideals of A in some fashion. Ideally, one would then find an explicit description of the Zariski topologies on spec A and prim A. It will often prove easier to describe pieces of these spaces, and we will shortly set up a framework that works particularly well for generic quantized coordinate rings.

SYMMETRY VIA GROUPS OF AUTOMORPHISMS

II.1.5. Terminology. Let H be a group acting by automorphisms on a ring R. This just means that we have specified a group homomorphism $\alpha : H \to \text{Aut } R$. However, we suppress the symbol α and for $\alpha(h)(a)$ we just write either $h.a$ (to

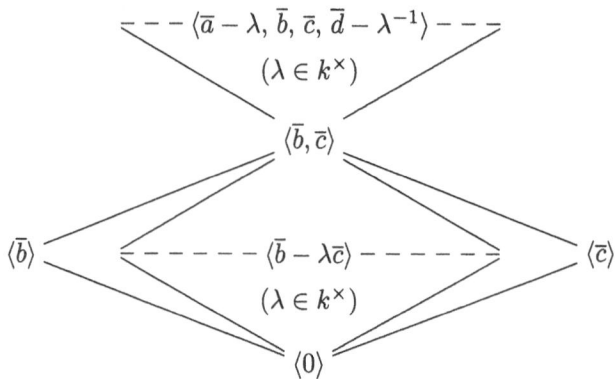

DIAGRAM II.1.3. spec $\mathcal{O}_q(SL_2(k))$

emphasize that this is an action) or $h(a)$ (to emphasize that the action operates by automorphisms). Then H permutes the prime, primitive, and maximal ideals of R, and so we obtain induced actions of H on spec R, on prim R, and on maxspec R. Observe that the action of any $h \in H$ on these spaces is by homeomorphisms.

We use the standard terminology of orbits and stabilizers for these induced actions. Thus, the *H-orbit* of a prime ideal P is the set

$$H.P = \{h(P) \mid h \in H\},$$

and the *stabilizer of P in H* is the subgroup

$$\mathrm{Stab}_H(P) = \{h \in H \mid h(P) = P\}.$$

In case H is a cyclic group, say $H = \langle \tau \rangle$, we abbreviate 'H-orbit' to 'τ-orbit', and similarly with other terminology involving the prefix 'H-'.

II.1.6. Examples. In all of our standard examples, there are useful actions of certain groups by automorphisms for which our standard generators are eigenvectors. Here we present the most basic examples; we leave it to the reader to check that in each case there do exist well-defined automorphisms and actions as described (Exercise II.1.E). For additional examples, see (II.1.14–18).

(a) First let $A = \mathcal{O}_q(k^2)$ and $H = (k^\times)^2$. There is a natural action of H on A by k-algebra automorphisms, such that

$$(\lambda, \mu).x^i y^j = \lambda^i \mu^j x^i y^j$$

for $(\lambda, \mu) \in H$ and $i, j \geq 0$. For instance, the H-orbit of the maximal ideal $\langle x, y-1 \rangle$ is the set $\{\langle x, y - \beta \rangle \mid \beta \in k^\times\}$, and the stabilizer of $\langle x, y - 1 \rangle$ is the subgroup $(k^\times) \times \{1\}$. On the other hand, the maximal ideal $\langle x, y \rangle$ forms a singleton H-orbit; its stabilizer is H.

Now assume that k is algebraically closed and q is not a root of unity, and recall the picture of spec A from (II.1.2). There are exactly six H-orbits in spec A:

$$\{\langle 0 \rangle\}, \qquad \{\langle x \rangle\}, \qquad \{\langle y \rangle\}, \qquad \{\langle x, y \rangle\},$$
$$\{\langle x, y - \beta \rangle \mid \beta \in k^\times\}, \qquad \{\langle x - \alpha, y \rangle \mid \alpha \in k^\times\}.$$

(b) Next, let $A = \mathcal{O}_q(SL_2(k))$ and again let $H = (k^\times)^2$. There is an action of H on A by k-algebra automorphisms such that

$$(\alpha, \beta).\bar{a} = \alpha\beta\bar{a} \qquad\qquad (\alpha, \beta).\bar{b} = \alpha\beta^{-1}\bar{b}$$
$$(\alpha, \beta).\bar{c} = \alpha^{-1}\beta\bar{c} \qquad\qquad (\alpha, \beta).\bar{d} = \alpha^{-1}\beta^{-1}\bar{d}$$

for $(\alpha, \beta) \in H$. If k is algebraically closed and q is not a root of unity, there are again six H-orbits in spec A, of which four are singletons.

We can also let a slightly larger group act on A, namely

$$H' = \{(\alpha_1, \alpha_2, \beta_1, \beta_2) \in (k^\times)^4 \mid \alpha_1\alpha_2\beta_1\beta_2 = 1\},$$

which acts on A by k-algebra automorphisms such that

$$(\alpha_1, \alpha_2, \beta_1, \beta_2).\bar{a} = \alpha_1\beta_1\bar{a} \qquad (\alpha_1, \alpha_2, \beta_1, \beta_2).\bar{b} = \alpha_1\beta_2\bar{b}$$
$$(\alpha_1, \alpha_2, \beta_1, \beta_2).\bar{c} = \alpha_2\beta_1\bar{c} \qquad (\alpha_1, \alpha_2, \beta_1, \beta_2).\bar{d} = \alpha_2\beta_2\bar{d}.$$

Observe that the H'-orbits in spec A coincide with the H-orbits.

(c) In the case $A = \mathcal{O}_q(M_2(k))$, we let $H = (k^\times)^4$ and observe that there is an action of H on A by k-algebra automorphisms such that

$$(\alpha_1, \alpha_2, \beta_1, \beta_2).a = \alpha_1\beta_1 a \qquad (\alpha_1, \alpha_2, \beta_1, \beta_2).b = \alpha_1\beta_2 b$$
$$(\alpha_1, \alpha_2, \beta_1, \beta_2).c = \alpha_2\beta_1 c \qquad (\alpha_1, \alpha_2, \beta_1, \beta_2).d = \alpha_2\beta_2 d.$$

There are many H-orbits in spec A in this case; for instance, if $q \neq \pm 1$ there are already six H-orbits consisting of maximal ideals of codimension 1 (Exercise II.1.F).

(d) Finally, consider the case $A = \mathcal{O}_q(GL_2(k))$, and observe that D_q is an eigenvector for all the automorphisms of $\mathcal{O}_q(M_2(k))$ considered in part (c). Hence, the action of $H = (k^\times)^4$ given there extends uniquely to an action of H on A by k-algebra automorphisms. We shall see later that this action is closely related to the actions of $(k^\times)^2$ and $(k^\times)^3$ on $\mathcal{O}_q(SL_2(k))$ described in part (b). For now, we just observe that if $q \neq \pm 1$, the maximal ideals of A with codimension 1 form a single H-orbit (Exercise II.1.G).

II.1.7. Definition. In all the examples just discussed, and others to be presented below, the 'natural' or 'obvious' choice of a group acting by k-algebra automorphisms on a quantized coordinate ring is a group of tuples of nonzero scalars. From the point of view of algebraic geometry, such groups are considered analogues of tori of arbitrary dimensions. For example, the standard 2-dimensional topological torus may be written as $S^1 \times S^1$, where S^1 is the group of complex numbers of modulus 1. Now S^1 is closed and compact in the Euclidean topology, whereas in the Zariski topology, it is dense in $\mathbb{C}$. The multiplicative group $\mathbb{C}^\times$ of nonzero complex numbers is viewed as a non-compact analogue of S^1, and thus $(\mathbb{C}^\times)^2$ becomes an analogue of $S^1 \times S^1$. This picture leads to the following standard terminology, which is used with respect to an arbitrary base field k.

An *algebraic torus over k*, or a *k-torus* for short, is a group of the form $(k^\times)^r$. We view $(k^\times)^r$ as an affine algebraic group in the naive sense; its structure as an affine algebraic variety comes from identifying it with the variety

$$\{(\alpha_1, \ldots, \alpha_r, \beta) \in k^{r+1} \mid \alpha_1 \alpha_2 \cdots \alpha_r \beta = 1\}.$$

Strictly speaking, we should refer to $(k^\times)^r$ as the group of k-rational points of the affine algebraic group $(\overline{k}^\times)^r$, but we shall not need that level of sophistication.

All of our standard examples of quantized coordinate rings admit natural actions of algebraic tori (acting via k-algebra automorphisms), analogous to the cases described in Examples II.1.6. We shall describe the general cases in (II.1.14–18).

H-PRIME IDEALS

II.1.8. Definitions. Let H be a group acting by automorphisms on a ring R. An ideal I of R is *H-stable* (or *H-invariant*) provided $h(I) = I$ for all $h \in H$. (Since H is a group, it suffices to check that $h(I) \subseteq I$ for $h \in H$.) By way of abbreviation, H-stable ideals are often called *H-ideals*. Given an arbitrary ideal I in R, we write $(I : H)$ for the largest H-ideal contained in I, that is,

$$(I : H) = \bigcap_{h \in H} h(I).$$

The ring R is said to be *H-simple* provided R is nonzero and the only H-ideals of R are 0 and R. For example, if k is infinite, $R = k[x^{\pm 1}]$, and $H = k^\times$ acts on R by k-algebra automorphisms such that $\alpha.x^i = \alpha^i x^i$ for $\alpha \in k^\times$ and $i \in \mathbb{Z}$, then R is H-simple.

II.1.9. Definitions. Let H be a group acting by automorphisms on a ring R. We say that R is an *H-prime ring* provided that R is nonzero and any product of nonzero H-ideals of R is nonzero. For example, any H-simple ring is H-prime. An

H-prime ideal of R is any (proper) H-ideal P of R such that R/P is an H-prime ring. Note that if Q is an ordinary prime ideal of R, then $(Q : H)$ is an H-prime ideal. Obviously any prime H-ideal is H-prime, but the converse typically fails. For instance, if $R = k \times k$ and $H = \langle \tau \rangle$ where τ is the flip (that is, $\tau(a, b) = (b, a)$ for $(a, b) \in R$), then 0 is an H-prime ideal of R but is not prime.

The set of all H-prime ideals of R will be denoted H-spec R. This set supports a topology analogous to the Zariski topology on spec R; see Exercise II.1.H.

II.1.10. Lemma. *Let R be a noetherian ring and H a group acting on R by automorphisms. If P is an H-prime ideal of R, then P is semiprime, the primes minimal over P form a single H-orbit, and $P = (Q : H)$ for any prime Q minimal over P.*

Proof. Recall that since R is noetherian, there are only finitely many primes minimal over P, and some product of these primes (with repetitions allowed) is contained in P.

Obviously H permutes the primes minimal over P, so if N is their intersection, then N is H-stable. Since P contains a power of N, it follows from H-primeness that $P \supseteq N$. Thus $P = N$ is semiprime.

Group the primes minimal over P into distinct H-orbits, say $X_1, \ldots, X_m$. Set $J_i = \bigcap X_i$ for each i. These J_i are H-ideals of R, and $J_1 \cap \cdots \cap J_m$ is the intersection of all the primes minimal over P. Since P is semiprime, $P = J_1 \cap \cdots \cap J_m \supseteq J_1 J_2 \cdots J_m$. By H-primeness, $P \supseteq J_l$ for some l, whence $P = J_l$. But then the product (in any order) of the primes in X_l is contained in P. Since these primes cannot be contained in any other primes minimal over P, the H-orbit X_l must exhaust the primes minimal over P. Therefore the primes minimal over P form a single H-orbit.

Finally, if Q is any prime minimal over P, then $(Q : H)$ is the intersection of all the primes minimal over P, whence $(Q : H) = N = P$. $\square$

II.1.11. Example. Let $A = \mathcal{O}_q(k^2)$ and $H = (k^\times)^2$ as in Example II.1.6(a), and assume that k is infinite. Then A has precisely four H-prime ideals, namely $\langle 0 \rangle$, $\langle x \rangle$, $\langle y \rangle$, $\langle x, y \rangle$ (see Exercise II.1.I).

The following corollary will be superseded, for most purposes, by Proposition II.2.9.

II.1.12. Corollary. *Let R be a noetherian ring, and suppose that a k-torus H acts on R by automorphisms. (Here R need not be a k-algebra.) If k is algebraically closed, then all H-prime ideals of R are prime.*

Proof. Let P be an H-prime ideal in R, and choose a prime Q minimal over P. By Lemma II.1.10, $(Q : H) = P$. Since P is H-stable, each $h \in H$ must send Q to a prime minimal over P. Thus, since there are only finitely many primes minimal over P, the H-orbit of Q must be finite. Consequently, $\mathrm{Stab}_H(Q)$ has finite index in H. It suffices to show that $\mathrm{Stab}_H(Q) = H$, for then we would have $P = (Q : H) = Q$.

Since k is algebraically closed, H is a divisible group, and so the factor $G = H/\mathrm{Stab}_H(Q)$ is a finite, divisible abelian group. Because G is finite, there is a positive integer m such that $g^m = 1$ for all $g \in G$. Therefore G cannot be divisible unless it is trivial, that is, $\mathrm{Stab}_H(Q) = H$, as desired. $\square$

II.1.13. Example. Let $A = \mathcal{O}_q(SL_2(k))$ and $H = (k^\times)^2$ as in Example II.1.6(b), and assume that k is algebraically closed. By Corollary II.1.12, the H-prime ideals of A are just the prime H-ideals. (Once we prove Proposition II.2.9, we shall not need to assume that k is algebraically closed, only that it is infinite.) If q is not a root of unity, then by Exercise II.1.K, A has precisely four H-prime ideals, namely $\langle 0 \rangle$, $\langle \bar{b} \rangle$, $\langle \bar{c} \rangle$, $\langle \bar{b}, \bar{c} \rangle$.

STANDARD EXAMPLES

As mentioned above, all our standard examples of quantized coordinate rings admit 'natural' or 'obvious' actions of algebraic tori (acting via k-algebra automorphisms). For later reference, we record these tori and actions below; they incorporate and generalise those discussed in Examples II.1.6 above. When referring to tori acting on these algebras, we shall mean the actions described here, unless specified otherwise.

II.1.14. The torus $H = (k^\times)^n$ acts on quantum affine n-spaces and on quantum tori of rank n in the same way, namely by automorphisms such that

$$(\alpha_1, \ldots, \alpha_n).x_i = \alpha_i x_i$$

for all i. This covers the single parameter algebras $\mathcal{O}_q(k^n)$ and $\mathcal{O}_q((k^\times)^n)$ as well as the multiparameter cases $\mathcal{O}_{\boldsymbol{q}}(k^n)$ and $\mathcal{O}_{\boldsymbol{q}}((k^\times)^n)$.

II.1.15. For quantum $n \times n$ matrices and quantum $GL_n(k)$, there is an action of the torus $H = (k^\times)^{2n}$ by automorphisms so that

$$(\alpha_1, \ldots, \alpha_n, \beta_1, \ldots, \beta_n).X_{ij} = \alpha_i \beta_j X_{ij}$$

for all i, j. Here actions by elements $(\alpha_1, \ldots, \alpha_n, 1, \ldots, 1)$ correspond to left winding automorphisms, while actions by elements $(1, \ldots, 1, \beta_1, \ldots, \beta_n)$ correspond to right winding automorphisms. (See (I.9.25) for the definitions of left and right winding automorphisms.) Again, the same actions apply to the single parameter algebras $\mathcal{O}_q(M_n(k))$ and $\mathcal{O}_q(GL_n(k))$ and to the multiparameter cases $\mathcal{O}_{\lambda,\boldsymbol{p}}(M_n(k))$ and $\mathcal{O}_{\lambda,\boldsymbol{p}}(GL_n(k))$.

II.1.16. The action of $(k^\times)^{2n}$ on $\mathcal{O}_q(M_n(k))$ and $\mathcal{O}_{\lambda,\boldsymbol{p}}(M_n(k))$ just described in (II.1.15) does not induce corresponding actions on $\mathcal{O}_q(SL_n(k))$ and $\mathcal{O}_{\lambda,\boldsymbol{p}}(SL_n(k))$. For these algebras, we take H to be the stabilizer of the quantum determinant, namely the subgroup

$$\{(\alpha_1,\ldots,\alpha_n,\beta_1,\ldots,\beta_n) \in (k^\times)^{2n} \mid \alpha_1\alpha_2\cdots\alpha_n\beta_1\beta_2\cdots\beta_n = 1\},$$

which is isomorphic to the torus $(k^\times)^{2n-1}$. This H fixes D_q and $D_{\lambda,\boldsymbol{p}}$, and so it induces actions by automorphisms of $\mathcal{O}_q(SL_n(k))$ and $\mathcal{O}_{\lambda,\boldsymbol{p}}(SL_n(k))$.

In case $n = 2$, the group H given here corresponds to the second of the two choices discussed in Example II.1.6(b). This choice will prove convenient with respect to some of the methods we introduce later.

II.1.17. The torus $H = (k^\times)^n$ acts on the quantized Weyl algebra $A_n^{Q,\Gamma}(k)$ by automorphisms so that

$$(\alpha_1,\ldots,\alpha_n).x_i = \alpha_i x_i \qquad \text{and} \qquad (\alpha_1,\ldots,\alpha_n).y_i = \alpha_i^{-1} y_i$$

for all i.

II.1.18. In the case of a quantized coordinate ring $\mathcal{O}_q(G)$, the tori that one expects to appear in the theory are the maximal tori of G. In the classical situation, a maximal torus of G with Lie algebra $\mathfrak{h}$ can be identified with the group $\mathrm{Hom}(L,\mathbb{C}^\times)$ (recall L from (I.7.1); the group operation on $\mathrm{Hom}(L,\mathbb{C}^\times)$ here is pointwise multiplication of functions). Thus, when working with the k-form $\mathcal{O}_q(G) = \mathcal{O}_q(G)_k$, we use the torus $H = \mathrm{Hom}(L,k^\times)$.

Recall from Lemma I.8.7 that $\mathcal{O}_q(G)$ is an $L \times L$ graded algebra. Hence, for each $h \in H$ we can define k-algebra automorphisms σ_h and τ_h on $\mathcal{O}_q(G)$ so that $\sigma_h(y) = h(\beta)y$ and $\tau_h(y) = h(\eta)y$ for $\beta,\eta \in L$ and $y \in \mathcal{O}_q(G)_{\beta,\eta}$ (Exercise II.1.L). In fact, σ_h (respectively, τ_h) can be exhibited as a left (respectively, right) winding automorphism of $\mathcal{O}_q(G)$ (Exercise II.1.L). Following the conventions in the literature, we concentrate on *right* winding automorphisms, and so we use the τ_h to define our standard action of H on $\mathcal{O}_q(G)$:

$$h.y = h(\eta)y \qquad \text{for } h \in H,\ \eta,\beta \in L,\ y \in \mathcal{O}_q(G)_{\beta,\eta}.$$

Recall from Theorem I.7.9 that in case $G = SL_n(\mathbb{C})$ and q is not a root of unity, we have $\mathcal{O}_q(G) \cong \mathcal{O}_q(SL_n(k))$. In this case, we have two distinct tori acting: the torus $H = \mathrm{Hom}(L,k^\times) \cong (k^\times)^{n-1}$ acting on $\mathcal{O}_q(G)$ as just defined, and the torus described in (II.1.16), call it $\widehat{H}$ (isomorphic to $(k^\times)^{2n-1}$), acting on $\mathcal{O}_q(SL_n(k))$. The isomorphism $\mathcal{O}_q(G) \cong \mathcal{O}_q(SL_n(k))$ identifies the action of H with the action of the subgroup

$$H_r = \{(1,\ldots,1,\beta_1,\ldots,\beta_n) \in (k^\times)^{2n} \mid \beta_1\beta_2\cdots\beta_n = 1\} \subseteq \widehat{H}.$$

It is known that for many purposes, these two actions have essentially the same effect. In particular, the H_r-primes in $\mathcal{O}_q(SL_n(k))$ are the same as the $\widehat{H}$-primes. This follows from Corollary II.1.12 (if k is algebraically closed) or Proposition II.2.9 (in general), once one has proved that $\mathcal{O}_q(SL_n(k))$ has only finitely many H_r-primes, since it is clear that $\widehat{H}$ permutes H_r-spec $\mathcal{O}_q(SL_n(k))$.

NOTES

The concepts introduced in this chapter are well established and standard; we shall not give any references.

EXERCISES

Exercise II.1.A. If R is a ring with ACC on ideals, show that spec R is compact, but not necessarily Hausdorff. For any ring R, show that the space maxspec R is T_1, that is, all its singleton subsets are closed. $\square$

Exercise II.1.B. Assuming that q is not a root of unity, show that $\mathcal{O}_q((k^\times)^2)$ is a simple ring. Use this fact to show that every nonzero prime ideal of $\mathcal{O}_q(k^2)$ must contain x or y. [Hint: Observe that $\left(x\mathcal{O}_q(k^2)\right)^i \left(y\mathcal{O}_q(k^2)\right)^j = x^i y^j \mathcal{O}_q(k^2)$ for all $i, j \geq 0$.] $\square$

Exercise II.1.C. If k is algebraically closed and q is not a root of unity, show that spec $\mathcal{O}_q(k^2)$ has no closed sets other than those described in (II.1.2). $\square$

Exercise II.1.D. Let $A = \mathcal{O}_q(SL_2(k))$ where q is not a root of unity. Compute spec A as follows.

(a) Observe that $A/\langle \bar{b}, \bar{c} \rangle \cong k[\bar{a}^{\pm 1}]$, and hence that the primes of A containing $\bar{b}$ and $\bar{c}$ consist of $\langle \bar{b}, \bar{c} \rangle$ together with the maximal ideals $\langle f(\bar{a}), \bar{b}, \bar{c} \rangle$ for irrreducible polynomials $f \in k[t^{\pm 1}]$. In case k is algebraically closed, these maximal ideals can be written in the form $\langle \bar{a} - \lambda, \bar{b}, \bar{c}, \bar{d} - \lambda^{-1} \rangle$ for $\lambda \in k^\times$.

(b) Show that the set $\{\bar{c}^i \mid i \geq 0\}$ is a denominator set in A, and that $A[\bar{c}^{-1}]/\langle \bar{b} \rangle \cong \mathcal{O}_q((k^\times)^2)$. Recall from Exercise II.1.B that $\mathcal{O}_q((k^\times)^2)$ is simple, and conclude that the only prime of A that contains $\bar{b}$ but not $\bar{c}$ is $\langle \bar{b} \rangle$. Similarly, show that the only prime of A that contains $\bar{c}$ but not $\bar{b}$ is $\langle \bar{c} \rangle$.

(c) Show that no prime of A can contain a power of $\bar{a}$. [Hint: If a prime P contains $\bar{a}^i$, then the H-prime $(P : H)$ also contains $\bar{a}^i$.]

(d) Show that in $\mathcal{O}_q(M_2(k))$, the powers of a form a denominator set. [Hint: One way is to use the iterated skew polynomial structure of $\mathcal{O}_q(M_2(k))$ discussed in Example I.1.16.] Hence, the powers of $\bar{a}$ form a denominator set in A.

(e) Show that the set $\{\bar{a}^i \bar{b}^j \bar{c}^l \mid i, j, l \geq 0\}$ is a denominator set in A. Let B denote the corresponding localization, and show that the element $\bar{b}\bar{c}^{-1}$ lies in the centre of B. Then show that $B \cong \mathcal{O}_q((k^\times)^2)[z^{\pm 1}]$.

(f) Use part (e) to show that every nonzero prime of B has the form $g(\bar{b}\bar{c}^{-1})B$ for some irreducible polynomial $g \in k[z^{\pm 1}]$. Conclude that the only nonzero primes of A not containing $\bar{b}$ or $\bar{c}$ are the ideals $A \cap g(\bar{b}\bar{c}^{-1})B$.

(g) For any $\lambda \in k$, show that $A/\langle \bar{b} - \lambda\bar{c}\rangle \cong \mathcal{O}_q(k^2)[x^{-1}]$, and conclude that $\langle \bar{b} - \lambda\bar{c}\rangle$ is a prime ideal of A.

(h) In case k is algebraically closed, show that the only nonzero primes of A not containing $\bar{b}$ or $\bar{c}$ are the ideals $\langle \bar{b} - \lambda\bar{c}\rangle$ for $\lambda \in k^\times$. □

Exercise II.1.E. In each of the cases described in Examples II.1.6, show that there exist k-algebra automorphisms and actions with the asserted properties. For instance, in case (a) this means showing that for each $\lambda, \mu \in k^\times$, there is a well-defined k-algebra automorphism $\tau_{\lambda,\mu}$ on $\mathcal{O}_q(k^2)$ such that $\tau_{\lambda,\mu}(x^iy^i) = \lambda^i\mu^jx^iy^j$ for all i, j, and that the map $(\lambda, \mu) \mapsto \tau_{\lambda,\mu}$ is a group homomorphism from $(k^\times)^2$ to $\operatorname{Aut}\mathcal{O}_q(k^2)$. □

Exercise II.1.F. Let $A = \mathcal{O}_q(M_2(k))$ with $q \neq \pm 1$. Show that the following is a complete list of the maximal ideals of A with codimension 1:

$\langle a - \alpha,\, b,\, c,\, d - \delta\rangle$	$(\alpha, \delta \in k)$
$\langle a,\, b - \beta,\, c,\, d\rangle$	$(\beta \in k)$
$\langle a,\, b,\, c - \gamma,\, d\rangle$	$(\gamma \in k).$

Now let the group $H = (k^\times)^4$ act on A as in Example II.1.6(c). Show that among the maximal ideals of A with codimension 1, there are precisely six H-orbits. □

Exercise II.1.G. Let $A = \mathcal{O}_q(GL_2(k))$ with $q \neq \pm 1$. Use the previous exercise to show that the maximal ideals of A with codimension 1 are just the ideals $\langle a - \alpha,\, b,\, c,\, d - \delta\rangle$ for $\alpha, \delta \in k^\times$. Now let the group $H = (k^\times)^4$ act on A as in Example II.1.6(d). Show that the maximal ideals of A with codimension 1 form a single H-orbit. □

Exercise II.1.H. Let H be a group acting by automorphisms on a ring R. Show that there is a topology on H-spec R whose closed sets are precisely the sets

$$\{P \in H\text{-spec}\,R \mid P \supseteq I\}$$

for ideals I of R, and that the rule $Q \mapsto (Q : H)$ gives a continuous map from spec R to H-spec R. □

Exercise II.1.I. Let $A = \mathcal{O}_q(k^2)$ and $H = (k^\times)^2$ as in Example II.1.6(a).

First assume that k is algebraically closed and q is not a root of unity. Using Example II.1.2, compute $(Q : H)$ for all primes Q of A, and conclude that the only H-prime ideals of A are $\langle 0\rangle$, $\langle x\rangle$, $\langle y\rangle$, and $\langle x, y\rangle$.

Now assume only that k is infinite, and show that any nonzero H-ideal of A contains a monomial x^iy^j for some i, j. Conclude again that the only H-prime ideals of A are $\langle 0\rangle$, $\langle x\rangle$, $\langle y\rangle$, and $\langle x, y\rangle$. □

Exercise II.1.J. Let H be a group acting by automorphisms on a ring R, and let X be an H-stable denominator set in R. Show that the action of H on R extends uniquely to an action on $R[X^{-1}]$ by automorphisms. Now assume that R is noetherian, and let $\phi : R \rightarrow R[X^{-1}]$ be the localization map. Show that there are bijections $Q \mapsto \phi^{-1}(Q)$ and $P \mapsto \phi(P)R[X^{-1}]$ between the H-primes of $R[X^{-1}]$ and those H-primes of R that are disjoint from X. In particular, if R is an H-prime ring and $0 \notin X$, then $R[X^{-1}]$ is an H-prime ring. $\square$

Exercise II.1.K. Let $A = \mathcal{O}_q(SL_2(k))$ and $H = (k^\times)^2$ as in Example II.1.6(b), and assume that q is not a root of unity. Either assume that k is algebraically closed and apply Corollary II.1.12, or else look ahead to Proposition II.2.9. In either case, it follows that the H-prime ideals of A are just the prime H-ideals. Use the description of the prime ideals of A from Exercise II.1.D to show that the only H-prime ideals of A are $\langle 0 \rangle$, $\langle \bar{b} \rangle$, $\langle \bar{c} \rangle$, and $\langle \bar{b}, \bar{c} \rangle$. $\square$

Exercise II.1.L. For $h \in H = \mathrm{Hom}(L, k^\times)$, verify that there exist k-algebra automorphisms σ_h and τ_h on $\mathcal{O}_q(G)$ as described in (II.1.18). Similarly, check that there are k-algebra homomorphisms $\phi_h, \psi_h : \mathcal{O}_q(G) \rightarrow k$ such that $\phi_h(y) = h(\beta)\varepsilon(y)$ and $\psi_h(y) = h(\eta)\varepsilon(y)$ for $y \in \mathcal{O}_q(G)_{\beta,\eta}$. Finally, show that σ_h coincides with the left winding automorphism corresponding to ϕ_h, while τ_h coincides with the right winding automorphism corresponding to ψ_h. [Hint: Evaluate on coordinate functions $c_{f,v}^V$ where f and v are weight vectors.] $\square$

Exercise II.1.M. Verify that the information given in (II.1.14–18) determines well-defined actions of tori via k-algebra automorphisms in each case. $\square$

STRATIFICATION

We continue the discussion of how an action of a group by automorphisms on a ring helps in analyzing the prime spectrum. If, say, H is a group acting on a ring R, then we can partition spec R into H-orbits. This partition is nice for theoretical purposes, but in practice it can be very difficult to deal with – there may be infinitely many H-orbits in spec R, and they can be hard to calculate. It proves useful to look for a coarser partition. Note that if P and P' are prime ideals of R with the same H-orbit, then

$$(P : H) = \bigcap \{Q \mid Q \in H.P\} = \bigcap \{Q' \mid Q' \in H.P'\} = (P' : H).$$

Following this hint, we partition spec R into "H-strata", where prime ideals P and P' belong to the same H-stratum if and only if $(P : H) = (P' : H)$.

STRATA

II.2.1. Definitions. Let H be a group acting by automorphisms on a ring R. Recall from (II.1.9) that $(P : H)$ is an H-prime ideal for any prime ideal P of R. For any H-prime ideal J, set

$$\text{spec}_J R = \{P \in \text{spec } R \mid (P : H) = J\}.$$

We will call $\text{spec}_J R$ the *H-stratum of* spec R *corresponding to* J. These H-strata give a partition of spec R, namely

$$\text{spec } R = \bigsqcup_{J \in H\text{-spec } R} \text{spec}_J R,$$

called the *H-stratification of* spec R. Similarly, there is an H-stratification of prim R by H-strata $\text{prim}_J R = (\text{prim } R) \cap (\text{spec}_J R)$.

The H-stratification of spec R meshes nicely with the Zariski topology in the sense that the closure of each H-stratum is a union of H-strata. Moreover, if there are only finitely many H-strata, then each H-stratum is an intersection of an open and a closed subset of spec R. (See Exercise II.2.A.)

II.2.2. An H-stratification of $\operatorname{spec} R$ as above offers an approach to describing $\operatorname{spec} R$, namely by finding and describing the individual H-strata. The most information is obtained when there are only finitely many H-strata – i.e., when H-$\operatorname{spec} R$ is finite – and when each H-stratum can be described with its relative topology.

We shall show later (Corollary II.4.12 and Theorems II.5.14, II.5.17) that under suitably generic conditions, when H is a torus acting on a quantized coordinate ring A as in (II.1.14–18), there are only finitely many H-strata in $\operatorname{spec} A$.

II.2.3. Examples. (a) Let $A = \mathcal{O}_q(k^2)$ where k is algebraically closed and q is not a root of unity, and let $H = (k^\times)^2$ act on A as in Example II.1.6(a). As noted in that example, there are six H-orbits in $\operatorname{spec} A$. On the other hand, A has just four H-prime ideals, namely $\langle 0 \rangle$, $\langle x \rangle$, $\langle y \rangle$, $\langle x, y \rangle$ (Example II.1.11). Consequently, there are four H-strata in $\operatorname{spec} A$:

$$\operatorname{spec}_{\langle 0 \rangle} A = \{ \langle 0 \rangle \}$$
$$\operatorname{spec}_{\langle x \rangle} A = \{ \langle x \rangle \} \cup \{ \langle x, y - \beta \rangle \mid \beta \in k^\times \}$$
$$\operatorname{spec}_{\langle y \rangle} A = \{ \langle y \rangle \} \cup \{ \langle x - \alpha, y \rangle \mid \alpha \in k^\times \}$$
$$\operatorname{spec}_{\langle x,y \rangle} A = \{ \langle x, y \rangle \}$$

(cf. Example II.1.2). Observe that the second and third of these strata look like prime spectra of Laurent polynomial rings; more precisely, there are homeomorphisms (with respect to the appropriate Zariski topologies)

$$\operatorname{spec}_{\langle x \rangle} A \approx \operatorname{spec} k[y^{\pm 1}] \qquad\qquad \operatorname{spec}_{\langle y \rangle} A \approx \operatorname{spec} k[x^{\pm 1}]$$

(Exercise II.2.B). As we shall see later, there are good reasons to complete this picture by viewing the singletons $\operatorname{spec}_{\langle 0 \rangle} A$ and $\operatorname{spec}_{\langle x,y \rangle} A$ as being homeomorphic to $\operatorname{spec} k$.

(b) Now let $A = \mathcal{O}_q(SL_2(k))$ where k is algebraically closed and q is not a root of unity, and let $H = (k^\times)^2$ act on A as in Example II.1.6(b). There are again four H-prime ideals (Example II.1.13), and thus four H-strata in $\operatorname{spec} A$:

$$\operatorname{spec}_{\langle 0 \rangle} A = \{ \langle 0 \rangle \} \cup \{ \langle \bar{b} - \lambda \bar{c} \rangle \mid \lambda \in k^\times \}$$
$$\operatorname{spec}_{\langle \bar{b} \rangle} A = \{ \langle \bar{b} \rangle \}$$
$$\operatorname{spec}_{\langle \bar{c} \rangle} A = \{ \langle \bar{c} \rangle \}$$
$$\operatorname{spec}_{\langle \bar{b},\bar{c} \rangle} A = \{ \langle \bar{b}, \bar{c} \rangle \} \cup \{ \langle \bar{a} - \lambda, \bar{b}, \bar{c}, \bar{d} - \lambda^{-1} \rangle \mid \lambda \in k^\times \}$$

(cf. Example II.1.3). Moreover, the first and last of these strata are homeomorphic to prime spectra of Laurent polynomial rings (Exercise II.2.C).

RATIONAL ACTIONS OF TORI

II.2.4. Observe that in Examples II.1.6 and in (II.1.14–17), we have a torus H acting on a k-algebra A in such a way that the standard generators for A are simultaneous eigenvectors for all $h \in H$. It follows that all monomials in these generators are simultaneous eigenvectors for H as well. For instance, we have $H = (k^\times)^4$ acting on $A = \mathcal{O}_q(M_2(k))$ so that

$$(\alpha_1, \alpha_2, \beta_1, \beta_2).(a^i b^j c^l d^m) = \alpha_1^{i+j} \alpha_2^{l+m} \beta_1^{i+l} \beta_2^{j+m} a^i b^j c^l d^m$$

for all i, j, l, m. Thus, the action of H on A is a *semisimple action*, i.e., the automorphisms corresponding to elements of H are simultaneously diagonalizable. We shall see that this special kind of action has strong and useful consequences for the structure of A and its prime spectrum.

II.2.5. Definitions. Let A be a k-algebra, and let H be a group acting on A by k-algebra automorphisms. A nonzero element $x \in A$ is an *H-eigenvector* provided $h(x) \in kx$ for all $h \in H$. Since H is acting via automorphisms, $h(x) \neq 0$ for all h, and so there must be a map $f : H \to k^\times$ such that $h(x) = f(h)x$ for all $h \in H$. The map f is the *H-eigenvalue* of the H-eigenvector x, and it is a group homomorphism.

Homomorphisms $H \to k^\times$ are called *characters* (or *k-characters*) of H. For any character f of H, set

$$A_f = \{x \in A \mid h(x) = f(h)x \text{ for all } h \in H\}.$$

If A_f is nonzero, it is called the *H-eigenspace of A with H-eigenvalue f*.

For example, let $A = \mathcal{O}_q((k^\times)^2)$ and let $H = (k^\times)^2$ act as in Example II.1.6(a). Each monomial $x^r y^s$ in A is an H-eigenvector; its H-eigenvalue is the character $(\alpha, \beta) \mapsto \alpha^r \beta^s$. In this case, provided k is infinite, the H-eigenspaces in A are all one-dimensional. On the other hand, if $H = (k^\times)^4$ acts on $A = \mathcal{O}_q(M_2(k))$ as in Example II.1.6(c), then each monomial $a^i b^j c^l d^m$ in A is an H-eigenvector with H-eigenvalue

$$(\alpha_1, \alpha_2, \beta_1, \beta_2) \mapsto \alpha_1^{i+j} \alpha_2^{l+m} \beta_1^{i+l} \beta_2^{j+m}.$$

Observe that in this case, the H-eigenspaces have arbitrarily large dimensions.

If H is an affine algebraic group over k, a *rational character* of H is any character $H \to k^\times$ which is also a morphism of algebraic varieties. (Typically, not all characters of H are rational – see Exercise II.2.D.) The set of rational characters, $X(H)$, forms an abelian group under pointwise multiplication, called the *(rational) character group of* H. (Warning: Many authors convert the operation in $X(H)$ to addition.) For example, if $H = (k^\times)^r$ and k is infinite, then $X(H)$

is free abelian of rank r, with a basis consisting of the r projections $(k^\times)^r \to k^\times$ (Exercise II.2.E).

Characters $g_1, \ldots, g_n \in X(H)$ are said to be *linearly independent* provided $g_1^{m_1} g_2^{m_2} \cdots g_n^{m_n} = 1$ only when all the $m_i = 0$. (This is just the condition of $\mathbb{Z}$-linear independence in $\mathbb{Z}$-modules, converted to multiplicative notation.)

II.2.6. Definitions. Let A be a k-algebra, let H be an affine algebraic group over k, and let H act on A by k-algebra automorphisms. The action of H on A is *rational* provided A is a directed union of finite dimensional H-invariant k-subspaces V_i such that the restriction maps $H \to \operatorname{Aut} A \to GL(V_i)$ are morphisms of algebraic varieties.

For example, suppose that H is a torus, and that its action on A is semisimple, that is, A is the direct sum of its H-eigenspaces. If all the characters that occur as H-eigenvalues for this action are rational, then the action is rational (Exercise II.2.F). In particular, all the actions of tori on quantized algebras discussed in (II.1.14–18) are rational (Exercise II.2.G).

The following theorem indicates that over an infinite field, a torus acts rationally precisely when it acts semisimply with rational eigenvalues. Thus for our purposes, it would suffice to take the latter conditions as our definition of a rational action.

II.2.7. Theorem. *Let A be a k-algebra, with k infinite, and let H be a k-torus acting on A by k-algebra automorphisms. The action of H on A is rational if and only if it is a semisimple action and the corresponding H-eigenvalues are all rational.*

Proof. We have just discussed the implication ($\Longleftarrow$). For the converse, see, e.g., [**173**, Ch. 5, Corollary to Theorem 36]. $\square$

One advantage of a rational action over an arbitrary one is that certain maps become Zariski-continuous, as in the following lemma.

II.2.8. Lemma. *Let A be a k-algebra, let H be an affine algebraic group over k, and assume that H acts rationally on A by k-algebra automorphisms. Let $Q \in \operatorname{spec} A$. Then the map $H \to \operatorname{spec} A$ given by $h \mapsto h(Q)$ is continuous. If A is noetherian, then $\operatorname{Stab}_H(Q)$ is closed in H.*

Proof. To see that the map $h \mapsto h(Q)$ is continuous, it suffices to show, for any ideal I of A, that the set

$$X = \{h \in H \mid h(Q) \supseteq I\}$$

is closed in H. Write $A = \bigcup_{j \in J} V_j$ for some finite dimensional H-invariant k-subspaces V_j. Then X equals the intersection of the sets

$$X_j = \{h \in H \mid h(Q \cap V_j) \supseteq I \cap V_j\},$$

and so it is enough to show that the X_j are all closed in H.

Fix $j \in J$, and choose idempotents $e_j, f_j \in \mathrm{End}_k(V_j)$ with images $Q \cap V_j$ and $I \cap V_j$ respectively. The set

$$Y_j = \{g \in GL(V_j) \mid (1 - e_j)g^{-1}f_j = 0\}$$

is clearly closed in $GL(V_j)$. Since the restriction map $\rho_j : H \to GL(V_j)$ is continuous by hypothesis, it follows that $X_j = \rho_j^{-1}(Y_j)$ is closed in H, as desired.

In particular, the set $Y = \{h \in H \mid h(Q) \supseteq Q\}$ is closed in H. If A is noetherian, then it follows from Exercise II.2.H that $Y = \mathrm{Stab}_H(Q)$, and the proof is complete. $\square$

II.2.9. Proposition. *Let A be a noetherian k-algebra, and suppose that a k-torus H acts rationally on A by k-algebra automorphisms. If k is infinite, then every prime ideal of A whose H-orbit is finite is H-stable. Consequently, every H-prime ideal of A is prime.*

Proof. As in the proof of Corollary II.1.12, the second conclusion will follow from the first. Thus, let Q be a prime ideal of A whose H-orbit is finite. Set $S = \mathrm{Stab}_H(Q)$; then S is a subgroup of H with finite index, and we just need to show that $S = H$. By Lemma II.2.8, S is closed in H.

For any $t \in H$, the translation map $h \mapsto th$ is a homeomorphism of H onto itself, and so the coset tS must be closed in H. Thus, if $t_1, \ldots, t_n$ is a transversal for S in H (that is, a complete, irredundant set of coset representatives), we have $H = t_1 S \sqcup \cdots \sqcup t_n S$, a finite disjoint union of closed subsets.

Recall that as an algebraic variety, H is isomorphic to the variety

$$\{(\alpha_1, \ldots, \alpha_n, \beta) \in k^{n+1} \mid \alpha_1 \alpha_2 \cdots \alpha_n \beta = 1\}$$

for some n. Hence, the coordinate ring of H has the form

$$R = k[x_1, \ldots, x_{n+1}]/\langle x_1 x_2 \cdots x_{n+1} - 1 \rangle,$$

where the x_i are independent indeterminates. (Here is where we need k to be infinite, to ensure that distinct elements of R determine distinct functions on H.) Note that R is isomorphic to the Laurent polynomial ring $k[x_1^{\pm 1}, \ldots, x_n^{\pm 1}]$, whence R is a domain.

Since the cosets $t_j S$ are closed in H, there exist ideals I_j in R such that

$$t_j S = \{h \in H \mid f(h) = 0 \text{ for all } f \in I_j\}$$

for all j. Any function in $I_1 \cap \cdots \cap I_n$ vanishes on all the $t_j S$ and hence on H. Thus $I_1 \cap \cdots \cap I_n = 0$. Since R is a domain, some $I_j = 0$, whence $t_j S = H$. Therefore we must have $S = H$, as desired. $\square$

RATIONAL $(k^\times)^r$-ACTIONS VERSUS $\mathbb{Z}^r$-GRADINGS

II.2.10. Let H be a k-torus acting rationally on a k-algebra A by k-algebra automorphisms, and assume that k is infinite. Then we have $A = \bigoplus_{f \in X(H)} A_f$ by Theorem II.2.7. If $a \in A_f$ and $b \in A_g$ for some $f, g \in X(H)$, then $h.a = f(h)a$ and $h.b = g(h)b$ for $h \in H$, whence

$$h.(ab) = (h.a)(h.b) = f(h)g(h)ab = (fg)(h)ab$$

for $h \in H$, that is, $ab \in A_{fg}$. Thus, we have $A_f A_g \subseteq A_{fg}$ for all $f, g \in X(H)$. This shows that the decomposition $A = \bigoplus_{f \in X(H)} A_f$ is an $X(H)$-grading on A.

See Definitions I.12.7 for the basic definitions and concepts associated with group-graded rings, in particular for the idea of a homogeneous ideal. We note for later use that with respect to the $X(H)$-grading of A above, an ideal I is homogeneous if and only if I is H-stable (Exercise II.2.I).

II.2.11. Lemma. *Let A be a k-algebra, where k is infinite, let H be a k-torus, and set $G = X(H)$. Then k-algebra G-gradings on A are equivalent to rational actions of H on A by k-algebra automorphisms. More precisely:*

(a) If $A = \bigoplus_{g \in G} A_g$ is a G-grading, then there is a rational action of H on A by k-algebra automorphisms such that $h.x = g(h)x$ for $h \in H$, $g \in G$, and $x \in A_g$. The H-eigenspace decomposition of A with respect to this H-action coincides with the given G-grading.

(b) If H acts rationally on A by k-algebra homomorphisms, then the H-eigenspace decomposition $A = \bigoplus_{g \in G} A_g$ is a G-grading. The action of H on A arising from this grading as in part (a) coincides with the given action.

Proof. See (II.2.10) and Exercise II.2.J. $\square$

In particular, if $H = (k^\times)^r$ for some r, then $X(H) \cong \mathbb{Z}^r$ by Exercise II.2.E, and so the lemma implies that rational $(k^\times)^r$-actions on k-algebras are equivalent to $\mathbb{Z}^r$-gradings. The gradings arising on our basic examples in this way can be described as follows.

II.2.12. Examples. Assume that k is infinite.

(a) Let A be either $\mathcal{O}_q(k^2)$ or $\mathcal{O}_q((k^\times)^2)$, and let $H = (k^\times)^2$ act on A as in Example II.1.6(a) or (II.1.14). Then $X(H)$ is free abelian of rank 2, with the two standard projections $H \to k^\times$ forming a basis. Hence, there is an isomorphism $\mathbb{Z}^2 \to X(H)$ under which each ordered pair (i, j) of integers corresponds to the character $(\lambda, \mu) \mapsto \lambda^i \mu^j$. Let us identify $X(H)$ with $\mathbb{Z}^2$ via this isomorphism. Hence, the H-action on A induces a $\mathbb{Z}^2$-grading as in Lemma II.2.11. Now each monomial $x^i y^j$ in A is an H-eigenvector with H-eigenvalue (i, j), and so with respect to the $\mathbb{Z}^2$-grading it is a homogeneous element with degree (i, j).

(b) Let A be either $\mathcal{O}_q(M_2(k))$ or $\mathcal{O}_q(GL_2(k))$, let $H = (k^\times)^4$ act on A as in Examples II.1.6(c)(d), and identify $X(H)$ with $\mathbb{Z}^4$. Then A obtains a $\mathbb{Z}^4$-grading where the generators are homogeneous elements with the following degrees:

$$\deg(a) = (1,0,1,0) \qquad\qquad \deg(b) = (1,0,0,1)$$
$$\deg(c) = (0,1,1,0) \qquad\qquad \deg(d) = (0,1,0,1).$$

Note that this grading accomodates the relation $ad - da = (q - q^{-1})bc$, since ad, da, and bc all have degree $(1,1,1,1)$.

(c) Recall that the standard action of $(k^\times)^4$ on $\mathcal{O}_q(M_2(k))$ above does not carry over to $\mathcal{O}_q(SL_2(k))$, because $ad - qbc - 1$ is not an H-eigenvector. Instead, $\mathcal{O}_q(SL_2(k))$ supports a rational action by $H = (k^\times)^2$, as in Example II.1.6(b). There is a corresponding $\mathbb{Z}^2$-grading on $\mathcal{O}_q(SL_2(k))$ where the generators are homogeneous elements with degrees as follows:

$$\deg(\bar{a}) = (1,1) \qquad\qquad \deg(\bar{b}) = (1,-1)$$
$$\deg(\bar{c}) = (-1,1) \qquad\qquad \deg(\bar{d}) = (-1,-1).$$

Example II.1.6(b) also describes an action on $\mathcal{O}_q(SL_2(k))$ by a larger torus, denoted H'. The character group of H' is naturally isomorphic to the quotient group $G = \mathbb{Z}^4/\mathbb{Z}(1,1,1,1)$, and the H'-grading on $\mathcal{O}_q(SL_2(k))$ corresponds to a G-grading very similar to that on $\mathcal{O}_q(M_2(k))$ in part (b); namely, the generators are homogeneous elements with degrees

$$\deg(\bar{a}) = \overline{(1,0,1,0)} \qquad\qquad \deg(\bar{b}) = \overline{(1,0,0,1)}$$
$$\deg(\bar{c}) = \overline{(0,1,1,0)} \qquad\qquad \deg(\bar{d}) = \overline{(0,1,0,1)}.$$

STRUCTURE OF STRATIFIED SPECTRA

We can now state the degree to which a rational action by a torus can help in analyzing a prime spectrum. Recall that a *regular* element in a ring is just any non-zero-divisor, and that when a group H acts on a ring R by automorphisms, the *fixed ring* R^H is the subring $\{r \in R \mid h(r) = r$ for all $h \in H\}$. As in the case of a noetherian domain (I.2.11), we denote the Goldie quotient ring of a prime noetherian ring R by Fract R. By Goldie's Theorem [**83**, Theorem 5.12; **158**, Theorem 2.3.6], Fract R always exists for such a ring R, and is simple artinian.

II.2.13. Stratification Theorem. *Let A be a noetherian k-algebra, with k infinite, let $H = (k^\times)^r$ be a torus acting rationally on A by k-algebra automorphisms, and let $J \in H$-spec A.*

(a) *J is a prime ideal of A.*

(b) *Let $\mathcal{E}_J$ denote the set of all regular H-eigenvectors in A/J. Then $\mathcal{E}_J$ is a denominator set, and the localization $A_J = (A/J)[\mathcal{E}_J^{-1}]$ is an H-simple ring (with respect to the induced H-action).*

(c) $\operatorname{spec}_J A$ is homeomorphic to $\operatorname{spec} A_J$ via localization and contraction.

(d) $\operatorname{spec} A_J$ is homeomorphic to $\operatorname{spec} Z(A_J)$ via contraction and extension.

(e) $Z(A_J)$ is a Laurent polynomial ring, in at most r indeterminates, over the fixed field $Z(A_J)^H = Z(\operatorname{Fract} A/J)^H$. The indeterminates can be chosen to be H-eigenvectors with linearly independent H-eigenvalues.

Proof. Part (a) is Proposition II.2.9. The remaining parts of the proof will be discussed in the following chapter; we deal with them by converting the H-action on A into an $X(H)$-grading as in Lemma II.2.11. Note that there will be no loss of generality in assuming $J = 0$. $\square$

II.2.14. Examples. (a) Let $A = \mathcal{O}_q(k^n)$ and $H = (k^\times)^n$ as in (II.1.14), and assume that k is infinite. For $w \subseteq \{1, \dots, n\}$, set $J_w = \langle x_i \mid i \in w \rangle$, and let E_w be the multiplicative set generated by the x_j for $j \notin w$. By Exercise II.2.K, the ideals J_w are precisely the H-primes of A. Now fix w. The image of $k^\times E_w$ in A/J_w coincides with $\mathcal{E}_{J_w}$ (in the notation of Theorem II.2.13), and the algebra $A_w = A_{J_w}$ is isomorphic to $\mathcal{O}_p((k^\times)^r)$ for a suitable p, where $r = n - \operatorname{card}(w)$. The centre $Z(A_w)$ has a basis of monomials in cosets of the x_j (for $j \notin w$) whose exponents satisfy certain equations in powers of the q_{ij}, equations which are particularly easy to solve when n is small and the q_{ij} are powers of some non-root of unity (see Exercise II.2.K).

(b) Let $A = \mathcal{O}_q(SL_2(k))$ and $H = (k^\times)^2$ as in Example II.1.6(b), with q not a root of unity. By Exercise II.1.K, there are four H-primes in A, namely $\langle 0 \rangle$, $\langle \bar{b} \rangle$, $\langle \bar{c} \rangle$, and $\langle \bar{b}, \bar{c} \rangle$.

If $J = \langle \bar{b}, \bar{c} \rangle$, then $A/J \cong k[\bar{a}^{\pm 1}]$. The H-eigenvectors in $k[\bar{a}^{\pm 1}]$ are just the monomials $\lambda \bar{a}^i$ for $\lambda \in k^\times$ and $i \in \mathbb{Z}$, and these are all invertible. Hence, $A_J = A/J$ in this case. Moreover, A_J is commutative, and so $Z(A_J) = A_J$. Of course, A_J already is a Laurent polynomial ring, as predicted by the theorem; note that the coefficient field $Z(A_J)^H$ equals k.

Next, let $J = \langle \bar{b} \rangle$, and note that $A/J \cong k[\bar{a}^{\pm 1}][\bar{c}; \tau]$ where τ is the k-algebra automorphism of $k[\bar{a}^{\pm 1}]$ such that $\tau(\bar{a}) = q^{-1}\bar{a}$. The H-eigenvectors here are the monomials $\lambda \bar{a}^i \bar{c}^j$ for $\lambda \in k^\times$, $i \in \mathbb{Z}$, and $j \in \mathbb{Z}^+$. Hence, $\mathcal{E}_J$ is generated by $k^\times \cup \{\bar{a}, \bar{c}\}$, and so $A_J \cong k[\bar{a}^{\pm 1}][\bar{c}^{\pm 1}; \tau]$, which is a copy of $\mathcal{O}_q((k^\times)^2)$. Since q is not a root of unity, $Z(A_J) = k$ (Exercise II.2.L), which we view as a Laurent polynomial ring in zero indeterminates.

Similarly, $A_{\langle \bar{c} \rangle} \cong k[\bar{a}^{\pm 1}][\bar{b}^{\pm 1}; \tau]$ and $Z(A_{\langle \bar{c} \rangle}) = k$.

Finally, consider the case $J = \langle 0 \rangle$; see Exercise II.2.L for hints and details. It can be shown that $\mathcal{E}_{\langle 0 \rangle}$ is generated by $\bar{a}, \bar{b}, \bar{c}, \bar{d}$ together with all nonzero polynomials in the element $\bar{b}\bar{c}$, and it follows that $A_{\langle 0 \rangle} = k(\bar{b}\bar{c})[\bar{b}^{\pm 1}][\bar{a}^{\pm 1}; \sigma]$ for a suitable automorphism σ. Some further calculation reveals that $Z(A_{\langle 0 \rangle}) = k[(\bar{b}\bar{c}^{-1})^{\pm 1}]$, and again $Z(A_{\langle 0 \rangle})^H = k$.

(c) For comparison with part (b), let $A = \mathcal{O}_q(GL_2(k))$ and $H = (k^\times)^4$ as in

Example II.1.6(d); assume that q is not a root of unity. The pattern of H-primes here matches the previous case: namely, the H-primes of A are $\langle 0 \rangle$, $\langle b \rangle$, $\langle c \rangle$, $\langle b,c \rangle$ (Exercise II.2.M). Thus, there are again four H-strata in spec A; the difference is that this time they are larger. For instance, $\mathrm{spec}_{\langle b,c \rangle} A$ contains maximal ideals $\langle a - \alpha, b, c, d - \delta \rangle$ for all $\alpha, \delta \in k^\times$. See Exercise II.2.N for further information.

   (d) Finally, consider the case where $A = \mathcal{O}_q(M_2(k))$ and $H = (k^\times)^4$ as in Example II.1.6(c), and assume that q is not a root of unity. Just as in the previous two cases, the ideals $\langle 0 \rangle$, $\langle b \rangle$, $\langle c \rangle$, $\langle b,c \rangle$ are H-primes of A, and it follows from part (c) that these are the only H-primes not containing D_q. There are 10 other H-primes in A (see Exercise II.2.O), and H-spec A can be pictured as in Diagram II.2.14 below. Thus, there are 14 H-strata in spec A. If k is algebraically closed and q is not a root of unity, these strata can be computed by the same techniques used in Exercises II.2.M-N; in fact, four of the H-strata can be read off from the H-strata of spec $\mathcal{O}_q(GL_2(k))$. We leave the details to the reader.

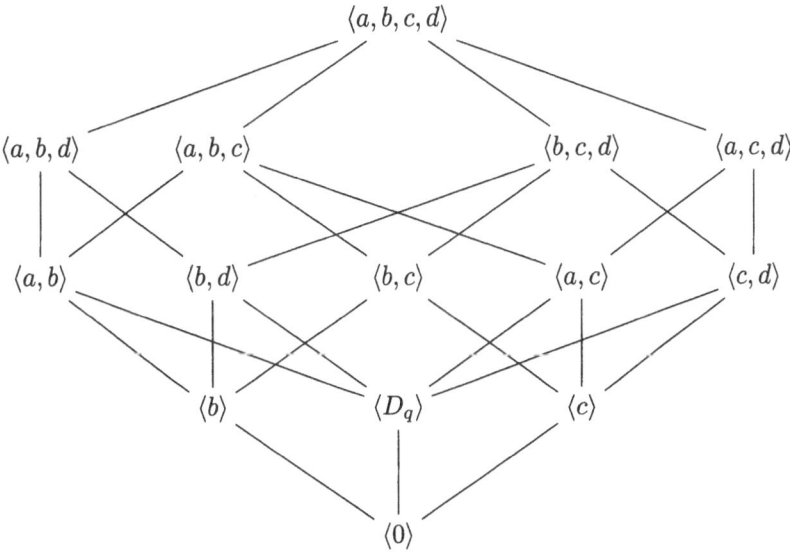

DIAGRAM II.2.14. H-spec $\mathcal{O}_q(M_2(k))$

NOTES

The H-stratifications discussed here were introduced by Goodearl and Letzter [80, 70]. Two versions of the Stratification Theorem were proved in those papers, each under an extra hypothesis – complete primeness of the H-primes in the case of [80], and a suitable supply of normal H-eigenvectors in the case of [70]. These extra hypotheses were removed when Goodearl and Stafford obtained a suitable graded version of Goldie's Theorem [82].

Exercises

Exercise II.2.A. Let H be a group acting by automorphisms on a ring R. Show that the closure of each H-stratum in spec R is a union of H-strata. If there are only finitely many H-strata, show that each H-stratum is an intersection of an open and a closed subset of spec R. $\square$

Exercise II.2.B. Let $A = \mathcal{O}_q(k^2)$ with k algebraically closed and q not a root of unity, and let $H = (k^\times)^2$ act on A as in Example II.1.6(a). Verify that the H-strata in spec A are correctly described in Example II.2.3(a), and establish the homeomorphisms $\operatorname{spec}_{\langle x \rangle} A \approx \operatorname{spec} k[y^{\pm 1}]$ and $\operatorname{spec}_{\langle y \rangle} A \approx \operatorname{spec} k[x^{\pm 1}]$. $\square$

Exercise II.2.C. Let $A = \mathcal{O}_q(SL_2(k))$ with k algebraically closed and q not a root of unity, and let $H = (k^\times)^2$ act on A as in Example II.1.6(b). Verify that the H-strata in spec A are correctly described in Example II.2.3(b), and establish the homeomorphisms $\operatorname{spec}_{\langle 0 \rangle} A \approx \operatorname{spec} k[z^{\pm 1}]$ and $\operatorname{spec}_{\langle \bar{b}, \bar{c} \rangle} A \approx \operatorname{spec} k[\bar{a}^{\pm 1}]$. $\square$

Exercise II.2.D. Let $H = (k^\times)^r$ be a torus over a field k which has a nontrivial automorphism. Show that not all characters of H are rational, i.e., $X(H)$ is properly contained in $\operatorname{Hom}(H, k^\times)$, even when $r = 1$. $\square$

Exercise II.2.E. Let $H = (k^\times)^r$ be a torus over an infinite field k. Show that $X(H)$ is free abelian with basis $p_1, \ldots, p_r$, where $p_j : H \to k^\times$ is the projection onto the j-th component. $\square$

Exercise II.2.F. Let $H = (k^\times)^r$ act on a k-algebra A by k-algebra automorphisms, and suppose that A is the direct sum of some H-eigenspaces corresponding to rational characters of H. Show that the action of H on A is rational. $\square$

Exercise II.2.G. Verify that all the actions of tori on quantized coordinate rings and quantized Weyl algebras given in (II.1.14–18) are rational. $\square$

Exercise II.2.H. Let R be a ring with ACC on ideals, ϕ an automorphism of R, and I an ideal of R. Show that if either $\phi(I) \subseteq I$ or $\phi(I) \supseteq I$, then $\phi(I) = I$. [Hint: In the first case, consider the ideals $\phi^{-n}(I)$ for $n \geq 0$.] $\square$

Exercise II.2.I. Let H be a k-torus acting rationally on a k-algebra A by k-algebra automorphisms, where k is infinite, and give A the corresponding $X(H)$-grading as in (II.2.10). Show that the homogeneous ideals of A (with respect to this grading) are precisely the H-ideals of A. $\square$

Exercise II.2.J. Complete the proof of Lemma II.2.11. $\square$

Exercise II.2.K. Let $A = \mathcal{O}_q(k^n)$ and $H = (k^\times)^n$ as in (II.1.14), and assume that k is infinite. For $w \subseteq \{1, \ldots, n\}$, define J_w and E_w as in Example II.2.14(a), and observe that J_w is an H-prime ideal of A. Observe also that E_w is a denominator set in A, and so its image is a denominator set in A/J_w; set $A_w = (A/J_w)[E_w^{-1}]$. Show that A_w is H-simple with respect to the natural induced action. [Hint: To simplify notation, note that $A_w \cong \mathcal{O}_p((k^\times)^r)$ for some multiplicatively antisymmetric matrix $p = (p_{ij})$, where $r = n - \operatorname{card}(w)$.] Use this information to conclude that the J_w are the only H-primes in A.

Show that $Z(A_w)$ is spanned by the cosets of the monomials $x_1^{s_1} x_2^{s_2} \cdots x_n^{s_n}$ such that $s_i = 0$ for $i \in w$ while $\prod_{j \notin w} q_{ij}^{s_j} = 1$ for $i \notin w$.

Now specialize to $n = 3$ and the single parameter case $A = \mathcal{O}_q(k^3)$. Assume that k is algebraically closed and q is not a root of unity. Compute $Z(A_J)$ and $\operatorname{spec}_J A$ for all eight H-primes J of A. $\square$

Exercise II.2.L. Let $A = \mathcal{O}_q(SL_2(k))$ and $H = (k^\times)^2$ as in Example II.1.6(b), with q not a root of unity. Fill in details for Example II.2.14(b) as follows.

(a) Check that $A_{\langle \overline{b} \rangle} \cong \mathcal{O}_q((k^\times)^2)$, and show that the centre of this algebra is just k. [Hint: An element $t \in \mathcal{O}_q((k^\times)^2)$ is central if and only if $xtx^{-1} = yty^{-1} = t$.]

(b) Show that $\mathcal{E}_{\langle 0 \rangle}$ is generated by $\overline{a}, \overline{b}, \overline{c}, \overline{d}$ together with all nonzero polynomials in the element $\overline{bc}$. [Hint: Observe that A is spanned by the products $\overline{b}^i \overline{a}^j \overline{c}^m$ and $\overline{b}^i \overline{d}^l \overline{c}^m$ for $i, j, l, m \geq 0$.] Then check that $A_{\langle 0 \rangle} = k(\overline{bc})[\overline{b}^{\pm 1}][\overline{a}^{\pm 1}; \sigma]$ for a suitable automorphism σ. Show that $Z(A_{\langle 0 \rangle}) = k[(\overline{bc}^{-1})^{\pm 1}]$ and $Z(A_{\langle 0 \rangle})^H = k$. $\square$

Exercise II.2.M. Let $A = \mathcal{O}_q(GL_2(k))$ and $H = (k^\times)^4$ as in Example II.1.6(d), with q not a root of unity.

(a) Observe that $A/\langle b, c \rangle \cong k[a^{\pm 1}, d^{\pm 1}]$. Compute the induced action of H on $A/\langle b, c \rangle$, and show that the latter algebra is H-simple. Hence, $\langle b, c \rangle$ is a maximal H-ideal of A.

(b) Observe that the powers of c form a denominator set in A, and show that $A[c^{-1}]/\langle b \rangle \cong \mathcal{O}_p((k^\times)^3)$ for a suitable p. Compute the induced action of H, and show that $A[c^{-1}]/\langle b \rangle$ is H-simple. Hence, the only H-prime of A that contains b but not c is $\langle b \rangle$. Similarly, the only H-prime of A that contains c but not b is $\langle c \rangle$.

(c) Show that no H-prime of A can contain a power of a.

(d) In view of Exercise II.1.D(d), the powers of a form a denominator set in $\mathcal{O}_q(M_2(k))$, and so also in A. Show that the set $\{a^i b^j c^l \mid i, j, l \geq 0\}$ is a denominator set in A, let B denote the corresponding localization, and observe that $d = a^{-1}(qbc + D_q)$ in B. Hence, B is generated by $a^{\pm 1}, b^{\pm 1}, c^{\pm 1}, D_q^{\pm 1}$. Use this observation to show that B is H-simple. [Hints: For any $\alpha \in k^\times$, the action of $(\alpha, 1, 1, \alpha^{-1}) \in H$ fixes b, c, D_q. Any ideal of B is stable under conjugation by a.] Conclude that every nonzero H-prime of A must contain b or c.

Therefore the only H-primes of A are $\langle 0 \rangle$, $\langle b \rangle$, $\langle c \rangle$, $\langle b, c \rangle$. $\square$

Exercise II.2.N. Let $A = \mathcal{O}_q(GL_2(k))$ and $H = (k^\times)^4$ as in Example II.1.6(d), with q not a root of unity. Show that the four H-strata of spec A are homeomorphic to prime spectra of Laurent polynomial rings as follows:

$$\mathrm{spec}_{\langle b,c \rangle}\, A \approx \mathrm{spec}\, k[a^{\pm 1}, d^{\pm 1}] \qquad \mathrm{spec}_{\langle b \rangle}\, A \approx \mathrm{spec}\, k[(ad)^{\pm 1}]$$

$$\mathrm{spec}_{\langle c \rangle}\, A \approx \mathrm{spec}\, k[(ad)^{\pm 1}] \qquad \mathrm{spec}_{\langle 0 \rangle}\, A \approx \mathrm{spec}\, k[(bc^{-1})^{\pm 1}, D_q^{\pm 1}]. \quad \square$$

Exercise II.2.O. Let $A = \mathcal{O}_q(M_2(k))$ with q not a root of unity, and let $H = (k^\times)^4$ act on A as in Example II.1.6(c).

(a) Check that each of the following ideals is an H-prime of A:

$$\langle 0 \rangle, \quad \langle b \rangle, \quad \langle c \rangle, \quad \langle b,c \rangle, \quad \langle a,b \rangle, \quad \langle a,c \rangle, \quad \langle b,d \rangle, \quad \langle c,d \rangle,$$
$$\langle a,b,c \rangle, \qquad \langle a,b,d \rangle, \qquad \langle a,c,d \rangle, \qquad \langle b,c,d \rangle, \qquad \langle a,b,c,d \rangle.$$

(b) Set $A_1 = A/\langle b,d \rangle$ and $A_2 = A/\langle c,d \rangle$, let $\pi_i : A \to A_i$ denote the quotient maps, and let θ be the composition

$$A \xrightarrow{\ \Delta\ } A \otimes A \xrightarrow{\ \pi_1 \otimes \pi_2\ } A_1 \otimes A_2$$

(where Δ is the standard bialgebra comultiplication on A). Show that $\ker(\theta) = \langle D_q \rangle$. [Hints: Check that $ad - q^2 da \in \langle D_q \rangle$, show that the cosets of the elements $a^i b^j d^m$ and $a^i c^l d^m$ span $A/\langle D_q \rangle$, and then show that θ maps these elements to linearly independent elements of $A_1 \otimes A_2$.] Conclude that $\langle D_q \rangle$ is an H-prime of A.

(c) Show that the 14 H-primes given in (a) and (b) are the only H-primes of A. [Hint: The H-primes not containing D_q correspond to H-primes of $\mathcal{O}_q(GL_2(k))$.] $\square$

PROOF OF THE STRATIFICATION THEOREM

Recall that part (a) of Theorem II.2.13 is already proved. To prove part (b), we need an 'H-form' of Goldie's Theorem: starting with a suitable H-prime k-algebra A, we want to show that the set $\mathcal{E}$ of regular H-eigenvectors in A forms a denominator set, and that the localization $A[\mathcal{E}^{-1}]$ is H-simple (as well as 'H-artinian' in an obvious sense). The action of H is, of course, extended to an action on $A[\mathcal{E}^{-1}]$ by k-algebra automorphisms in the obvious manner. In view of Lemma II.2.11, it suffices to prove an analogous graded version of Goldie's Theorem in the context of $X(H)$-graded rings, and we proceed in that direction.

A GRADED GOLDIE THEOREM

II.3.1. Definitions. Let R be a ring graded by a group G. Many standard ring- and module-theoretic concepts are converted to graded versions by placing the prefix 'gr-' in front of the usual names, as follows (cf. (I.12.7)).

We say that R is *gr-prime* provided R is nonzero and the product of any two nonzero homogeneous ideals of R is nonzero; *gr-semiprime* provided R has no nonzero nilpotent homogeneous ideals; and *gr-simple* provided R is nonzero and its only homogeneous ideals are 0 and R. The adjectives *gr-artinian* and *gr-noetherian* refer to chain conditions on homogeneous one-sided ideals; in particular, R is gr-artinian provided it satisfies the DCC for homogeneous right and left ideals, while it is right gr-noetherian if it has the ACC on homogeneous right ideals. Finally, R is *right gr-Goldie* provided R contains no infinite direct sums of nonzero homogeneous right ideals and R has the ACC on right annihilators of sets of homogeneous elements. In particular, if R is right gr-noetherian, then R is right gr-Goldie.

II.3.2. Example. A complete generalisation of Goldie's Theorem to the gr-world would say that if R is a gr-semiprime, right gr-Goldie ring, then the set of homogeneous regular elements in R forms a right denominator set and the corresponding localization of R is a finite direct sum of gr-simple, gr-artinian rings. Unfortunately, this does not always hold. For example, let

$$R = k\langle x, y \mid xy = yx = 0\rangle,$$

and observe that the powers of x together with the powers of y form a k-basis for R. The ideals $\langle x \rangle$ and $\langle y \rangle$ are prime, since $R/\langle x \rangle \cong k[y]$ and $R/\langle y \rangle \cong k[x]$, and, moreover, $\langle x \rangle \cap \langle y \rangle = 0$. Thus R is a commutative, noetherian, semiprime ring.

Now give R the $\mathbb{Z}$-grading such that $R_n = kx^n$ for $n \geq 0$ while $R_n = ky^{-n}$ for $n < 0$. The only homogeneous regular elements of R are the nonzero scalars, so they obviously form a denominator set, and the corresponding localization is just R. However, R is neither gr-artinian nor a direct sum of gr-simple rings.

Fortunately for our purposes, the gr-prime case behaves much better, at least for rings graded by abelian groups. In addition to this case of a graded Goldie theorem, we shall also need the corresponding graded analogue of Goldie's lemma concerning the existence of regular elements.

II.3.3. Proposition. *Let G be an abelian group and R a G-graded, gr-prime, right gr-Goldie ring. Then every essential homogeneous right ideal of R contains a homogeneous regular element.*

Proof. [**82**, Theorem 4]. □

II.3.4. Theorem. *Let G be an abelian group and R a G-graded, gr-prime, right gr-Goldie ring. Then the set $\mathcal{E}$ of homogeneous regular elements of R is a right denominator set, the G-grading on R extends uniquely to $R[\mathcal{E}^{-1}]$, and $R[\mathcal{E}^{-1}]$ is a gr-simple, gr-artinian ring.*

Proof. [**82**, Theorem 1]. □

For our purposes, it suffices to translate the noetherian cases of these results into the context of torus actions, as follows.

II.3.5. Corollary. *Let A be a noetherian k-algebra, with k infinite, let H be a k-torus acting rationally on A by k-algebra automorphisms, and assume that A is an H-prime ring.*

(a) Every H-stable essential right or left ideal of A contains a regular H-eigenvector.

(b) The set $\mathcal{E}$ of all regular H-eigenvectors in A is a denominator set.

(c) The action of H on A extends uniquely to a rational action of H on $A[\mathcal{E}^{-1}]$ by k-algebra automorphisms, and $A[\mathcal{E}^{-1}]$ is an H-simple ring. Further, $A[\mathcal{E}^{-1}]$ satisfies the DCC on H-stable right and left ideals.

Proof. Most of this follows from Lemma II.2.11, Proposition II.3.3, and Theorem II.3.4. For the rationality of the H-action on $A[\mathcal{E}^{-1}]$, see Exercise II.3.A. □

II.3.6. Proof of Theorem II.2.13(b) and (c). Part (b) is immediate from Corollary II.3.5(b)(c). Part (c) will follow by standard localization theory, once we show that the prime ideals of A/J disjoint from $\mathcal{E}_J$ are precisely the factors P/J for $P \in \mathrm{spec}_J A$. It is clear that if $P \in \mathrm{spec}_J A$, then P/J is a prime of A/J and that P/J contains no nonzero H-eigenvectors, whence $(P/J) \cap \mathcal{E}_J = \varnothing$. Conversely, any prime of A/J disjoint from $\mathcal{E}_J$ has the form P/J for some $P \in \operatorname{spec} A$ containing J, whence $(P : H) \supseteq J$ and $(P : H)/J$ contains no regular H-eigenvectors. By Corollary II.3.5(a), $(P : H)/J$ is not essential as a right (or left) ideal of A/J. Since A/J is a prime ring (Theorem II.2.13(a)), it follows that $(P : H)/J = 0$, and therefore $P \in \operatorname{spec}_J A$. $\square$

CENTRES OF GR-SIMPLE RINGS

To prove parts (d) and (e) of Theorem II.2.13, we need to analyze the centres of the H-simple rings A_J. We again transfer the problem to group-graded rings, and analyze the centres of suitable gr-simple rings.

II.3.7. Lemma. *Let G be an abelian group and R a G-graded, gr-simple ring.*

(a) The centre $Z(R)$ is a homogeneous subring of R, with homogeneous components $Z(R)_g = Z(R) \cap R_g$.

(b) The set $G_Z = \{g \in G \mid Z(R)_g \neq 0\}$ is a subgroup of G, and $Z(R)$ is strongly graded by G_Z.

(c) Every nonzero homogeneous element of $Z(R)$ is invertible. In particular, $Z(R)_1$ is a field.

(d) As $Z(R)$-modules, $Z(R)$ is a direct summand of R.

(e) Suppose that G_Z is a free abelian group of finite rank, say rank t. Choose a basis $\{g_1, \ldots, g_t\}$ for G_Z, and choose a nonzero element $z_j \in Z(R)_{g_j}$ for each j. Then $Z(R) = Z(R)_1[z_1^{\pm 1}, \ldots, z_t^{\pm 1}]$, a Laurent polynomial ring over the field $Z(R)_1$.

Proof. Part (a) is easily checked (Exercise II.3.B). Part (c) follows from the graded-simplicity of R (Exercise II.3.B), and then part (b) is clear.

(d) Note that $S = \bigoplus_{g \in G_Z} R_g$ is a homogeneous subring of R containing $Z(R)$. Further, S is a left S-module direct summand of R, with complement $\bigoplus_{g \in G \setminus G_Z} R_g$. Hence, we just need to show that $Z(R)$ is a $Z(R)$-module direct summand of S.

Choose nonzero elements $z_g \in Z(R)_g$ for $g \in G_Z$. Note that $S_g = S_1 z_g$ and $Z(R)_g = Z(R)_1 z_g$ for all $g \in G_Z$. Since $Z(R)_1$ is a field, we can choose a basis $\mathcal{B}$ for S_1 as a vector space over $Z(R)_1$, with $1 \in \mathcal{B}$. Hence,

$$S = \bigoplus_{g \in G_Z} S_1 z_g = \bigoplus_{\beta \in \mathcal{B}} \bigoplus_{g \in G_Z} Z(R)_1 \beta z_g = \bigoplus_{\beta \in \mathcal{B}} Z(R) \beta.$$

Since $1 \in \mathcal{B}$, this shows that $Z(R)$ is a direct summand of S, as desired.

(e) Since the monomials $z_1^{i_1} z_2^{i_2} \cdots z_t^{i_t}$ are homogeneous with distinct degrees $g_1^{i_1} g_2^{i_2} \cdots g_t^{i_t}$, it is clear that $z_1, \ldots, z_t$ are algebraically independent over $Z(R)_1$. For each $g \in G_Z$, there exist unique $m_1, \ldots, m_t \in \mathbb{Z}$ such that $g = g_1^{m_1} g_2^{m_2} \cdots g_t^{m_t}$, and $z_g = z_1^{m_1} z_2^{m_2} \cdots z_t^{m_t}$ is a nonzero element of $Z(R)_g$. Therefore $Z(R) = \bigoplus_{g \in G_Z} Z(R)_1 z_g = Z(R)_1[z_1^{\pm 1}, \ldots, z_t^{\pm 1}]$. $\square$

II.3.8. Proposition. *Let G be an abelian group and R a G-graded, gr-simple ring. Then there exist bijections between the sets of ideals of R and $Z(R)$, given by contraction and extension, that is,*

$$I \mapsto I \cap Z(R) \qquad \text{and} \qquad J \mapsto JR.$$

Proof. It is clear from Lemma II.3.7(d) that $JR \cap Z(R) = J$ for every ideal J of $Z(R)$. It remains to show that $(I \cap Z(R))R = I$ for any ideal I of R. Set $J = I \cap Z(R)$, and suppose that $I \neq JR$. Pick an element $x \in I \setminus JR$ of minimal length, say length n. (By the *length* of a nonzero element of R is meant the number of nonzero homogeneous components it has.) Then $x = x_1 + \cdots + x_n$ for some nonzero elements $x_i \in R_{g_i}$, where $g_1, \ldots, g_n$ are distinct elements of G.

Now Rx_1R is a nonzero homogeneous ideal of R, so $Rx_1R = R$ by graded-simplicity and $\sum_j a_j x_1 b_j = 1$ for some $a_j, b_j \in R$. Express each a_j, b_j as a sum of homogeneous elements, and substitute these expressions in the terms $a_j x_1 b_j$. This expands the sum $\sum_j a_j x_1 b_j$ in the form $\sum_t c_t x_1 d_t$ with all c_t, d_t homogeneous. Hence, after relabelling, we may assume that the a_j and b_j are all homogeneous, say of degrees e_j and f_j, respectively.

Comparing identity components in the equation $\sum_j a_j x_1 b_j = 1$, we obtain

$$\sum_{e_j g_1 f_j = 1} a_j x_1 b_j = 1.$$

Thus, after deleting all other $a_j x_1 b_j$ terms, we may assume that $e_j g_1 f_j = 1$ for all j. Since G is abelian, it follows that $e_j f_j = g_1^{-1}$ for all j.

Set $x' = \sum_j a_j x b_j$, and note that $x' \in I$. Further,

$$x' = \sum_{i=1}^n \sum_j a_j x_i b_j$$

where $\sum_j a_j x_i b_j \in R_{g_i^{-1} g_i}$ for each i. As a result, x' is an element whose support is contained in $\{1, g_1^{-1} g_2, \ldots, g_1^{-1} g_n\}$ and whose identity component is 1. Hence, $x_1 x'$ is an element of I whose support is contained in $\{g_1, \ldots, g_n\}$ and whose g_1-component is x_1. Comparing this element with x, we see that $x - x_1 x'$ is an element of I whose support is contained in $\{g_2, \ldots, g_n\}$. By the minimality of n, we must have $x - x_1 x' \in JR$, and so $x' \notin JR$. Therefore, after replacing x by x', there is no loss of generality in assuming that $g_1 = 1$ and $x_1 = 1$.

$\underline{\text{Claim:}}$ There do not exist $g \in G$ and a nonzero element $y \in I$ whose support is properly contained in the set $\{gg_1, \ldots, gg_n\}$.

Suppose there do exist such g and y. Say the gg_s-component of y is nonzero. As above, there exists an element of the form $y' = \sum_j a_j y b_j$ whose support is properly contained in $\{g_s^{-1}g_1, \ldots, g_s^{-1}g_n\}$ and whose identity component is 1. Then $y' \in I$, and $x - x_s y'$ is an element of I whose support is properly contained in $\{g_1, \ldots, g_n\}$. Since y' and $x - x_s y'$ are elements of I of length less than n, they must lie in JR, by the minimality of n. But then $x \in JR$, contradicting our assumptions. Therefore the claim is proved.

Finally, consider any homogeneous element $r \in R$, say $r \in R_g$. Then $rx - xr$ is an element of I with support contained in $\{gg_2, \ldots, gg_n\}$, and so $rx - xr = 0$ by the claim. It follows that $x \in Z(R)$ and so $x \in J$, contradicting our assumption that $x \notin JR$.

Therefore $I = JR$. $\quad\square$

II.3.9. Corollary. *Let A be a noetherian k-algebra, with k infinite, let $H = (k^\times)^r$ be a torus acting rationally on A by k-algebra automorphisms, and assume that A is an H-simple ring.*

(a) $Z(A)$ is a Laurent polynomial ring, in at most r indeterminates, over the fixed field $Z(A)^H = Z(\text{Fract } A)^H$. The indeterminates can be chosen to be H-eigenvectors with linearly independent H-eigenvalues.

(b) Every ideal of A is generated by its intersection with $Z(A)$.

(c) Contraction and extension provide mutually inverse homeomorphisms between spec A *and* spec $Z(A)$.

Proof. Most of parts (a) and (b), except for the equality $Z(A)^H = Z(\text{Fract } A)^H$ in part (a), is given by Lemmas II.2.11 and II.3.7 together with Proposition II.3.8. Part (c) follows (Exercise II.3.C).

The inclusion $Z(A)^H \subseteq Z(\text{Fract } A)^H$ is clear. For the reverse inclusion, consider an element $u \in Z(\text{Fract } A)^H$, and observe that the set

$$I = \{a \in A \mid au \in A\}$$

is a nonzero H-ideal of A. Since A is H-simple, $I = A$, whence $u \in A$. Therefore $u \in Z(A)^H$, as desired. $\quad\square$

II.3.10. Proof of Theorem II.2.13(d) and (e). Corollary II.3.9. $\quad\square$

II.3.11. In the previous chapter (Examples II.2.14), we have seen how the Stratification Theorem applies to $\mathcal{O}_q(k^n)$, $\mathcal{O}_q(SL_2(k))$, $\mathcal{O}_q(GL_2(k))$, and $\mathcal{O}_q(M_2(k))$. It applies similarly to other quantized coordinate algebras. Now if in a given situation there are infinitely many H-strata, then likely these strata are relatively small, perhaps even singletons, and so the H-stratification may not contain much information. Thus, it is preferable if there are only finitely many H-strata. In Chapter II.5, we shall develop some techniques to show that H-spec is finite in many iterated skew polynomial rings, and apply these results to our collection of quantized algebras.

In studying an individual H-stratum $\text{spec}_J A$, we are led to a Laurent polynomial ring over a field $Z(A_J)^H$, and we need to identify this field. We shall also develop techniques to show that in many cases, $Z(A_J)^H$ is just the base field k. That is the best possible outcome, since then $\mathrm{spec}_J A$ is homeomorphic to an affine scheme over k.

NOTES

This proof of the Stratification Theorem was taken mainly from [80], with a few modifications from [70]. A different proof of Proposition II.3.8 can be found in [109]. The graded Goldie Theorem (II.3.4) was proved by Goodearl and Stafford in [82].

EXERCISES

Exercise II.3.A. Let A be a noetherian k-algebra, with k infinite, let H be a k-torus acting rationally on A by k-algebra automorphisms, and let $\mathcal{E}$ be a denominator set in A consisting of H-eigenvectors. Show that the action of H on A extends uniquely to an action on $A[\mathcal{E}^{-1}]$ by k-algebra automorphisms, and that the latter action is rational. $\square$

Exercise II.3.B. Let G be an abelian group and R a G-graded, gr-simple ring. Show that $Z(R)$ is homogeneous, and that all nonzero homogeneous elements of $Z(R)$ are invertible. $\square$

Exercise II.3.C. Let $A \supseteq Z(A) \supseteq Z$ be rings such that $I = A(I \cap Z)$ and $J = AJ \cap Z$ for all ideals I of A and J of Z. Show that the rules $P \mapsto P \cap Z$ and $Q \mapsto AQ$ provide mutually inverse homeomorphisms between $\mathrm{spec}\,A$ and $\mathrm{spec}\,Z$. $\square$

CHAPTER II.4

PRIME IDEALS IN $\mathcal{O}_q(G)$

In this chapter, we outline a picture of spec $\mathcal{O}_q(G)$ originating in work of Soibelman and Vaksman [196, 197, 205] as extended by Hodges-Levasseur [94, 95], Joseph [115, 116], and Hodges-Levasseur-Toro [96]. We then use this information to determine the H-stratification of spec $\mathcal{O}_q(G)$. The original results were proved under the assumptions that either $k = \mathbb{C}$ and $q \in \mathbb{C}^\times$ is not a root of unity, or char$(k) = 0$ and $q \in k^\times$ is transcendental over $\mathbb{Q}$, but most of the proofs carry over to the general situation. Since a great deal of calculation is involved, we mainly establish notation here and cite results without presenting proofs. In choice of notation and formulation, we follow the single parameter (untwisted) case of [96], but with cross-references to [116].

Throughout the chapter, let G be a fixed connected, complex, semisimple algebraic group with Lie algebra $\mathfrak{g}$, and fix associated data, such as the lattice L, as in Chapters I.7 and I.8. In particular, q is a non-root of unity in the base field k. Various items will be indexed by the double Weyl group, $W \times W$, whose elements we will typically write in the form $w = (w_+, w_-)$. The torus $H = \mathrm{Hom}(L, k^\times)$ is assumed to act on $\mathcal{O}_q(G)$ by right winding automorphisms as in (II.1.18), and we recall that this action is rational (Exercise II.2.G).

We begin with some consequences of Theorem I.8.15.

II.4.1. Definitions. Recall that an element x in a ring R is *normal* provided $xR = Rx$. In case we just have that a coset $x + I$ is normal in a factor ring R/I, we say that x is *normal modulo I*.

A *polynormal sequence* in R is any sequence $x_1, \ldots, x_m$ of elements of R such that each x_j is normal modulo $\langle x_1, \ldots, x_{j-1} \rangle$. (In particular, this entails that x_1 is normal in R.) An ideal of R is called *polynormal* provided it has a polynormal sequence of generators.

II.4.2. Proposition. *Let $\lambda \in L^+$.*

(a) *If $\eta, \beta \in L$, $f \in V(\lambda)^*_\beta$ and $u \in V(\lambda)_\eta$, then the element $c^{V(\lambda)}_{f,u}$ in $\mathcal{O}_q(G)$*

is normal modulo each of the ideals

$$I_{f,u}^+ = \langle c_{f,t}^{V(\lambda)} \mid t \in \bigcup_{\substack{\nu \in Q^+ \\ \nu \neq 0}} U_q^+(\mathfrak{g})_\nu u \rangle + \langle c_{s,u}^{V(\lambda)} \mid s \in \bigcup_{\substack{\nu \in Q^+ \\ \nu \neq 0}} f U_q^+(\mathfrak{g})_\nu \rangle$$

$$I_{f,u}^- = \langle c_{f,t}^{V(\lambda)} \mid t \in \bigcup_{\substack{\nu \in Q^+ \\ \nu \neq 0}} U_q^-(\mathfrak{g})_{-\nu} u \rangle + \langle c_{s,u}^{V(\lambda)} \mid s \in \bigcup_{\substack{\nu \in Q^+ \\ \nu \neq 0}} f U_q^-(\mathfrak{g})_{-\nu} \rangle.$$

(b) Let $v_1, \ldots, v_m$ and $f_1, \ldots, f_m$ be bases for $V(\lambda)$ and $V(\lambda)^*$, respectively, consisting of weight vectors. Then the elements $c_{f_j,u_i}^{V(\lambda)}$ can be arranged in a poly-normal sequence in $\mathcal{O}_q(G)$, which generates the ideal

$$\langle c_{f,u}^{V(\lambda)} \mid f \in V(\lambda)^* \text{ and } u \in V(\lambda) \rangle.$$

Proof. (a) For any $\mu \in L^+$, $\rho, \gamma \in L$, $v \in V(\mu)_\rho$ and $g \in V(\mu)_\gamma^*$, Theorem I.8.15 implies that

$$q^{-(\eta,\rho)} c_{g,v}^{V(\mu)} c_{f,u}^{V(\lambda)} - q^{-(\beta,\gamma)} c_{f,u}^{V(\lambda)} c_{g,v}^{V(\mu)} \in I_{f,u}^+.$$

Since the coordinate functions $c_{g,v}^{V(\mu)}$ generate $\mathcal{O}_q(G)$ as a k-algebra, it follows that $c_{f,u}^{V(\lambda)}$ is normal modulo $I_{f,u}^+$, as desired.

Normality of $c_{f,u}^{V(\lambda)}$ modulo $I_{f,u}^-$ is proved in the same way.

(b) Exercise II.4.A. $\square$

A PARTITION OF $\operatorname{spec} \mathcal{O}_q(G)$ INDEXED BY $W \times W$

The general theory of H-stratifications presented in Chapter II.2 was inspired by the observed properties of a specific stratification of $\operatorname{spec} \mathcal{O}_q(G)$ constructed in the literature. Since the latter stratification provides the detail needed to determine the H-primes and H-strata for $\mathcal{O}_q(G)$, we indicate how it is formed.

II.4.3. Definitions. Given a $U_q(\mathfrak{g})$-module M and a subset $X \subseteq M$, write $X^\perp$ for the *orthogonal of X in M^**, that is,

$$X^\perp = \{ f \in M^* \mid \ker(f) \supseteq X \}.$$

For $y \in W$ and $\lambda \in L^+$, set

$$V_y^+(\lambda)^\perp = \{ c_{f,u}^{V(\lambda)} \mid u \in V(\lambda)_\lambda \text{ and } f \in (U_q^{\geq 0}(\mathfrak{g}) V(\lambda)_{y\lambda})^\perp \}$$

$$V_y^-(\lambda)^\perp = \{ c_{f,u}^{V(\lambda)} \mid u \in V(\lambda)_{w_0\lambda} \text{ and } f \in (U_q^{\leq 0}(\mathfrak{g}) V(\lambda)_{yw_0\lambda})^\perp \}$$

and then define ideals $J_y^{\pm} = \left\langle \bigcup_{\lambda \in L^+} V_y^{\pm}(\lambda)^{\perp} \right\rangle$ in $\mathcal{O}_q(G)$. Finally, for ordered pairs $w = (w_+, w_-) \in W \times W$, set

$$J_w = J_{w_+}^+ + J_{w_-}^-.$$

It follows from Proposition II.4.2 that the ideals $J_y^{\pm}$ are polynormal (Exercise II.4.B). Hence, each J_w is a polynormal ideal of $\mathcal{O}_q(G)$.

Furthermore, the coordinate functions $c_{f,u}^{V(\lambda)}$ used to generate J_w are H-eigenvectors (Exercise II.4.C), and so J_w is an H-ideal.

II.4.4. Definitions. For $y \in W$, let E_y^+ and E_y^- denote the multiplicative sets in $\mathcal{O}_q(G)$ generated by $k^{\times}$ together with

$$\{c_{f,u}^{V(\lambda)} \mid \lambda \in L^+, \ 0 \neq f \in V(\lambda)_{-y\lambda}^* \text{ and } 0 \neq u \in V(\lambda)_\lambda\} \qquad \text{and}$$

$$\{c_{f,u}^{V(-w_0\lambda)} \mid \lambda \in L^+, \ 0 \neq f \in V(-w_0\lambda)_{y\lambda}^* \text{ and } 0 \neq u \in V(-w_0\lambda)_{-\lambda}\},$$

respectively. (The existence of nonzero vectors f and u of the appropriate weights in the display above follows from Proposition I.6.12, which also shows that each f and u is unique up to a scalar multiple.) It follows from Proposition II.4.2 that the elements of $E_y^{\pm}$ are normal modulo $J_y^{\pm}$ (Exercise II.4.B).

For $w = (w_+, w_-) \in W \times W$, let $\mathcal{E}_w$ denote the multiplicative set in $\mathcal{O}_q(G)/J_w$ generated by the image of $E_{w_+}^+ \cup E_{w_-}^-$. Then $\mathcal{E}_w$ consists of normal elements, and so there exists an Ore localization

$$A_w = \left(\mathcal{O}_q(G)/J_w\right)[\mathcal{E}_w^{-1}].$$

Since J_w is an H-ideal and $\mathcal{E}_w$ is generated by H-eigenvectors, there is an induced (rational) action of H on A_w by k-algebra automorphisms (Exercise II.4.D).

II.4.5. Theorem. *The algebra A_w is nonzero for all $w \in W \times W$.*

Proof. [**96**, Proposition 4.3] (cf. [**116**, §10.3.2]). $\square$

II.4.6. Definition. For $w \in W \times W$, set

$$\mathrm{spec}_w\, \mathcal{O}_q(G) = \{P \in \mathrm{spec}\, \mathcal{O}_q(G) \mid P \supseteq J_w \text{ and } (P/J_w) \cap \mathcal{E}_w = \varnothing\}.$$

(It can be shown that this set coincides with the set $\mathbf{X}(w)$ defined in [**116**, §9.3.9].) Localization provides a bijection of $\mathrm{spec}_w\, \mathcal{O}_q(G)$ onto $\mathrm{spec}\, A_w$.

II.4.7. Theorem. *The prime spectrum $\mathrm{spec}\, \mathcal{O}_q(G)$ is the disjoint union of the subsets $\mathrm{spec}_w\, \mathcal{O}_q(G)$ for $w \in W \times W$.*

Proof. [**96**, Theorem 4.4] (cf. [**116**, Corollary 9.3.9]). $\square$

It will turn out that the sets $\mathrm{spec}_w\, \mathcal{O}_q(G)$ coincide with the H-strata in $\mathrm{spec}\, \mathcal{O}_q(G)$. However, further information about the structure of the localizations A_w must be developed in order to show this.

II.4.8. Since the action of H on each A_w is rational (Exercise II.4.D), A_w is the direct sum of its H-eigenspaces. In addition, we can identify $X(H)$ with L (Exercise II.4.E), which allows us to index the H-eigenspaces of A_w by elements of L, as follows:

$$(A_w)_\lambda = \{a \in A_w \mid h.a = h(\lambda)a \text{ for all } h \in H\}$$

for $\lambda \in L$. Thus, we have $A_w = \bigoplus_{\lambda \in L}(A_w)_\lambda$. (This decomposition can also be proved directly, as in Exercise II.4.F.) Note that the zero element of L corresponds to the trivial character $h \mapsto 1$ on H, whence $(A_w)_0$ coincides with the fixed ring A_w^H.

II.4.9. Theorem. *Let $w \in W \times W$ and $\lambda \in L$. Then $(A_w)_\lambda$ is invariant under the adjoint action of $\mathcal{O}_q(G)$, and $(A_w)_\lambda = A_w^H c_\lambda$ for some unit c_λ. Moreover, the only nonzero ideal of A_w^H invariant under $\operatorname{ad}\mathcal{O}_q(G)$ is A_w^H itself.*

Proof. [**96**, Theorem 4.7(3) and Corollary 4.13] (cf. [**116**, §10.3.3]). □

II.4.10. Corollary. *If $w \in W \times W$, then A_w is H-simple.*

Proof. Remember that the sets $(A_w)_\lambda$ are the eigenspaces for the action of H on A_w. Hence, if I is a nonzero H-stable ideal of A_w, we must have

$$I = \bigoplus_{\lambda \in L}\big(I \cap (A_w)_\lambda\big) = \bigoplus_{\lambda \in L}(I \cap A_w^H c_\lambda).$$

Consequently, $I \cap A_w^H c_\lambda \neq 0$ for some $\lambda \in L$, and since c_λ is a unit, it follows that $I \cap A_w^H \neq 0$.

Now $I \cap A_w^H$ is a nonzero ideal of A_w^H. Since I is an ideal of A_w, it is invariant under $\operatorname{ad}\mathcal{O}_q(G)$, whence $I \cap A_w^H$ is invariant under $\operatorname{ad}\mathcal{O}_q(G)$. Theorem II.4.9 then shows that $I \cap A_w^H = A_w^H$, and therefore $I = A_w$. □

THE H-STRATIFICATION OF $\operatorname{spec}\mathcal{O}_q(G)$

II.4.11. Proposition. *Let $w \in W \times W$, and let K_w denote the kernel of the localization map $\mathcal{O}_q(G) \to A_w$. Then K_w is an H-prime ideal of $\mathcal{O}_q(G)$, and $\operatorname{spec}_{K_w}\mathcal{O}_q(G) = \operatorname{spec}_w\mathcal{O}_q(G)$.*

Proof. Let $\phi_w : \mathcal{O}_q(G) \to A_w$ be the localization map, and note that ϕ_w is H-equivariant. Hence, K_w is an H-ideal of $\mathcal{O}_q(G)$. Since A_w is H-simple (Corollary II.4.10), its zero ideal is an H-prime ideal, and thus a prime ideal by Proposition II.2.9. Hence, K_w/J_w is a prime ideal of $\mathcal{O}_q(G)/J_w$, and therefore K_w is an H-prime ideal of $\mathcal{O}_q(G)$.

By localization theory, the rule $Q \mapsto \phi_w^{-1}(Q)/J_w$ gives a bijection between the primes of A_w and the primes P/J_w of $\mathcal{O}_q(G)/J_w$ which are disjoint from $\mathcal{E}_w$. Since A_w is H-simple, any such Q satisfies $(Q : H) = 0$, from which it follows that $(P/J_w : H) = K_w/J_w$ for the corresponding P, that is, $(P : H) = K_w$. Thus $(P : H) = K_w$ for all $P \in \mathrm{spec}_w \, \mathcal{O}_q(G)$, and so $\mathrm{spec}_w \, \mathcal{O}_q(G) \subseteq \mathrm{spec}_{K_w} \, \mathcal{O}_q(G)$.

Conversely, if $P \in \mathrm{spec}_{K_w} \, \mathcal{O}_q(G)$, then P/J_w is a prime of $\mathcal{O}_q(G)/J_w$ satisfying $(P/J_w : H) = K_w/J_w$. In particular, any H-eigenvector contained in P/J_w must lie in K_w/J_w. Since the elements of $\mathcal{E}_w$ are H-eigenvectors which map to units in A_w, and since A_w is nonzero (Theorem II.4.5), $\mathcal{E}_w$ is disjoint from K_w/J_w. Thus P/J_w is disjoint from $\mathcal{E}_w$, and therefore $P \in \mathrm{spec}_w \, \mathcal{O}_q(G)$. $\square$

II.4.12. Corollary. *The ideals K_w for $w \in W \times W$ are precisely the H-primes of $\mathcal{O}_q(G)$, and the partition $\mathrm{spec} \, \mathcal{O}_q(G) = \bigsqcup_{w \in W \times W} \mathrm{spec}_w \, \mathcal{O}_q(G)$ coincides with the H-stratification of $\mathrm{spec} \, \mathcal{O}_q(G)$.*

Proof. Each $\mathrm{spec}_w \, \mathcal{O}_q(G)$ is an H-stratum of $\mathrm{spec} \, \mathcal{O}_q(G)$ by Proposition II.4.11. Since $\mathrm{spec} \, \mathcal{O}_q(G)$ is the disjoint union of these strata (Theorem II.4.7), there are no other H-strata. Therefore the H-strata in $\mathrm{spec} \, \mathcal{O}_q(G)$ are precisely the sets $\mathrm{spec}_w \, \mathcal{O}_q(G)$. Since the H-primes in $\mathcal{O}_q(G)$ are the minimal elements of the H-strata of $\mathrm{spec} \, \mathcal{O}_q(G)$, we thus conclude that the H-primes of $\mathcal{O}_q(G)$ are precisely the ideals K_w. $\square$

II.4.13. Definition. Let $w = (w_+, w_-) \in W \times W$. Recall that W can be viewed as a group of linear transformations on $\mathfrak{h}^*$. Thus, view $w_+ - w_-$ as such a linear transformation, and set

$$t(w) = \dim_{\mathbb{C}} \ker(w_+ - w_-).$$

In [96] and [116], many formulas are given in terms of the value $s(w) = n - t(w)$, where $n = \dim \mathfrak{h} = \mathrm{rank}(G)$. As in [116], s can also be defined on single elements of W via the rule $s(y) = n - \dim_{\mathbb{C}} \ker(\mathrm{id} - y)$, and then the two-variable version is given by $s(w_+, w_-) = s(w_+^{-1} w_-)$. It is known that $s(y)$ equals the minimum number of factors needed to express y as a product of (not necessarily simple) reflections [116, Lemma A.1.18]; thus $s(y) \leq \ell(y)$.

II.4.14. Theorem. *Let $w \in W \times W$. If k is algebraically closed, then $Z(A_w)$ is a Laurent polynomial ring of the form $k[z_1^{\pm 1}, \ldots, z_{t(w)}^{\pm 1}]$.*

Proof. Recall the eigenspace decomposition $A_w = \bigoplus_{\lambda \in L} (A_w)_\lambda$ from (II.4.8), and set $Z_\lambda = Z(A_w) \cap (A_w)_\lambda$ for $\lambda \in L$. By [96, Theorem 4.14], $\dim(Z_\lambda) \leq 1$. In particular, $Z_0 = k \cdot 1$. As in Lemma II.3.7, the set $L_w = \{\lambda \in L \mid Z_\lambda \neq 0\}$ is a subgroup of L, and $Z(A_w)$ is a Laurent polynomial ring over k in $\mathrm{rank}(L_w)$ indeterminates (Exercise II.4.G). In the proof of [96, Proposition 4.17], it is shown that $\mathrm{rank}(L_w) = n - s(w) = t(w)$. $\square$

The calculations in [116, §10.3.3] make it appear likely that Theorem II.4.14 also holds if k is not algebraically closed.

II.4.15. Corollary. *If $w \in W \times W$ and k is algebraically closed, then $Z(A_{K_w})$ is a Laurent polynomial ring of the form $k[z_1^{\pm 1}, \ldots, z_{t(w)}^{\pm 1}]$.*

Proof. It suffices to show that $Z(A_{K_w}) \cong Z(A_w)$.

Let B_w denote the image of $\mathcal{O}_q(G)$ under the localization map $\mathcal{O}_q(G) \to A_w$, and identify B_w with $\mathcal{O}_q(G)/K_w$. Since K_w is a prime ideal by Propositions II.4.11 and II.2.9, B_w is a prime ring. The image of $\mathcal{E}_w$ in B_w consists of normal H-eigenvectors, all of which must be nonzero because $A_w \neq 0$ (Theorem II.4.5). Recall that nonzero normal elements in prime rings are regular (Exercise II.4.H). Therefore the image of $\mathcal{E}_w$ is contained in $\mathcal{E}_{K_w}$, and hence $B_w \subseteq A_w \subseteq A_{K_w}$.

Since A_{K_w} is a localization of A_w and A_w is H-simple (Corollary II.4.10), A_{K_w} is H-simple and $Z(A_w) \subseteq Z(A_{K_w})$. To prove the reverse inclusion, we may concentrate on H-eigenvectors, since $Z(A_{K_w})$ is a homogeneous subring of A_{K_w} (Lemma II.3.7). If $z \in Z(A_{K_w})$ is an H-eigenvector, then the set

$$I = \{a \in A_w \mid az \in A_w\}$$

is an H-ideal of A_w, and $I \neq 0$ because A_{K_w} is a localization of A_w. Since A_w is H-simple, we must have $I = A_w$, whence $z \in A_w$. Therefore $Z(A_{K_w}) = Z(A_w)$, as desired. $\square$

PRIMITIVE IDEALS AND SYMPLECTIC LEAVES

One of the major results of Soibelman's work [**196, 197**] was the construction of a bijection between primitive ideals and symplectic leaves in the setting of compact groups. (We refer ahead to Chapter III.5 for the concepts of Poisson structures and symplectic leaves.) More precisely, if K is a maximal compact subgroup of a complex semisimple Lie group, and if $\mathbb{C}[K]_q$ is the standard single parameter quantization of the function algebra on K, then there is an associated Poisson structure on K, and Soibelman constructed a bijection between the (isomorphism classes of) irreducible *-representations of $\mathbb{C}[K]_q$ and the symplectic leaves in K (cf. [**126**, Chapter 3, Corollary 6.2.8]).

Hodges and Levasseur then conjectured that a similar result should hold for prim $\mathcal{O}_q(G)_{\mathbb{C}}$, relative to the associated Poisson structure on G (see [**94**, Appendix A] for a description of this structure). Let us write $\mathrm{Symp}\, G$ for the set of symplectic leaves in G. In the case where $G = SL_2(\mathbb{C})$, the Poisson structure is obtained as described in Example III.5.5(2). A comparison of that example with Example II.8.6 reveals a natural bijection between prim $\mathcal{O}_q(SL_2(\mathbb{C}))$ and $\mathrm{Symp}\, SL_2(\mathbb{C})$. Moreover, this bijection can be made equivariant with respect to the standard maximal torus H of $SL_2(\mathbb{C})$, which acts on $\mathcal{O}_q(SL_2(\mathbb{C}))$ in the usual way and on $\mathrm{Symp}\, SL_2(\mathbb{C})$ by left translation (Exercise II.4.I).

II.4.16. Theorem. *There is an H-equivariant bijection*

$$\mathrm{prim}\, \mathcal{O}_q(G)_{\mathbb{C}} \longrightarrow \mathrm{Symp}\, G.$$

Proof. This was first proved by Hodges and Levasseur for $G = SL_3(\mathbb{C})$ in [**94**, Theorem 4.4.1] and for $G = SL_n(\mathbb{C})$ in [**95**, Theorem 4.2 ff.]. The general case follows from work of Joseph in [**115**, **116**], and is given explicitly as the single parameter case of [**96**, Theorems 4.16 and 4.18]. As shown in the latter paper, this theorem does not hold for all of the multiparameter quantized coordinate rings $\mathcal{O}_{q,p}(G)_{\mathbb{C}}$. $\square$

NOTES

Most of the computations and results outlined here are taken from the work of Hodges-Levasseur [**94**, **95**], Joseph [**115**, **116**], and Hodges-Levasseur-Toro [**96**]. That the partitions of spec $\mathcal{O}_q(G)$ obtained in those papers could be reinterpreted as H-stratifications was observed by Brown and Goodearl [**19**].

EXERCISES

Exercise II.4.A. Prove part (b) of Proposition II.4.2. $\square$

Exercise II.4.B. For $y \in W$ and each choice of sign $(\pm)$, show that the ideal $J_y^{\pm}$ in Definition II.4.3 is polynormal, and that the elements of the set $E_y^{\pm}$ in Definition II.4.4 are normal modulo $J_y^{\pm}$. $\square$

Exercise II.4.C. Let V be a module in $\mathcal{C}_q(\mathfrak{g}, L)$, let $u \in V_\eta$ for some $\eta \in L$, and let $f \in V^*$. Show that $c_{f,u}^V$ is an H-eigenvector. $\square$

Exercise II.4.D. Show that the action of H on $\mathcal{O}_q(G)$ induces (uniquely) actions by k-algebra automorphisms on the localizations A_w, and that these actions are rational. $\square$

Exercise II.4.E. For each $\lambda \in L$, evaluation at λ defines a character $e_\lambda : H \to k^\times$ (that is, $e_\lambda(h) = h(\lambda)$ for $h \in H$). Show that these characters are rational, and then show that the rule $\lambda \mapsto e_\lambda$ defines an isomorphism of L onto $X(H)$. [Hint: Choose a basis $\lambda_1, \ldots, \lambda_t$ for L and show that the rule $h \mapsto (h(\lambda_1), \ldots, h(\lambda_t))$ defines an isomorphism of H onto $(k^\times)^t$ as algebraic groups.] $\square$

Exercise II.4.F. Let $w \in W \times W$, and decompose A_w as in (II.4.8) without using rationality and Exercise II.4.E, as follows. First, define

$$(A_w)_\lambda = \{a \in A_w \mid h.a = h(\lambda)a \text{ for all } h \in H\}$$

for $\lambda \in L$, and show that the $(A_w)_\lambda$ are independent k-subspaces of A_w. Then show that the coset of $c_{f,v}^{V(\mu)}$ lies in $(A_w)_\lambda$ whenever $\mu \in L^+$, $f \in V(\mu)^*$, and $v \in V(\mu)_\lambda$, and use this to see that $\bigoplus_{\lambda \in L}(A_w)_\lambda$ contains the image of $\mathcal{O}_q(G)$. Finally, show that each of the generators of $\mathcal{E}_w$ lies in $(A_w)_{\pm\lambda}$ for a suitable $\lambda \in L^+$ and choice of sign, and conclude that $\bigoplus_{\lambda \in L}(A_w)_\lambda = A_w$. $\square$

Exercise II.4.G. In the proof of Theorem II.4.14, check that L_w is a subgroup of L and that $Z(A_w)$ is a Laurent polynomial ring over k in $\mathrm{rank}(L_w)$ indeterminates. $\square$

Exercise II.4.H. If R is a prime ring and $x \in R$ is a nonzero normal element, show that x is regular. $\square$

Exercise II.4.I. Let $G = SL_2(\mathbb{C})$ and $A = \mathcal{O}_q(G)$, and let $H \subset G$ be the subgroup of diagonal matrices. Let H act on A by right winding automorphisms as in (II.1.18), and on $\mathrm{Symp}\,G$ by left translation. Use Examples II.8.6 and III.5.5(2) to construct an H-equivariant bijection between $\mathrm{prim}\,A$ and $\mathrm{Symp}\,G$. $\square$

H-PRIMES IN ITERATED
SKEW POLYNOMIAL ALGEBRAS

In order for the Stratification Theorem to give the clearest picture of the prime spectrum of a k-algebra A, we would want H-spec A to be finite, so that there are only finitely many H-strata in spec A, and we would want the coefficient fields $Z(A_J)^H$ to be equal to k. Further, in specific examples, we would want to compute H-spec A in detail, and to find such information as the number of variables in each Laurent polynomial ring $Z(A_J)$.

Many of the standard quantized coordinate rings are built from iterated skew polynomial algebras, and the standard tori acting as automorphisms admit the skew polynomial variables as eigenvectors. A number of results, such as a finiteness theorem for H-spec, can be proved in such a context. We discuss the requisite skew polynomial technology in this chapter, and utilize it to prove a finiteness theorem which we can apply to many of our examples. Further results on iterated skew polynomial algebras will be considered in the following chapter.

BASIC SETUP

II.5.1. To avoid repetition, we label here the precise iterated skew polynomial setup that we will work with:

(a) Let $A = k[x_1][x_2; \tau_2, \delta_2] \cdots [x_n; \tau_n, \delta_n]$ be an iterated skew polynomial algebra over k. (By convention, the τ_i are k-algebra automorphisms and the δ_i are k-linear.)

(b) Let H be a group acting on A by k-algebra automorphisms.

(c) Assume that $x_1, \ldots, x_n$ are H-eigenvectors.

(d) Assume that there exist $h_1, \ldots, h_n \in H$ such that $h_i(x_j) = \tau_i(x_j)$ for $i > j$ and such that the h_i-eigenvalue of x_i is not a root of unity for any i.

Note that condition (d) is nontrivial even for $i = 1$, since it requires the existence of $h_1 \in H$ such that the h_1-eigenvalue of x_1 is not a root of unity. The first part of the condition says that the action of h_i on the subalgebra $k\langle x_1, \ldots, x_{i-1} \rangle$ agrees with τ_i, for each $i = 2, \ldots, n$.

II.5.2. Example. Let $A = \mathcal{O}_q(M_2(k))$ with q not a root of unity. Recall from Example I.1.16 that A is an iterated skew polynomial algebra of the form

$$A = k[a][b; \tau_2][c; \tau_3][d; \tau_4, \delta_4],$$

where the τ_i are k-algebra automorphisms such that

$$\tau_2(a) = q^{-1}a \qquad \tau_3(a) = q^{-1}a \qquad \tau_3(b) = b$$
$$\tau_4(a) = a \qquad \tau_4(b) = q^{-1}b \qquad \tau_4(c) = q^{-1}c,$$

while δ_4 is a k-linear τ_4-derivation such that

$$\delta_4(a) = (q^{-1} - q)bc \qquad\qquad \delta_4(b) = \delta_4(c) = 0.$$

In particular, condition (a) of (II.5.1) is satisfied. As in Example II.1.6(c), we let the torus $H = (k^\times)^4$ act on A by k-algebra automorphisms such that

$$(\alpha_1, \alpha_2, \beta_1, \beta_2).a = \alpha_1\beta_1 a \qquad\qquad (\alpha_1, \alpha_2, \beta_1, \beta_2).b = \alpha_1\beta_2 b$$
$$(\alpha_1, \alpha_2, \beta_1, \beta_2).c = \alpha_2\beta_1 c \qquad\qquad (\alpha_1, \alpha_2, \beta_1, \beta_2).d = \alpha_2\beta_2 d.$$

Conditions (b) and (c) of (II.5.1) are immediate.

There are many choices of $h_1, h_2, h_3, h_4 \in H$ satisfying condition (d). For instance, we may take

$$h_1 = (q, 1, 1, 1) \qquad\qquad h_2 = (q^{-1}, 1, 1, 1)$$
$$h_3 = (q^{-1}, q, 1, q) \qquad\qquad h_4 = (1, q^{-1}, 1, q^{-1}).$$

II.5.3. In working with the setup of (II.5.1), we of course proceed one variable at a time, i.e., by induction on n. An appropriate induction setup is the following:

(a) Let $S = R[x; \tau, \delta]$ be a skew polynomial algebra over a noetherian k-algebra R.

(b) Let H be a group acting on S by k-algebra automorphisms.

(c) Assume that R is H-stable and that x is an H-eigenvector.

(d) Assume that there exists $h_0 \in H$ such that $h_0|_R = \tau$ and such that the h_0-eigenvalue of x is not a root of unity.

It is convenient to let $q \in k^\times$ denote the *inverse* of the h_0-eigenvalue of x, so that $h_0(x) = q^{-1}x$. By Exercise II.5.A, $\delta\tau = q\tau\delta$. Thus (τ, δ) is a q-skew derivation, with q not a root of unity. This allows us to apply the q-Leibniz Rules, which we recall from (I.8.4):

$$\delta^n(ab) = \sum_{i=0}^{n} \binom{n}{i}_q \tau^{n-i}\delta^i(a)\delta^{n-i}(b)$$

$$x^n a = \sum_{i=0}^{n} \binom{n}{i}_q \tau^{n-i}\delta^i(a)x^{n-i}$$

for all $a, b \in R$ and $n \geq 0$.

Skew polynomial simplifications

We recall a few general facts which are useful for simplifying certain skew polynomial rings.

II.5.4. Definitions. Let τ be an automorphism of a ring R. The *inner τ-derivation* corresponding to an element $d \in R$ is the map $R \to R$ sending $r \mapsto dr - \tau(r)d$.

Now let $T = R[x; \tau, \delta]$, where δ is an arbitrary τ-derivation (i.e., not necessarily inner). A *(τ, δ)-ideal* of R is any ideal I such that $\tau(I) = I$ and $\delta(I) \subseteq I$. Given any such I, note that τ induces an automorphism of R/I (which we continue to denote τ), and that δ induces a τ-derivation on R/I (again denoted δ). Moreover, $IT = TI$ is an ideal of T, and $T/IT = (R/I)[y; \tau, \delta]$, where $y = x + IT$.

II.5.5. Lemma. *Let $R[x; \tau, \delta]$ be a skew polynomial ring.*

(a) The map $-\delta\tau^{-1}$ is a τ^{-1}-derivation on R^{op}, and $R[x; \tau, \delta]^{\mathrm{op}}$ can be presented in the form $R[x; \tau, \delta]^{\mathrm{op}} = R^{\mathrm{op}}[x; \tau^{-1}, -\delta\tau^{-1}]$.

(b) Suppose that τ is an inner automorphism, say there is a unit $u \in R$ such that $\tau(r) = u^{-1}ru$ for $r \in R$. Then $u\delta$ is an ordinary derivation on R, and $R[x; \tau, \delta] = R[ux; u\delta]$.

(c) Suppose that δ is an inner τ-derivation, say there is some $d \in R$ such that $\delta(r) = dr - \tau(r)d$ for $r \in R$. Then $R[x; \tau, \delta] = R[x - d; \tau]$.

Proof. See [**68**, Lemma 1.5]. □

II.5.6. Lemma. *Let $R[x; \tau, \delta]$ be a skew polynomial ring, and let S be a right denominator set in R such that $\tau(S) = S$.*

(a) The automorphism τ extends uniquely to an automorphism of $R[S^{-1}]$, and then δ extends uniquely to a τ-derivation on $R[S^{-1}]$. Moreover, $\delta(rs^{-1}) = \delta(r)s^{-1} - \tau(rs^{-1})\delta(s)s^{-1}$ for all $r \in R$ and $s \in S$.

(b) The set S is a right denominator set in $R[x; \tau, \delta]$, and $R[x; \tau, \delta][S^{-1}] = R[S^{-1}][x1^{-1}; \tau, \delta]$.

Proof. See [**68**, Lemmas 1.3, 1.4]. □

Some *q*-skew calculations

Given a skew derivation (τ, δ) on a ring R, we shall need to consider both *τ-prime* and *(τ, δ)-prime* ideals in R. The definitions are routine – just replace ideals by τ-stable ideals (respectively, (τ, δ)-ideals) in the definition of a prime ideal. In fact, τ-prime ideals are the same as $\langle \tau \rangle$-prime ideals. If I is an ideal of R, we write $(I : \tau)$ for the largest τ-stable ideal contained in I, that is, $(I : \tau) = (I : \langle \tau \rangle)$.

II.5.7. Lemma. *Let (τ, δ) be a q-skew derivation on a k-algebra R, where $q \in k^{\times}$ is not a root of unity.*

(a) Every minimal τ-prime ideal of R is δ-invariant.

(b) If R is noetherian, then every (τ, δ)-prime ideal of R is also τ-prime.

Proof. (a) Let I be a minimal τ-prime of R, set

$$J = \{r \in R \mid \delta^i(r) \in I \text{ for all } i \geq 0\},$$

and observe that J is a δ-invariant, τ-stable ideal contained in I. We claim that J is τ-prime. Once this is verified, $J = I$ by the minimality of I, and the δ-invariance of I follows.

To prove that J is τ-prime, we apply Exercise II.5.B (with $H = \langle \tau \rangle$). Thus, it suffices to show that for any $u, w \in R \setminus J$, there exist $v \in R$ and $t \in \mathbb{Z}$ such that $uv\tau^t(w) \notin J$. Choose nonnegative integers m and n minimal such that $\delta^m(u), \delta^n(w) \notin I$. By Exercise II.5.B, there exist $v \in R$ and $s \in \mathbb{Z}$ such that $\delta^m(u)v\tau^s\delta^n(w) \notin I$. Set $a = \tau^{-n}(u)$ and $c = \tau^s(w)$, and observe that $\delta^i(a) = q^{-in}\tau^{-n}\delta^i(u) \in I$ for $0 \leq i < m$, and similarly $\delta^j(c) \in I$ for $0 \leq j < n$. Now set $b = \tau^{-n}(v)$. Then

$$\tau^n\delta^m(a)\tau^n(b)\delta^n(c) = q^{-mn+ns}\delta^m(u)v\tau^s\delta^n(w) \notin I.$$

Now expand $\delta^{m+n}(abc)$ using a double application of the q-Leibniz Rule:

$$\delta^{m+n}(abc) = \sum_{i=0}^{m+n} \binom{m+n}{i}_q \tau^{m+n-i}\delta^i(ab)\delta^{m+n-i}(c)$$

$$= \sum_{i=0}^{m+n}\sum_{j=0}^{i} \binom{m+n}{i}_q \binom{i}{j}_q \tau^{m+n-i}\left(\tau^{i-j}\delta^j(a)\delta^{i-j}(b)\right)\delta^{m+n-i}(c)$$

$$= \sum_{i=0}^{m+n}\sum_{j=0}^{i} \binom{m+n}{i}_q \binom{i}{j}_q \tau^{m+n-j}\delta^j(a)\tau^{m+n-i}\delta^{i-j}(b)\delta^{m+n-i}(c).$$

All but one term of the last sum lie in I because $\delta^i(a)$ and $\delta^j(c)$ are in I for $i < m$ and $j < n$. The remaining term is $\binom{m+n}{m}_q \tau^n\delta^m(a)\tau^n(b)\delta^n(c)$, which is not in I because the q-binomial coefficients are invertible in R. Hence, $\delta^{m+n}(abc) \notin I$, and so $abc \notin J$.

Finally, $uv\tau^{n+s}(w) = \tau^n(abc) \notin J$, and therefore J is τ-prime, as desired.

(b) It suffices to show that if R is a (τ, δ)-prime ring, then it is also a τ-prime ring. Choose minimal primes $Q_1, \ldots, Q_t$ of R such that $Q_1Q_2 \cdots Q_t = 0$. Then each $(Q_j : \tau)$ is a minimal τ-prime of R (use Lemma II.1.10), and so $(Q_j : \tau)$ is δ-invariant by part (a). Since $(Q_1 : \tau)(Q_2 : \tau) \cdots (Q_t : \tau) = 0$ and R is (τ, δ)-prime, some $(Q_j : \tau) = 0$, and therefore R is τ-prime. $\square$

II.5.8. Lemma. *Let $S = R[x; \tau, \delta]$ be a q-skew polynomial algebra over k, where R is noetherian and $q \in k^\times$ is not a root of unity. Let $P \in \operatorname{spec} S$, and suppose that all primes of R minimal over $P \cap R$ have finite τ-orbits. Then there exist a*

δ-invariant τ-prime ideal I of R and a prime Q of R such that $I \subseteq P \cap R \subseteq Q$ and Q is minimal over I.

Proof. Set $I = (P \cap R : \tau)$. Since R is noetherian, it follows from Exercise II.2.H that

$$I = \{r \in R \mid \tau^n(r) \in P \text{ for all } n \geq 0\}.$$

Given $r \in I$, we have

$$\tau^n \delta(r) = q^{-n} \delta \tau^n(r) = q^{-n} [x \tau^n(r) - \tau^{n+1}(r) x] \in P$$

for all $n \geq 0$, whence $\delta(r) \in I$. Thus I is a (τ, δ) ideal.

We next claim that I is (τ, δ)-prime. If A and B are (τ, δ)-ideals of R such that $AB \subseteq I$, then AS and BS are ideals of S such that $(AS)(BS) = ABS \subseteq IS \subseteq P$. Now either $AS \subseteq P$ or $BS \subseteq P$, say the former. Then $A \subseteq P \cap R$, and consequently $A \subseteq I$ because A is a τ-ideal. This verifies the claim. Thus, by Lemma II.5.7, I must be τ-prime.

There are primes $Q_1, \ldots, Q_t$ minimal over $P \cap R$ such that $Q_1 Q_2 \cdots Q_t \subseteq P \cap R$. The product $(Q_1 : \tau)(Q_2 : \tau) \cdots (Q_t : \tau)$ is a τ-ideal contained in $P \cap R$, and so it must also be contained in I. Since I is τ-prime, some $(Q_j : \tau) \subseteq I$. By hypothesis, the τ-orbit of Q_j is finite, and it follows from Exercise II.5.C that Q_j is minimal over $(Q_j : \tau)$. Therefore Q_j is minimal over I. $\square$

FINITENESS OF H-spec

II.5.9. Lemma. *Assume the setup of* (II.5.3). *Then every H-prime of S contracts to a δ-invariant H-prime of R.*

Proof. Let J be an H-prime of S, and observe that $J \cap R$ is δ-invariant, since $\delta(r) = xr - h_0(r)x \in J$ for all $r \in J \cap R$. Let P be a prime of S minimal over J. By Lemma II.1.10, the H-orbit of P is finite and $J = (P : H)$. Thus $h_0^n(P) = P$ for some $n > 0$, whence $\tau^n(P \cap R) = P \cap R$. If Q is any prime minimal over $P \cap R$, then $\tau^i(Q)$ is minimal over $\tau^i(P \cap R)$ for all i. It follows that the τ-orbit of Q must be finite.

By Lemma II.5.8, there exist a δ-invariant τ-prime $I \subseteq P \cap R$ and a prime $Q \supseteq P \cap R$ such that Q is minimal over I. By Lemma II.1.10, $(Q : h_0) = (Q : \tau) = I$. Consequently, $(Q : H) = (I : H)$. Since $I \subseteq P \cap R \subseteq Q$, it follows that

$$(Q : H) = (P \cap R : H) = (P : H) \cap R = J \cap R.$$

Therefore $J \cap R$ is an H-prime ideal of R. $\square$

II.5.10. Lemma. *Assume the setup of* (II.5.3), *and suppose that R is H-simple but S is not. Let $\lambda : H \to k^\times$ denote the H-eigenvalue of x.*

(a) *There is a unique element $d \in R$ such that $h(d) = \lambda(h)d$ for all $h \in H$ and $\delta(r) = dr - \tau(r)d$ for all $r \in R$.*

(b) *There are precisely two H-prime ideals in S, namely 0 and $(x - d)S$.*

Proof. (a) Let I be a proper nonzero H-ideal in S, and let n be the minimum degree for nonzero elements of I. (Necessarily $n > 0$, since $I \neq S$.) The set consisting of 0 and the leading coefficients of the elements in I of degree n is then a nonzero H-ideal of R and so equals R. Hence, there exists a monic polynomial $s \in I$ with degree n, say

$$s = x^n + cx^{n-1} + [\text{lower terms}].$$

Set $d = -\binom{n}{1}_q^{-1} q^{n-1} c$. By Exercise II.5.D, $\delta(r) = dr - \tau(r)d$ for all $r \in R$. For $h \in H$, we have

$$h(s) = \lambda(h)^n x^n + \lambda(h)^{n-1} h(c)x^{n-1} + [\text{lower terms}].$$

Then $h(s) - \lambda(h)^n s$ is an element of I of degree less than n, whence $h(s) = \lambda(h)^n s$, and consequently $h(c) = \lambda(h)c$. Thus $h(d) = \lambda(h)d$ for all $h \in H$

Suppose also that $e \in R$ satisfies $h(e) = \lambda(h)e$ for all $h \in H$ and $\delta(r) = er - \tau(r)e$ for all $r \in R$. Set $f = d - e$. Then $h(f) = \lambda(h)f$ for all $h \in H$, and $fa = \tau(a)f$ for all $a \in R$. Hence, $fR = Rf$ is an H-ideal of R, and so f is either zero or a unit. However, $\tau(f) = q^{-1}f$, and so $f^2 = q^{-1}f^2$, whence $f^2 = 0$. Therefore $f = 0$, and so $e = d$.

(b) Set $z = x - d$; then z is an H-eigenvector with H-eigenvalue λ and $S = R[z; \tau]$. Hence, zS is an H-ideal of S, and there is an H-equivariant ring isomorphism $S/zS \cong R$. Therefore, S/zS is an H-simple ring, and so zS is an H-prime ideal of S. As in the proof of part (a), every nonzero H-ideal of S contains a monic polynomial. Since the product of two monic polynomials is always nonzero, S is an H-prime ring; that is, 0 is an H-prime ideal of S.

Now let P be a nonzero H-prime ideal of S, and let n be the minimum degree for nonzero elements of P. As in part (a), there is a monic (in z) polynomial $p \in P$ of degree n, say $p = z^n + p_{n-1}z^{n-1} + \cdots + p_1 z + p_0$. Since p is monic, the Division Algorithm applies, showing that $P = pS = Sp$. Note that $n > 0$ because $P \neq S$.

For $a \in R$, observe that $pa - \tau^n(a)p$ is an element of P with degree less than n, whence $pa = \tau^n(a)p$. Thus $p_{n-i}a = \tau^i(a)p_{n-i}$ for all $a \in R$ and all $i = 1, \ldots, n$. Similarly, given $h \in H$, observe that $h(p) - \lambda(h)^n p$ is an element of P with degree less than n, whence $h(p) = \lambda(h)^n p$. Thus $h(p_{n-i}) = \lambda(h)^i p_{n-i}$ for all $h \in H$ and all $i = 1, \ldots, n$. Now each $p_{n-i}R = Rp_{n-i}$ is an H-ideal of R, and therefore p_{n-i} is either zero or a unit.

We next compute that

$$zp = z^{n+1} + \tau(p_{n-1})z^n + \cdots + \tau(p_1)z^2 + \tau(p_0)z$$
$$= z^{n+1} + q^{-1}p_{n-1}z^n + \cdots + q^{-n+1}p_1 z^2 + q^{-n}p_0 z,$$

and so $zp - pz = (q^{-1} - 1)p_{n-1}z^n + \cdots + (q^{-n+1} - 1)p_1z^2 + (q^{-n} - 1)p_0z$. Since $zp - pz$ is an element in P of degree at most n, it follows that $zp - pz = ap$, where $a = (q^{-1} - 1)p_{n-1}$. Thus $(q^{-i} - 1)p_{n-i} = ap_{n-i+1}$ for $i > 0$, and $ap_0 = 0$.

Suppose that p_0 is a unit. Then $a = 0$, and since q is not a root of unity, it follows that $p_{n-i} = 0$ for all i. Thus $p = z^n$, and so $(zS)^n \subseteq P$. Furthermore, because zS is an H-ideal and P is H-prime, $zS \subseteq P$. Therefore, $n = 1$ in this case, whence $p = z$ and $P = zS$.

Now assume that $p_0 = 0$, and set $f = z^{n-1} + p_{n-1}z^{n-2} + \cdots + p_1$, so that $p = fz$. By the minimality of n, we see that $f \notin P$. Since zS is an H-ideal, it follows that $z \in P$, and we conclude – as in the previous case – that $P = zS$.   □

II.5.11. Proposition. *Assume the setup of* (II.5.3).

(a) *There are at most twice as many H-prime ideals in S as in R.*

(b) *If all H-prime ideals of R are completely prime, then the same is true for S.*

Proof. (a) By Lemma II.5.9, every H-prime of S contracts to a δ-invariant H-prime of R. Hence, it suffices to show that at most two H-primes of S can contract to any given δ-invariant H-prime Q in R. In view of Exercise II.5.E, we may assume, after replacing R by Fract R/Q, that R is an H-simple artinian ring. If S is H-simple, then S has just one H-prime, namely 0. If S is not H-simple, then by Lemma II.5.10, S has just two H-primes.

(b) As in part (a), we can reduce to the case that R is H-simple artinian. Then R is a domain, and so is S. If there exists a nonzero H-prime P in S, then by Lemma II.5.10, P is of the form $(x - d)S$, and $S/P \cong R$. Thus in this case too, S/P is a domain.   □

II.5.12. Theorem. *Let $A = k[x_1][x_2; \tau_2, \delta_2] \cdots [x_n; \tau_n, \delta_n]$ and H be as in* (II.5.1). *Then all H-prime ideals of A are completely prime, and there are at most 2^n of them.*

Proof. By induction, using Proposition II.5.11.   □

For instance, if $A = \mathcal{O}_q(M_2(k))$ with q not a root of unity and $H = (k^\times)^4$ acts on A as in Example II.1.6(c), Theorem II.5.12 implies that A has at most 16 H-primes. In fact, as we saw in Example II.2.14(d), there are just 14 H-primes in A. For 13 of these H-prime ideals J, it is clear that A/J is an iterated skew polynomial algebra over k and hence a domain. In the remaining case, the method of Exercise II.2.O shows that $A/\langle D_q \rangle$ is a domain. Thus, there is a direct confirmation of the conclusion that all H-primes of A are completely prime.

APPLICATIONS TO QUANTIZED COORDINATE RINGS

II.5.13. In order to apply Theorem II.5.12 to some of our standard examples, we need specific iterated skew polynomial presentations, whose existence was previously left to Exercise I.2.I.

(a) If A is either $\mathcal{O}_q(k^n)$ or $\mathcal{O}_q(k^n)$, then $A = k[x_1][x_2; \tau_2] \cdots [x_n; \tau_n]$ for suitable automorphisms τ_i. If $A = \mathcal{O}_q(k^n)$, then $\tau_i(x_j) = q^{-1}x_j$ for $i > j$, while if $A = \mathcal{O}_q(k^n)$, then $\tau_i(x_j) = q_{ij}x_j$ for $i > j$.

(b) Now let A be either $\mathcal{O}_q(M_n(k))$ or $\mathcal{O}_{\lambda,p}(M_n(k))$. If the generators X_{ij} are listed in lexicographic order, that is,

$$X_{11}, X_{12}, \ldots, X_{1n}, X_{21}, X_{22}, \ldots, X_{2n}, \ldots, X_{n1}, X_{n2}, \ldots, X_{nn},$$

then A can be presented as an iterated skew polynomial algebra

$$k[X_{11}][X_{12}; \tau_{12}] \cdots [X_{ij}; \tau_{ij}, \delta_{ij}] \cdots [X_{nn}; \tau_{nn}, \delta_{nn}].$$

In the case $A = \mathcal{O}_q(M_n(k))$, we have

$\tau_{lm}(X_{ij}) = q^{-1}X_{ij}$	$\delta_{lm}(X_{ij}) = 0$	$(i < l, \; j = m)$
$\tau_{lm}(X_{ij}) = q^{-1}X_{ij}$	$\delta_{lm}(X_{ij}) = 0$	$(i = l, \; j < m)$
$\tau_{lm}(X_{ij}) = X_{ij}$	$\delta_{lm}(X_{ij}) = 0$	$(i < l, \; j > m)$
$\tau_{lm}(X_{ij}) = X_{ij}$	$\delta_{lm}(X_{ij}) = (q^{-1} - q)X_{im}X_{lj}$	$(i < l, \; j < m).$

We leave the case $A = \mathcal{O}_{\lambda,p}(M_n(k))$ to the reader.

(c) Finally, $A_n^{Q,\Gamma}(k)$ can be written as an iterated skew polynomial algebra of the form $k[y_1][x_1; \tau_1, \delta_1][y_2; \sigma_2][x_2; \tau_2, \delta_2] \cdots [y_n; \sigma_n][x_n; \tau_n, \delta_n]$.

II.5.14. Theorem. Let A be one of the following algebras:

(a) $\mathcal{O}_q(k^n)$ or $\mathcal{O}_q(k^n)$. No restrictions on q or (q_{ij}) are needed, but assume that k contains a non-root of unity.

(b) $\mathcal{O}_q(M_n(k))$ or $\mathcal{O}_q(GL_n(k))$, where q is not a root of unity.

(c) $\mathcal{O}_{\lambda,p}(M_n(k))$ or $\mathcal{O}_{\lambda,p}(GL_n(k))$, where λ is not a root of unity.

(d) $A_n^{Q,\Gamma}(k)$, where $q_1, \ldots, q_n$ are non-roots of unity.

Let H be the k-torus acting on A as in (II.1.14–17).

Then H-spec A is finite, and so there are only finitely many H-strata in spec A. Further, all H-prime ideals of A are completely prime.

Proof. First consider the algebras $\mathcal{O}_\bullet(k^n)$, $\mathcal{O}_\bullet(M_n(k))$, and $A_n^{Q,\Gamma}(k)$. The iterated skew polynomial presentations of these algebras discussed in (II.5.13) clearly satisfy conditions (a),(b),(c) of (II.5.1), so only condition (d) remains to be verified in order to apply Theorem II.5.12 in these cases.

(a) Let $A = \mathcal{O}_q(k^n)$. Choose an element $\lambda \in k^\times$ which is not a root of unity, and set

$$h_i = (q_{i1}, q_{i2}, \ldots, q_{i,i-1}, \lambda, 1, 1, \ldots, 1) \in H.$$

Then $h_i(x_j) = \tau_i(x_j)$ for $i > j$ and $h_i(x_i) = \lambda x_i$.

(b) First let $A = \mathcal{O}_q(M_n(k))$. Then take, for instance,

$$h_{lm} = (\alpha_{lm1}, \ldots, \alpha_{lmn}, \beta_{lm1}, \ldots, \beta_{lmn})$$
$$= (q, \ldots, q, 1, *, \ldots, *, q^{-1}, \ldots, q^{-1}, q^{-2}, q^{-1}, \ldots, q^{-1}),$$

where $\alpha_{lmi} = q$ for $i < l$ and $\alpha_{lml} = 1$, while $\beta_{lmj} = q^{-1}$ for $j \neq m$ and $\beta_{lmm} = q^{-2}$ (the scalars α_{lmi} for $i > l$ can be arbitrary.)

The conclusions of the corollary for the case $A = \mathcal{O}_q(GL_n(k))$ follow from the case of $\mathcal{O}_q(M_n(k))$ by localization (see Exercise II.1.J).

(c)(d) Exercise II.5.F. $\square$

PASSAGE FROM $\mathcal{O}_q(GL_n(k))$ TO $\mathcal{O}_q(SL_n(k))$

One might also expect to see the case $A = \mathcal{O}_q(SL_n(k))$ in Theorem II.5.14, since this algebra is a quotient of $\mathcal{O}_q(M_n(k))$. However, we do not have the same torus H acting on both algebras – case (b) of Theorem II.5.14 refers to the action of $(k^\times)^{2n}$ on $\mathcal{O}_q(M_n(k))$ given in (II.1.15), and this action does not induce an action on $\mathcal{O}_q(SL_n(k))$. Instead, we have a torus H of rank $2n-1$ acting on $\mathcal{O}_q(SL_n(k))$ as in (II.1.16), and so what we have proved about $\mathcal{O}_q(M_n(k))$ does not yield information about H-primes in $\mathcal{O}_q(SL_n(k))$ for this H. To avoid this difficulty, we return to the method we used in Chapter I.2 to show that $\mathcal{O}_q(SL_n(k))$ is a domain. Specifically, recall the isomorphism constructed in Lemma I.2.9:

II.5.15. Lemma. *There exists a k-algebra isomorphism*

$$\theta : \mathcal{O}_q(SL_n(k))[z^{\pm 1}] \to \mathcal{O}_q(GL_n(k)),$$

where z is a central indeterminate, such that

$$\begin{aligned}
\theta(\overline{X}_{1j}) &= D_q^{-1} X_{1j} && \text{(all } j) \\
\theta(\overline{X}_{ij}) &= X_{ij} && (i > 1; \text{ all } j) \\
\theta(z) &= D_q. && \square
\end{aligned}$$

II.5.16. Lemma. *Set $A = \mathcal{O}_q(SL_n(k))$ and $B = \mathcal{O}_q(GL_n(k))$, and let $A[z^{\pm 1}]$ be a Laurent polynomial ring over A. Let $H^+ = (k^\times)^{2n}$ act on B and the subgroup*

$$H = \{(\alpha_1, \ldots, \alpha_n, \beta_1, \ldots, \beta_n) \in H^+ \mid \alpha_1 \alpha_2 \cdots \alpha_n \beta_1 \beta_2 \cdots \beta_n = 1\}$$

on A as in (II.1.15–16). Let $\theta : A[z^{\pm 1}] \to B$ be the isomorphism given in Lemma II.5.15.

(a) *Conjugation of the action of H^+ on B by θ gives an action of H^+ by k-algebra automorphisms on $A[z^{\pm 1}]$ such that*

$$(\alpha_1, \ldots, \alpha_n, \beta_1, \ldots, \beta_n).\overline{X}_{ij} = \beta_j(\alpha_2\alpha_3 \cdots \alpha_n\beta_1\beta_2 \cdots \beta_n)^{-1}\overline{X}_{ij} \qquad (i = 1)$$
$$(\alpha_1, \ldots, \alpha_n, \beta_1, \ldots, \beta_n).\overline{X}_{ij} = \alpha_i\beta_j\overline{X}_{ij} \qquad (i > 1)$$
$$(\alpha_1, \ldots, \alpha_n, \beta_1, \ldots, \beta_n).z = \alpha_1\alpha_2 \cdots \alpha_n\beta_1\beta_2 \cdots \beta_n z.$$

(b) *The group homomorphisms from H and H^+ to $\operatorname{Aut} A$ given by the above actions have the same image.*

(c) *If k contains a non-root of unity, the rule $P \mapsto \theta(P[z^{\pm 1}])$ yields a bijection between H-spec A and H^+-spec B.*

Proof. (a)(b) Exercise II.5.H.

(c) In view of (b), H-spec $A = H^+$-spec A. By Exercise II.5.I, the rule $P \mapsto P[z^{\pm 1}]$ gives a bijection from H^+-spec A onto H^+-spec $A[z^{\pm 1}]$. (If $\lambda \in k^\times$ is not a root of unity, the element $(\lambda, 1, 1, \ldots, 1) \in H^+$ can take the role of g_0 in the exercise.) The proof is finished by applying θ, which has been arranged to be H^+-equivariant. $\square$

II.5.17. Theorem. *Let A be either $\mathcal{O}_q(SL_n(k))$ where q is not a root of unity, or $\mathcal{O}_{\lambda,\mathbf{p}}(SL_n(k))$ where λ is not a root of unity, and let H be the k-torus (isomorphic to $(k^\times)^{2n-1}$) acting on A described in (II.1.16). Then H-spec A is finite, and so there are only finitely many H-strata in $\operatorname{spec} A$. Further, all H-primes of A are completely prime.*

Proof. Apply Theorem II.5.14 together with Lemma II.5.16 or Exercise II.5.J. $\square$

NOTES

The development in this chapter is taken from work of Goodearl and Letzter in [**80**], using skew polynomial technology from [**77, 69**].

EXERCISES

Exercise II.5.A. In the setup of (II.5.3), show that $\delta\tau = q\tau\delta$. [Hint: $\delta(r) = xr - h_0(r)x$ for $r \in R$.] $\square$

Exercise II.5.B. Let H be a group acting by automorphisms on a ring R, and let P be a proper H-ideal of R. Show that P is H-prime if and only if for any $u, w \in R \setminus P$, there exist $v \in R$ and $h \in H$ such that $uvh(w) \notin P$. $\square$

Exercise II.5.C. Let τ be an automorphism of a ring R and Q a prime of R. If the τ-orbit of Q is finite, show that Q is minimal over $(Q : \tau)$. $\square$

Exercise II.5.D. Let $S = R[x; \tau, \delta]$ be a q-skew polynomial algebra over k, where $q \in k^\times$ is not a root of unity. Let I be a proper nonzero ideal of S, let n be the minimum degree for nonzero elements of I, and suppose that I contains a monic polynomial s of degree n, say

$$s = x^n + cx^{n-1} + [\text{lower terms}].$$

(Note that $n > 0$ because $I \neq S$.) Set $d = -\binom{n}{1}_q^{-1} q^{n-1} c$, and show that $\delta(r) = dr - \tau(r)d$ for all $r \in R$. □

Exercise II.5.E. Let H be a group acting by automorphisms on a noetherian ring R, and assume that R is an H-prime ring. Then R is semiprime by Lemma II.1.10. Show that Fract R is H-simple (with respect to the induced H-action). □

Exercise II.5.F. Finish the proof of Theorem II.5.14. □

Exercise II.5.G. Verify the conclusions of Theorem II.5.14 directly for the case $A = \mathcal{O}_q(GL_2(k))$. □

Exercise II.5.H. Prove parts (a) and (b) of Lemma II.5.16. □

Exercise II.5.I. Let $A[z^{\pm 1}]$ be a Laurent polynomial ring over a noetherian k-algebra A. Let G be a group acting on $A[z^{\pm 1}]$ by k-algebra automorphisms, such that A is G-stable and z is a G-eigenvector. Suppose there exists $g_0 \in G$ such that $g_0|_A$ is the identity on A while the g_0-eigenvalue of z is not a root of unity. Show that the rule $P \mapsto P[z^{\pm 1}]$ gives a bijection from G-spec A onto G-spec $A[z^{\pm 1}]$. □

Exercise II.5.J. State and prove versions of Lemmas II.5.15 and II.5.16 for $\mathcal{O}_{\lambda, p}(SL_n(k))$. □

MORE ON ITERATED SKEW POLYNOMIAL ALGEBRAS

We continue our analysis of iterated skew polynomial algebras of the form described in (II.5.1) and applications to quantized coordinate rings. Our first main goal is to show that in the context of the Stratification Theorem, the coefficient fields $Z(A_J)^H$ for many standard examples must equal the base field k. We also discuss a theorem which shows that in many standard examples, when suitable parameters are non-roots of unity, all prime ideals are completely prime (i.e., all prime factors are domains).

COEFFICIENT FIELDS

Recall that the coefficient fields for the Laurent polynomial rings in part (e) of the Stratification Theorem (II.2.13) have the form $Z(A_J)^H = Z(\text{Fract } A/J)^H$, where H is the torus acting on the noetherian k-algebra A and J is an H-prime ideal of A. We investigate circumstances under which we can conclude that $Z(\text{Fract } A/J)^H = k$, since in that case the theorem tells us that $\text{spec}_J A$, which is homeomorphic to $\text{spec } Z(A_J)$, is homeomorphic to the scheme of irreducible subvarieties of a k-torus.

II.6.1. Definitions. Let A be a noetherian k-algebra, H a group acting on A by k-algebra automorphisms, and P an H-prime ideal of A. We will say that P is H-*rational* provided the fixed field $Z(\text{Fract } A/P)^H$ is algebraic over k, and *strongly H-rational* provided $Z(\text{Fract } A/P)^H = k$.

II.6.2. Lemma. *Assume the setup of (II.5.3), and let Q be a strongly H-rational, δ-invariant, H-prime ideal of R. Then every H-prime of S contracting to Q is strongly H-rational.*

Proof. Since Q is a semiprime, δ-invariant, τ-stable ideal, we may pass from R to $\text{Fract } R/Q$. Hence, there is no loss of generality in assuming that $Q = 0$ and R is an H-simple artinian ring (see Exercise II.5.E). Now $\text{Fract } R = R$, and so $Z(R)^H = k$ by hypothesis.

Assume first that S is H-simple. Then 0 is the only H-prime of S, and we just need to show that $Z(\text{Fract } S)^H = k$.

Consider a nonzero element $u \in Z(\text{Fract } S)^H$, set $I = \{s \in S \mid su \in S\}$, and observe that I is a nonzero H-ideal of S. By H-simplicity, $I = S$, and so $u \in S$. Similarly, $u^{-1} \in S$. Now let $u_n x^n$ be the leading term of u. Since u commutes with elements of R, we have $u_n \tau^n(r) = r u_n$ for $r \in R$, and so u_n is a normal element of R. Further, since u is fixed by H, we see that u_n is an H-eigenvector. Hence, $Ru_n = u_nR$ is a nonzero H-ideal of R, and a second application of H-simplicity implies that u_n is a unit in R. Consequently, the equation $uu^{-1} = 1$ in S forces $n = 0$, whence $u \in R$, and therefore $u \in Z(R)^H = k$.

Now assume that S is not H-simple, let $d \in R$ be the element given in Lemma II.5.10, and set $z = x - d$. Then $S = R[z; \tau]$, and the lemma shows that the H-primes of S are just 0 and zS. There is an H-equivariant k-algebra isomorphism $R \cong S/zS$, and so $Z(\text{Fract } R)^H = k$ implies $Z(\text{Fract } S/zS)^H = k$. Thus, zS is strongly H-rational, and it only remains to show that $Z(\text{Fract } S)^H = k$.

Observe that the prime radical of any proper, nonzero H-ideal I of S is an intersection of H-primes and so must equal zS. Hence, I must contain a power of z.

Now consider an element $u \in Z(\text{Fract } S)^H$. The set $I = \{s \in S \mid us \in S\}$ is a nonzero H-ideal of S, and so $z^n \in I$ for some $n \geq 0$. Hence, $u = vz^{-n}$ for some $v \in S$. Write $v = v_0 + v_1z + \cdots + v_tz^t$ for some $v_i \in R$. Since u commutes with z, so does v, from which we see that $v_i = \tau(v_i) = h_0(v_i)$ for all i. Moreover, u is fixed by h_0 and $h_0(z^n) = q^{-n}z^n$, whence $h_0(v) = q^{-n}v$. This implies that $q^{-i}h_0(v_i) = q^{-n}v_i$ for all i. Since q is not a root of unity, it follows that $v_i = 0$ for all $i \neq n$. Thus $u = v_n \in R$, and therefore $u \in Z(R)^H = k$. $\square$

II.6.3. Corollary. *Assume the setup of (II.5.3). If all H-primes of R are strongly H-rational, then the same holds for S.*

Proof. Lemmas II.5.9 and II.6.2. $\square$

II.6.4. Theorem. *Let $A = k[x_1][x_2; \tau_2, \delta_2] \cdots [x_n; \tau_n, \delta_n]$ and H be as in (II.5.1). Then all H-prime ideals of A are strongly H-rational.*

Proof. By induction, using Corollary II.6.3. $\square$

II.6.5. Corollary. *Let A be one of the algebras*

$$\mathcal{O}_q(k^n) \qquad \mathcal{O}_q(k^n) \qquad \mathcal{O}_q(M_n(k)) \qquad \mathcal{O}_{\lambda,p}(M_n(k))$$
$$\mathcal{O}_q(GL_n(k)) \qquad \mathcal{O}_{\lambda,p}(GL_n(k)) \qquad A_n^{Q,\Gamma}(k),$$

with non-root of unity restrictions as in Theorem II.5.14. Let H be the k-torus acting on A as in (II.1.14–17).

Then $Z(\text{Fract } A/J)^H = k$ for any H-prime ideal J of A, and hence $\text{spec}_J A$ is homeomorphic to the prime spectrum of some Laurent polynomial ring over k.

Proof. Theorems II.6.4 and II.2.13. $\square$

II.6.6. Corollary. *Let A be either $\mathcal{O}_q(SL_n(k))$ where q is not a root of unity, or $\mathcal{O}_{\lambda,p}(SL_n(k))$ where λ is not a root of unity, and let H be the k-torus acting on A as in (II.1.16). Then $Z(\text{Fract } A/J)^H = k$ for any H-prime ideal J of A, and hence*

$\operatorname{spec}_J A$ *is homeomorphic to the prime spectrum of some Laurent polynomial ring over* k.

Proof. Apply Corollary II.6.5 together with Lemma II.5.16 or Exercise II.5.J to get $Z(\operatorname{Fract} A/J)^H = k$. Then apply Theorem II.2.13. $\square$

COMPLETE PRIMENESS

II.6.7. A classical theorem of Dixmier [48], as extended by Gabriel [58], states that if $\mathfrak{g}$ is a solvable finite dimensional Lie algebra over a field k of characteristic zero, then all prime factor rings of $U(\mathfrak{g})$ are domains, i.e., all prime ideals of $U(\mathfrak{g})$ are completely prime. In particular, this result incorporates Lie's Theorem: in case k is algebraically closed, all finite dimensional irreducible $U(\mathfrak{g})$-modules are 1-dimensional. A key ingredient in proving Dixmier's theorem is that when k is algebraically closed, $U(\mathfrak{g})$ has the structure of an iterated differential operator ring over k. Sigurdsson [195] developed a proof entirely within the latter context – he proved that if

$$S = R[x_1; \delta_1][x_2; \delta_2] \cdots [x_n; \delta_n]$$

is an iterated differential operator ring over a commutative noetherian $\mathbb{Q}$-algebra R, then all primes of S are completely prime.

For applications to quantized coordinate rings, an analogue of Sigurdsson's theorem for iterated q-skew polynomial rings was developed by Goodearl and Letzter [78]. We do not prove the most general result here, but give a version for the setup of (II.5.1), which allows us to take advantage of some of the machinery developed in the previous chaper.

In Sigurdsson's theorem, the hypothesis of characteristic zero is essential to ensure the non-vanishing of the binomial coefficients that appear when Leibniz' Rule is used. Similarly, when dealing with q-skew polynomial rings, one needs q to be a non-root of unity so that the q-binomial coefficients in the q-Leibniz Rule do not vanish.

II.6.8. Example. Consider $A = \mathcal{O}_q(k^2)$ where k is algebraically closed, and assume first that q is not a root of unity. In view of Example II.1.2, any prime factor of A is isomorphic to k or to a polynomial ring $k[t]$ or to A itself. All of these algebras are domains, and so all primes of A are completely prime.

Now suppose that q is a primitive l-th root of unity for some $l > 1$. By Exercise II.6.B, A has many primes which are not completely prime. For instance, $A/\langle x^l - 1, y^l - 1 \rangle \cong M_l(k)$.

II.6.9. Theorem. *Let* $A = k[x_1][x_2; \tau_2, \delta_2] \cdots [x_n; \tau_n, \delta_n]$ *and* H *be as in* (II.5.1). *Assume also that* H *is a* k-torus *and that the* H-action on A is rational. There are scalars $\lambda_{ij} \in k^\times$ such that $\tau_i(x_j) = \lambda_{ij} x_j$ for all $i > j$. If the subgroup $\langle \lambda_{ij} \rangle \subseteq k^\times$ is torsionfree, then all prime ideals of A are completely prime.*

Proof. We discuss the proof of this theorem in the following section. $\square$

II.6.10. Corollary. *Let A be one of the following algebras:*

(a) $\mathcal{O}_q(k^n)$, $\mathcal{O}_q((k^\times)^n)$, $\mathcal{O}_q(M_n(k))$, $\mathcal{O}_q(GL_n(k))$, or $\mathcal{O}_q(SL_n(k))$, *where q is not a root of unity.*

(b) $\mathcal{O}_q(k^n)$ *or* $\mathcal{O}_q((k^\times)^n)$, *where the group $\langle q_{ij} \rangle$ is torsionfree.*

(c) $\mathcal{O}_{\lambda,\boldsymbol{p}}(M_n(k))$, $\mathcal{O}_{\lambda,\boldsymbol{p}}(GL_n(k))$, *or* $\mathcal{O}_{\lambda,\boldsymbol{p}}(SL_n(k))$, *where the group $\langle \lambda, p_{ij} \rangle$ is torsionfree.*

(d) $A_n^{Q,\Gamma}(k)$, *where the group $\langle q_i, \gamma_{ij} \rangle$ is torsionfree.*

Then all prime ideals of A are completely prime.

Proof. In the cases $\mathcal{O}_\bullet(k^n)$, $\mathcal{O}_\bullet(M_n(k))$, and $A_n^{Q,\Gamma}(k)$, there is a k-torus H acting by k-algebra automorphisms on A as in (II.1.14–17), and the conditions of (II.5.1) are satisfied (see the proof of Theorem II.5.14). Let $\lambda_{ij} \in k^\times$ be the scalars such that $\tau_i(x_j) = \lambda_{ij} x_j$ for $i > j$. The group $\langle \lambda_{ij} \rangle$ is contained in the group $\langle q \rangle$, $\langle q_{ij} \rangle$, $\langle \lambda, p_{ij} \rangle$, or $\langle q_i, \gamma_{ij} \rangle$ in cases (a), (b), (c), (d), respectively. Thus, by Theorem II.6.9, all primes of A are completely prime in these cases.

The cases $\mathcal{O}_\bullet(SL_n(k))$ follow directly from those above, while the cases $\mathcal{O}_\bullet((k^\times)^n)$ and $\mathcal{O}_\bullet(GL_n(k))$ follow by localization. $\square$

Complete primeness has also been proved for the prime ideals in $\mathcal{O}_q(G)$ when $q \in \mathbb{C}^\times$ is not a root of unity [**115**, Theorem 11.4].

PROOF OF THE COMPLETE PRIMENESS THEOREM

II.6.11. Lemma. *Let $S = R[x; \tau, \delta]$ be a q-skew polynomial algebra over k, where R is noetherian and $q \in k^\times$ is not a root of unity. Assume that R is a τ-prime, artinian ring. If either R or S is not a simple ring, then δ is an inner τ-derivation.*

Proof. By Lemma II.1.10, R is semiprime, and its minimal primes form a single τ-orbit. Then R is semisimple artinian, and if $e_1, \ldots, e_n$ are its distinct primitive central idempotents, the minimal primes of R are the ideals $R(1 - e_i)$. Since the cyclic group $\langle \tau \rangle$ acts transitively on the ideals $R(1-e_i)$, it must also act transitively on the idempotents $1 - e_i$, and hence on the e_i. Thus, after renumbering the e_i if necessary, we may assume that $\tau(e_i) = e_{i+1}$ for $i = 1, \ldots, n-1$ and $\tau(e_n) = e_1$.

If $n \geq 2$, then δ is inner by Exercise II.6.C. Thus, we may assume that $n = 1$, that is, R is a simple ring.

Now by hypothesis, S is not simple, and so it contains a proper nonzero ideal I. Let s be a nonzero element of I with minimal degree, say degree n and leading coefficient s_n. Since R is simple, there exist elements $a_i, b_i \in R$ such that $\sum_i a_i s_n b_i = 1$. Then $\sum_i a_i s \tau^{-n}(b_i)$ is an element of I with degree n and leading coefficient 1. Hence, there is no loss of generality in assuming that s is monic, and then Exercise II.5.D shows that δ is inner. $\square$

II.6.12. Corollary. *Let $S = R[x; \tau, \delta]$ be a q-skew polynomial algebra over k, where R is noetherian and $q \in k^\times$ is not a root of unity. Let P be a prime* ideal of S such that $P \cap R = 0$. If either $P \neq 0$ or R is not a prime ring, then $\mathrm{Fract}(S/P) \cong \mathrm{Fract}(R[z;\tau]/P')$ for some $P' \in \mathrm{spec}\, R[z;\tau]$.

Proof. Note that all the minimal primes of R must have finite τ-orbits. Lemma II.5.8 implies that R must be a τ-prime ring, and so, in particular, R is semiprime. Now the set X of regular elements of R is a τ-stable denominator set in R, hence also a denominator set in S. Set $\overline{R} = R[X^{-1}]$ and $\overline{S} = S[X^{-1}] = \overline{R}[x;\tau,\delta]$. Observe that τ-primeness passes from R to $\overline{R}$ (Exercise II.1.J). Since P is disjoint from X, the set $P\overline{S}$ is a prime of $\overline{S}$, and Exercise II.6.D shows that $\mathrm{Fract}(S/P) \cong \mathrm{Fract}(\overline{S}/P\overline{S})$.

If $P \neq 0$, then $P\overline{S} \neq 0$ and so $\overline{S}$ is not simple. On the other hand, if R is not prime, then $\overline{R}$ is not simple. Hence, Lemma II.6.11 shows that δ is inner on $\overline{R}$. Consequently, $\overline{S} = \overline{R}[z;\tau]$ where $z = x - d$ for a suitable element $d \in R$, and so $\overline{S} = R[z;\tau][X^{-1}]$. Now $P\overline{S} = P'\overline{S}$ for some $P' \in \mathrm{spec}\, R[z;\tau]$ disjoint from X, and by Exercise II.6.D, $\mathrm{Fract}(\overline{S}/P'\overline{S}) \cong \mathrm{Fract}(R[z;\tau]/P')$. This establishes the corollary. $\square$

II.6.13. Lemma. *Let (τ,δ) be a skew derivation on a ring R. If there exists $c \in Z(R)$ such that $c - \tau(c)$ is invertible in R, then δ is an inner τ-derivation.*

Proof. Exercise II.6.E. $\square$

II.6.14. Theorem. *Assume the setup of (II.5.3), and suppose that H is a k-torus whose action, at least on R, is rational. Let $P \in \mathrm{spec}\, S$. Then either $P = QS$ for some δ-invariant, τ-stable prime ideal Q of R, or $\mathrm{Fract}(S/P) \cong \mathrm{Fract}(R[z;\tau]/P')$ for some $P' \in \mathrm{spec}\, R[z;\tau]$.*

Proof. Note that assumption (d) of (II.5.3) entails that k is infinite.

Since P is prime, $(P:H)$ is H-prime. By Lemma II.5.9, $(P:H) \cap R$ is a δ-invariant H-prime ideal of R. In particular, $(P:H) \cap R$ is τ-stable, and so we may factor it out. Thus, there is no loss of generality in assuming that $(P:H) \cap R = 0$ and that R is an H-prime ring.

By Corollary II.3.5, the set $\mathcal{E}$ of regular H-eigenvectors in R is a denominator set, and $R[\mathcal{E}^{-1}]$ is H-simple (with respect to the induced, rational H-action). Then $\tau(\mathcal{E}) = \mathcal{E}$, and so we have $S[\mathcal{E}^{-1}] = R[\mathcal{E}^{-1}][x;\tau,\delta]$. Since

$$(P \cap R : H) = (P : H) \cap R = 0,$$

the ideal $P \cap R$ contains no nonzero H-eigenvectors, and so P is disjoint from $\mathcal{E}$. By Exercise II.6.D, the prime $\overline{P} = P[\mathcal{E}^{-1}]$ in $S[\mathcal{E}^{-1}]$ satisfies $\mathrm{Fract}(S/P) \cong \mathrm{Fract}(S[\mathcal{E}^{-1}]/\overline{P})$. Moreover, if $\overline{P} = \overline{Q}S[\mathcal{E}^{-1}]$ for a δ-invariant, τ-stable prime $\overline{Q}$ of $R[\mathcal{E}^{-1}]$, then $Q = \overline{Q} \cap R$ is a δ-invariant, τ-stable prime of R and $P = \overline{P} \cap S = QS$. Thus there is no loss of generality in passing to $S[\mathcal{E}^{-1}]$, that is, we may now assume that R is H-simple.

At this point, the proof splits into two cases, depending on whether τ restricts to the identity on $Z(R)$ or not.

First suppose that τ is not trivial on $Z(R)$. Since $Z(R)$ is homogeneous with respect to the $X(H)$-grading on R (Lemma II.3.7(a)), it is spanned by its H-eigenvectors. Hence, there is an H-eigenvector $c \in Z(R)$ such that $\tau(c) \neq c$. Since H is abelian, $\tau(c)$ is an H-eigenvector with the same H-eigenvalue as c, and so $c - \tau(c)$ is a nonzero H-eigenvector in $Z(R)$. By Lemma II.3.7(c), $c - \tau(c)$ is invertible, and hence δ is inner by Lemma II.6.13. Thus $S = R[z; \tau]$ for some z, and we have $S/P = R[z; \tau]/P$ in this case.

Now assume that τ is trivial on $Z(R)$. Since all ideals of R are centrally generated (Corollary II.3.9(b)), it follows that all ideals of R are τ-stable. In particular, the ideal $Q = P \cap R$ is τ-stable, and then it must also be δ-invariant, since $\delta(r) = xr - \tau(r)x \in P$ for all $r \in Q$. Consequently, QS is an ideal of S, and we can identify S/QS as a q-skew polynomial ring $S' = R'[x; \tau, \delta]$, where $R' = R/Q$. Set $P' = P/QS$, which is a prime of S' such that $P' \cap R' = 0$. If either $P' \neq 0$ or R' is not a prime ring, then Corollary II.6.12 shows that $\mathrm{Fract}(S'/P') \cong \mathrm{Fract}(R'[z; \tau]/P'')$ for some $P'' \in \mathrm{spec}\, R'[z; \tau]$, and in this case we are done. On the other hand, if $P' = 0$ and R' is a prime ring, then $P = QS$ and Q is a δ-invariant, τ-stable prime ideal of R. $\square$

II.6.15. Corollary. *Assume the setup of* (II.5.3), *and suppose that H is a k-torus whose action, at least on R, is rational. If all primes of $R[z; \tau]$ are completely prime, then the same holds for S.*

Proof. Note that since $R[z; \tau]/\langle z \rangle \cong R$, all primes of R must be completely prime. Hence, if Q is a δ-invariant, τ-stable prime of R, then $S/QS \cong (R/Q)[x; \tau, \delta]$ is a domain, and so QS is a completely prime ideal of S. Now apply Theorem II.6.14. $\square$

II.6.16. Theorem. *Let $\boldsymbol{\lambda} = (\lambda_{ij})$ be an $n \times n$ multiplicatively antisymmetric matrix over k. If the group $\langle \lambda_{ij} \rangle \subseteq k^\times$ is torsionfree, then all prime ideals of $\mathcal{O}_{\boldsymbol{\lambda}}(k^n)$ are completely prime.*

Sketch of proof. (See [**78**, Theorem 2.1] for full details.)

Proceed by induction on n, the case $n = 1$ being clear. If we have a prime ideal of $\mathcal{O}_{\boldsymbol{\lambda}}(k^n)$ that contains a generator x_i, then we can pass to $\mathcal{O}_{\boldsymbol{\lambda}}(k^n)/\langle x_i \rangle$, which is isomorphic to $\mathcal{O}_{\boldsymbol{\mu}}(k^{n-1})$ for a suitable $\boldsymbol{\mu}$. This case thus follows from the induction hypothesis. To deal with primes containing none of the x_i, we can pass to the algebra $A = \mathcal{O}_{\boldsymbol{\lambda}}((k^\times)^n)$ by localization. Thus, we need only show that any prime P of A is completely prime.

Let G be the subgroup of the group of units of A generated by $x_1, \ldots, x_n$. The commutator subgroup of G is $\langle \lambda_{ij} \rangle$, and $G/\langle \lambda_{ij} \rangle$ is free abelian of rank n. Set

$$H = \{z \in G \mid z + P \in Z(A/P)\},$$

and observe that H is a normal subgroup of G containing $\langle \lambda_{ij} \rangle$. If T is the k-subalgebra of A generated by H, then $T/(T \cap P)$ is a commutative domain, since it embeds in $Z(A/P)$. Let $K = \mathrm{Fract}\big(T/(T \cap P)\big)$, let X be the set of nonzero elements of $T/(T \cap P)$, and set $B = (A/P)[X^{-1}]$. Then B is a K-algebra.

One now checks that the group G/H is torsionfree, and thus free abelian of rank at most n. Let $y_1 H, \ldots, y_m H$ be a basis for G/H, and check that the elements $q_{ij} = y_i y_j y_i^{-1} y_j^{-1} + P$ lie in K for all i, j. Then $\boldsymbol{q} = (q_{ij})$ is a multiplicatively antisymmetric matrix over K, and there exists a K-algebra surjection $\mathcal{O}_{\boldsymbol{q}}(K^m) \rightarrow B$.

Finally, one applies a criterion of McConnell and Pettit [**157**, 1.3] to show that $\mathcal{O}_{\boldsymbol{q}}(K^m)$ is a simple ring. Thus $B \cong \mathcal{O}_{\boldsymbol{q}}(K^m)$, whence B is a domain, and therefore A/P is a domain. $\square$

II.6.17. Proof of Theorem II.6.9. Set $\lambda_{ii} = 1$ for all i and $\lambda_{ji} = \lambda_{ij}^{-1}$ for $i > j$, so that $\boldsymbol{\lambda} = (\lambda_{ij})$ is a multiplicatively antisymmetric matrix. By Theorem II.6.16, all primes of $\mathcal{O}_{\boldsymbol{\lambda}}(k^n)$ are completely prime.

Let $m \in \{1, \ldots, n\}$ be minimal such that $\delta_i = 0$ for all $i > m$; we proceed by induction on m. If $m = 1$, then all $\delta_i = 0$ and $A = \mathcal{O}_{\boldsymbol{\lambda}}(k^n)$, in which case we are done.

Now assume that $m > 1$, and that the theorem holds in situations with smaller values of m. Thus $\delta_i = 0$ for $i = m+1, \ldots, n$, and so $x_i x_j = \lambda_{ij} x_j x_i$ whenever $i > m, j$. Let R denote the k-subalgebra of A generated by the x_i with $i \neq m$, and observe that R is an iterated skew polynomial algebra of the form

$$R = k[x_1][x_2; \tau_2, \delta_2] \cdots [x_{m-1}; \tau_{m-1}, \delta_{m-1}][x_{m+1}; \sigma_{m+1}] \cdots [x_n; \sigma_n].$$

By Exercise II.6.H, $A = R[x_m; \tau, \delta]$ for a suitable k-algebra automorphism τ and k-linear τ-derivation δ. (If $m = n$, then $\tau = \tau_n$ and $\delta = \delta_n$.)

The algebra $R[z; \tau]$ can be written as an iterated skew polynomial algebra

$$R[z; \tau] = k[y_1][y_2; \tau_2', \delta_2'] \cdots [y_n; \tau_n', \delta_n']$$

where the y_i are the variables $x_1, \ldots, x_{m-1}, x_{m+1}, \ldots, x_n, z$ (in that order) and $\delta_i' = 0$ for $i \geq m$. By Exercise II.6.I, there is a rational action of a k-torus H^+ on $R[z; \tau]$ by k-algebra automorphisms such that $R[z; \tau]$ and H^+ fit the setup of (II.5.1). Further, $\tau_s'(y_t) = \lambda_{st}' y_t$ for $s > t$ where each λ_{st}' equals some λ_{ij}. Hence, the group $\langle \lambda_{st}' \rangle$ is torsionfree. Our induction hypothesis thus applies to $R[z; \tau]$, and we conclude that all primes of $R[z; \tau]$ are completely prime.

Finally, Corollary II.6.15 implies that all primes of A are completely prime, and the induction step is established. $\square$

NOTES

The strong H-rationality theorem (II.6.4) was developed by Goodearl using the methods of [**80**]. The complete primeness theorem (II.6.9) is a special case of the theorem proved by Goodearl and Letzter in [**78**] (see also [**69**]).

Exercises

Exercise II.6.A. Verify the conclusions of Corollary II.6.5 directly for the algebra $\mathcal{O}_q(GL_2(k))$, and identify the number of indeterminates in each Laurent polynomial ring. □

Exercise II.6.B. Let $A = \mathcal{O}_q(k^2)$ where q is a primitive l-th root of unity for some $l > 1$. Show that $A/\langle x^l - \alpha^l, \, y^l - \beta^l \rangle \cong M_l(k)$ for any $\alpha, \beta \in k^\times$. [Hint: Show that the matrices

$$\begin{bmatrix} 0 & \alpha & 0 & \cdots & 0 \\ 0 & 0 & \alpha & \cdots & 0 \\ \vdots & & \ddots & & \vdots \\ 0 & 0 & 0 & \cdots & \alpha \\ \alpha & 0 & 0 & \cdots & 0 \end{bmatrix} \qquad \text{and} \qquad \begin{bmatrix} \beta & 0 & 0 & \cdots & 0 \\ 0 & q\beta & 0 & \cdots & 0 \\ 0 & 0 & q^2\beta & \cdots & 0 \\ \vdots & & & \ddots & \vdots \\ 0 & 0 & 0 & \cdots & q^{l-1}\beta \end{bmatrix}$$

generate $M_l(k)$.] □

Exercise II.6.C. Let (τ, δ) be a skew derivation on a ring R. Suppose that there exist orthogonal central idempotents $e_1, \ldots, e_n \in R$ such that $e_1 + \cdots + e_n = 1$ and $\tau(e_i) = e_{i+1}$ for all i, where indices are interpreted modulo n. If $n \geq 2$, show that δ is an inner τ-derivation. [Hint: Show that there exist elements $a_i \in e_i R$ for all i such that each $\delta(e_i) = a_i - a_{i+1}$, and consider the element $d = a_1 + \cdots + a_n$.] □

Exercise II.6.D. Let S be a noetherian ring, X a denominator set in S, and P a prime ideal of S disjoint from X. Recall that $P[X^{-1}]$ is a prime ideal of $S[X^{-1}]$, and show that $\mathrm{Fract}(S[X^{-1}]/P[X^{-1}]) \cong \mathrm{Fract}(S/P)$. □

Exercise II.6.E. Prove Lemma II.6.13. [Hint: Apply δ to the equations $rc = cr$, for $r \in R$.] □

Exercise II.6.F. If $S = R[x; \tau, \delta]$ is a skew polynomial ring over a semiprime right Goldie ring R, show that R_R and S_S have the same uniform ranks, as follows. First, after localizing, we may assume that R is semisimple. Observe that it now suffices to show that IS is a uniform right ideal of S for any simple right ideal I of R. Choose a uniform right ideal $U \subseteq IS$ and a nonzero element $u \in U$ with minimum degree n and leading coefficient c. Show that $J = \tau^{-n}(\mathrm{r.ann}_R(c))$ is a maximal right ideal of R and that $\mathrm{r.ann}_S(u) = JS$. Conclude that $uS \cong IS$, and therefore that IS is uniform. □

Exercise II.6.G. Assume the hypotheses of Theorem II.6.14, and suppose there is a finite upper bound, say b, for the uniform ranks of the factors $R[z; \tau]/P'$ for $P' \in \mathrm{spec}\, R[z; \tau]$. Show that $\mathrm{rank}(S/P) \leq b$ for all $P \in \mathrm{spec}\, S$. □

Exercise II.6.H. Let $A = k[x_1][x_2; \tau_2, \delta_2] \cdots [x_n; \tau_n, \delta_n]$ be an iterated skew polynomial algebra over k, and assume there exist $\lambda_{ij} \in k^\times$ such that $\tau_i(x_j) = \lambda_{ij} x_j$

for $i > j$. Suppose there is some $m < n$ such that $\delta_i = 0$ for $i > m$. Show that A can be rewritten as an iterated skew polynomial algebra of the form

$$k[x_1][x_2; \tau_2, \delta_2] \cdots [x_{m-1}; \tau_{m-1}, \delta_{m-1}][x_{m+1}; \sigma_{m+1}] \cdots [x_n; \sigma_n][x_m; \tau, \delta],$$

where each σ_i is the restriction of τ_i to the subalgebra

$$k\langle x_1, \ldots, x_{m-1}, x_{m+1}, \ldots, x_{i-1}\rangle,$$

while τ and δ are determined by

$$
\begin{aligned}
\tau(x_j) &= \lambda_{mj} x_j, & \delta(x_j) &= \delta_m(x_j) & (1 \leq j \leq m-1) \\
\tau(x_j) &= \lambda_{jm}^{-1} x_j, & \delta(x_j) &= 0 & (m+1 \leq j \leq n). \quad \square
\end{aligned}
$$

Exercise II.6.I. Let $R[z; \tau] = k[y_1][y_2; \tau_2', \delta_2'] \cdots [y_n; \tau_n', \delta_n']$ as in (II.6.17). Set $H^+ = H \times (k^\times)^{n-m+1}$, and show that there is a rational action of H^+ on $R[z; \tau]$ by k-algebra automorphisms such that

$$
(h, \alpha_1, \ldots, \alpha_{n-m+1})(y_i) = \begin{cases} h(y_i) & (i < m) \\ \alpha_{i-m+1} y_i & (i \geq m) \end{cases}
$$

for all i and all $(h, \alpha_1, \ldots, \alpha_{n-m+1}) \in H^+$. Show that $R[z; \tau]$ and H^+ fit the setup of (II.5.1). $\square$

THE PRIMITIVE SPECTRUM

For many purposes, especially in representation theory, the primitive ideals of a ring R are more important than the general prime ideals. In other words, one may be more interested in prim R than in spec R, although knowledge of spec R may be very helpful in determining prim R. In this chapter and the next, we discuss the structure of prim R and its relationship to spec R.

IDENTIFYING PRIMITIVE IDEALS

II.7.1. A fundamental goal in the representation theory of an algebra A is to find and classify the irreducible representations of A (that is, the simple A-modules). For many infinite dimensional algebras, however, such as the enveloping algebras of semisimple Lie algebras, finding the irreducible representations appears to be an impossible problem. Dixmier has promulgated the following programme as a substitute, and also as an approach towards the complete goal: Start by finding the primitive ideals of A, and then for each primitive ideal P, find at least one irreducible representation with annihilator P.

To follow this programme, one needs ways of recognizing primitive ideals without first finding irreducible representations. For commutative or PI rings, this is easy – the primitive ideals are just the maximal ideals, using Kaplansky's Theorem (I.13.3) for the latter case. Maximal ideals are always primitive, of course, but in general primitive ideals need not be maximal. Typically, generic quantized coordinate rings will contain non-maximal primitive ideals, as in the example below.

II.7.2. Example. Let $A = \mathcal{O}_q(k^2)$ with k algebraically closed and q not a root of unity. From Example II.1.2 we know all the prime ideals of A; let us see which ones are primitive.

First, the maximal ideals $\langle x - \alpha, y \rangle$ and $\langle x, y - \beta \rangle$, for $\alpha, \beta \in k$, are automatically primitive. Second, note that $A/\langle x \rangle \cong k[y]$, a commutative polynomial ring. Since $k[y]$ is not a primitive ring, $\langle x \rangle$ is not a primitive ideal of A. Similarly, $\langle y \rangle$ is not primitive. Finally, Exercise II.7.A shows that A is a primitive ring, whence $\langle 0 \rangle$ is a primitive ideal of A. In particular, this provides an example of a primitive ideal which is not maximal.

To summarize: All prime ideals of A except for $\langle x \rangle$ and $\langle y \rangle$ are primitive.

II.7.3. The most basic step in Dixmier's programme is easy: every primitive ideal is prime. Thus, what we are really asking is how to tell which prime ideals are primitive. Dixmier proposed two different criteria – one algebraic, one topological – which, as he and Moeglin verified, characterize the primitive ideals in enveloping algebras. We discuss these criteria separately, beginning with the algebraic one.

II.7.4. Definition. Let A be a noetherian k-algebra. A prime ideal P of A is *rational* provided the field $Z(\text{Fract } A/P)$ is algebraic over k. For example, if A is a commutative affine k-algebra, then the rational primes of A are precisely the maximal ideals (Exercise II.7.B). (The notion of H-rational H-primes (II.6.1) is modelled on this concept.)

II.7.5. Example. Let $A = \mathcal{O}_q(k^2)$ with k algebraically closed and q not a root of unity. Note that any maximal ideal P of A has codimension 1, that is, $A/P \cong k$. Any such prime P is clearly rational. On the other hand, the field $\text{Fract } A/\langle x \rangle \cong k(x)$ is transcendental over k, and so $\langle x \rangle$ is not rational. Similarly, $\langle y \rangle$ is not rational. Finally, $Z(\text{Fract } A) = k$ (Exercise II.7.C), and so $\langle 0 \rangle$ is rational.

To summarize: All prime ideals of A except for $\langle x \rangle$ and $\langle y \rangle$ are rational. Therefore the rational prime ideals in A are precisely the primitive ideals.

The topological criterion which we discuss next has the advantage that it can be defined in (the prime spectra of) arbitrary rings, not just in algebras.

II.7.6. Definitions. A subset L of a topological space X is *locally closed* provided there exists an open set $U \supseteq L$ such that L is closed in U. Equivalently, L is locally closed if and only if it is an intersection of an open set and a closed set in X. A *locally closed point* in X is any point $x \in X$ such that the singleton $\{x\}$ is locally closed; this may be rephrased as saying that $\{x\}$ is closed in some neighborhood of x.

In a ring R, we shall say that a prime ideal P is *locally closed* provided P is a locally closed point of spec R, where spec R is equipped, as usual, with the Zariski topology. Note, in particular, that any maximal ideal M of R is locally closed in spec R, since the singleton $\{M\}$ is closed.

It is helpful to translate the above condition into more ideal-theoretic terms, as follows:

II.7.7. Lemma. *A prime ideal P in a ring R is locally closed in spec R if and only if the intersection of all prime ideals properly containing P is an ideal properly containing P.*

Proof. Let J be the intersection of all the prime ideals of R properly containing P. If $J \supsetneq P$, then $W(J) \cap V(P) = \{P\}$, and so P is locally closed. Conversely, if P is locally closed, then there are ideals I and L in R such that $V(I) \cap W(L) = \{P\}$. In this case, $J \supseteq L + P \supsetneq P$. $\square$

II.7.8. Example. Once again, let $A = \mathcal{O}_q(k^2)$ with k algebraically closed and q not a root of unity. As just noted, the maximal ideals of A are all locally closed. On the other hand,

$$\bigcap \{\text{primes} \supsetneq \langle x \rangle \} = \bigcap_{\beta \in k^\times} \langle x, \, y - \beta \rangle = \langle x \rangle,$$

and so $\langle x \rangle$ is not locally closed. Similarly, $\langle y \rangle$ is not locally closed. Finally, the intersection of the nonzero primes of A contains $\langle xy \rangle$ and so is nonzero, whence $\langle 0 \rangle$ is locally closed.

To summarize: All prime ideals of A except for $\langle x \rangle$ and $\langle y \rangle$ are locally closed. Therefore the locally closed primes of A are precisely the primitive ideals.

II.7.9. Definition. Dixmier [49] and Moeglin [161] proved that in the enveloping algebra of any finite dimensional complex Lie algebra, the sets of primitive, locally closed, and rational prime ideals coincide; Irving and Small [104] extended this result to finite dimensional Lie algebras over arbitrary fields of characteristic zero. Any algebra in which the above sets coincide is said to satisfy the *Dixmier-Moeglin equivalence*. We shall see that many quantized coordinate rings satisfy this equivalence. For instance, Examples II.7.2,5,8 show that when k is algebraically closed and q is not a root of unity, the Dixmier-Moeglin equivalence holds in $\mathcal{O}_q(k^2)$. On the other hand, this equivalence fails, for example, in local rings such as $k[x]_{(x)}$, where (0) is locally closed but not primitive.

JACOBSON RINGS

It turns out that some of the implications involved in the Dixmier-Moeglin equivalence hold for fairly general reasons. We discuss one of these implications in this section, and another in the following section.

II.7.10. Definition. A *Jacobson ring* is a ring R in which all prime ideals P are semiprimitive, that is, $J(R/P) = 0$.

Recall that one advantage of studying semiprimitive over primitive ideals is that there are several characterizations of the Jacobson radical which do not involve irreducible modules. One version of Hilbert's Nullstellensatz states that any commutative affine algebra over a field is a Jacobson ring. Another example is $\mathcal{O}_q(k^2)$ – see Exercise II.7.D.

II.7.11. Lemma. *In a Jacobson ring R, every locally closed prime is primitive.*

Proof. Let P be a locally closed prime of R, and let $\{P_i \mid i \in I\}$ be the set of all (left) primitive ideals of R containing P. Then $\bigcap_{i \in I} P_i = P$ because $J(R/P) = 0$. By Lemma II.7.7, the P_i cannot all properly contain P. Thus some $P_i = P$, whence P is (left) primitive. (Note that the same argument shows that P must also be right primitive.) $\square$

It is sometimes very easy to see that certain algebras are Jacobson rings:

II.7.12. Proposition. *Let A be a noetherian k-algebra such that $\dim_k A <$ card k. (In particular, this holds if k is uncountable and A is a countably generated k-algebra.) Then A is a Jacobson ring.*

Proof. If k is finite, then so is A, and the conclusion is clear. Hence, we may assume that k is infinite. In particular, card k = card $k^\times$.

We first show that $J(A)$ is a nil ideal (i.e., all its elements are nilpotent). Consider $c \in J(A)$. Then $c - \alpha = -\alpha(1 - \alpha^{-1}c)$ is invertible for all $\alpha \in k^\times$. Thus, we obtain card $k^\times$ elements $(c - \alpha)^{-1}$ in A. Since $\dim_k A <$ card $k^\times$, the above elements must be linearly dependent over k, from which it follows that c is algebraic over k. Now

$$c^n + \beta_{n-1}c^{n-1} + \cdots + \beta_{m+1}c^{m+1} + \beta_m c^m = 0$$

for some $n \geq m \geq 0$ and some $\beta_i \in k$ with $\beta_m \neq 0$. Hence, $(1 + d)c^m = 0$ where $d = \beta_m^{-1}(c^{n-m} + \beta_{n-1}c^{n-m-1} + \cdots + \beta_{m+1}c)$. Since $d \in J(A)$, it follows that $c^m = 0$.

Now let P be a prime ideal of A. If $J(A/P)$ were nonzero, it would be essential as a right ideal of the prime noetherian ring A/P, and so it would contain a regular element. On the other hand, the previous paragraph (applied to the ring A/P) shows that $J(A/P)$ is nil, and so we have a contradiction. Therefore $J(A/P) = 0$, as desired. $\square$

The proposition shows that in case k is uncountable, all our standard quantized coordinate rings (and quantized enveloping algebras) are Jacobson rings. In fact, this conclusion remains valid when k is countable, but takes more effort to prove. We will discuss some results that apply to this case in the following section.

THE NONCOMMUTATIVE NULLSTELLENSATZ

The next step is to find some general conditions which ensure that primitive ideals are rational. To aid this task, the following lemma lets us reduce from centres of Goldie quotient rings of prime factors to centres of endomorphism rings of simple modules. Note that in the situation of the lemma, it is rather trivial to embed Fract $Z(R/P)$ into $Z(\text{End}_R M)$. The usefulness of the lemma lies in its control over $Z(\text{Fract } R/P)$, since, in general, the centre of the Goldie quotient ring of a prime noetherian ring is larger than the quotient field of the centre of the ring. For example, let $A = \mathcal{O}_q(k^3)$ with q not a root of unity. It is easily checked that $Z(A) = k$, whereas $x_1 x_2^{-1} x_3 \in Z(\text{Fract } A)$; thus, $Z(\text{Fract } A)$ is a transcendental extension of Fract $Z(A)$.

II.7.13. Lemma. *Let R be a left noetherian ring and P the annihilator of a simple left R-module M. Then there exists a $Z(R/P)$-algebra embedding*

$$\phi : Z(\text{Fract } R/P) \longrightarrow Z(\text{End}_R M).$$

In particular, if R is a k-algebra, then ϕ is a k-algebra embedding.

Proof. We may assume that $P = 0$. Set $Q = \operatorname{Fract} R$ and $E = \operatorname{End}_R M$, and let $\mathcal{Z}$ denote the set of all pairs $(r, s) \in R \times R$ such that s is regular and $s^{-1}r \in Z(Q)$. We first observe that if $(r, s) \in \mathcal{Z}$, then

$$sar = sas(s^{-1}r) = s(s^{-1}r)as = ras$$

for all $a \in R$. In particular, $sr = rs$.

We next claim that for any $(r, s) \in \mathcal{Z}$, there is a unique $f \in Z(E)$ such that $f(sm) = rm$ for all $m \in M$. Since M is faithful, we may choose $m_0 \in M$ such that $sm_0 \neq 0$. Then sm_0 generates M, and so there can be at most one $f \in E$ satisfying $f(sm_0) = rm_0$.

Our first task is to find some $f \in E$ such that $f(sm_0) = rm_0$. Since sm_0 generates M, such an f will exist provided $\operatorname{ann}_R(sm_0) \subseteq \operatorname{ann}_R(rm_0)$. If not, there is some $a \in R$ with $asm_0 = 0$ but $arm_0 \neq 0$. Then arm_0 generates M, and so $barm_0 = m_0$ for some $b \in R$. Now $x = ba$ is an element of R such that $xsm_0 = 0$ and $xrm_0 = m_0$. But then $sm_0 = sxrm_0 = rxsm_0 = 0$, a contradiction. Thus there does exist $f \in E$ such that $f(sm_0) = rm_0$, as desired.

Now any $m \in M$ has the form $m = csm_0$ for some $c \in R$, whence

$$f(sm) = sf(csm_0) = scf(sm_0) = scrm_0 = rcsm_0 = rm.$$

Further, for any $g \in E$ we have

$$gf(sm_0) = g(rm_0) = rg(m_0) = f(sg(m_0)) = fg(sm_0),$$

and so $gf = fg$. Thus $f \in Z(E)$, and the claim is proved.

Next, consider (r_1, s_1) and (r_2, s_2) in $\mathcal{Z}$ such that $s_1^{-1}r_1 = s_2^{-1}r_2$, and let $f_1, f_2 \in Z(E)$ correspond to these pairs as in the claim. Since r_1 and s_1 commute, we have $r_1 s_1^{-1} = s_2^{-1}r_2$, and so $s_2 r_1 = r_2 s_1$. Hence,

$$f_1(s_2 s_1 m) = s_2 f_1(s_1 m) = s_2 r_1 m = r_2 s_1 m = f_2(s_2 s_1 m)$$

for all $m \in M$. Since $s_2 s_1 \neq 0$, there is some $m' \in M$ such that $s_2 s_1 m' \neq 0$, and so we conclude that $f_1 = f_2$.

Therefore there exists a well-defined function $\phi : Z(Q) \to Z(E)$ such that $\phi(s^{-1}r)(sm) = rm$ for all $(r, s) \in \mathcal{Z}$ and $m \in M$. It is easily checked that ϕ is a unital homomorphism of $Z(R)$-algebras. Since $Z(Q)$ is a field, ϕ must thus be an embedding. $\square$

II.7.14. Definition. Let A be a noetherian k-algebra. We say that A *satisfies the Nullstellensatz over k* provided

(a) A is a Jacobson ring.

(b) The endomorphism ring of every irreducible A-module is algebraic over k.

(The definition of this concept for non-noetherian k-algebras is slightly different – see [**158**, 9.1.4].)

II.7.15. Lemma. *Let A be a noetherian k-algebra satisfying the Nullstellensatz over k. Then for all prime ideals of A, the following implications hold:*

$$\text{locally closed} \quad \Longrightarrow \quad \text{primitive} \quad \Longrightarrow \quad \text{rational.}$$

Proof. The first implication is proved in Lemma II.7.11, and the second follows immediately from Lemma II.7.13. □

As before, cardinality restrictions are the easiest way to ensure the Nullstellensatz:

II.7.16. Proposition. *Let A be a noetherian k-algebra such that $\dim_k A <$ card k. Then A satisfies the Nullstellensatz over k.*

Proof. Proposition II.7.12 shows that A is Jacobson.

Now let M be an irreducible A-module and set $E = \text{End}_A M$. Choose a nonzero element $m \in M$, and note that since $M = Am$, the map $f \mapsto f(m)$ provides a k-linear embedding of E into M. Again using $M = Am$, we thus have

$$\dim_k E \leq \dim_k M \leq \dim_k A < \text{card } k.$$

Now consider any $f \in E$. Obviously if $f \in k \cdot 1$, then f is algebraic over k. Otherwise, $f - \alpha$ is invertible for all $\alpha \in k$. Since $\dim_k E < \text{card } k$, the elements $(f - \alpha)^{-1}$ for $\alpha \in k$ must be linearly dependent over k, from which it follows that f is algebraic over k. Therefore E is algebraic over k. □

The proposition above applies to quantized coordinate rings and quantized enveloping algebras over uncountable fields. To allow for smaller fields, the following theorems are useful.

II.7.17. Theorem. *Let A be a k-algebra containing a sequence of subalgebras $A_0 = k \subseteq A_1 \subseteq \cdots \subseteq A_t = A$ such that for all $i > 0$, either A_i is a finitely generated A_{i-1}-module on each side, or A_i is generated by A_{i-1} together with an element z_i such that $A_{i-1}z_i + A_{i-1} = z_i A_{i-1} + A_{i-1}$.*

Then A is noetherian and satisfies the Nullstellensatz over k.

Proof. Noetherianness follows from [**158**, Theorem 1.2.10]. The Nullstellensatz is a special case of [**158**, Theorem 9.4.21]. □

II.7.18. Corollary. *The algebras*

$\mathcal{O}_q(k^n)$	$\mathcal{O}_q((k^\times)^n)$	$\mathcal{O}_q(M_n(k))$	$\mathcal{O}_q(GL_n(k))$
$\mathcal{O}_q(SL_n(k))$	$\mathcal{O}_q(k^n)$	$\mathcal{O}_q((k^\times)^n)$	$\mathcal{O}_{\lambda,\boldsymbol{p}}(M_n(k))$
$\mathcal{O}_{\lambda,\boldsymbol{p}}(GL_n(k))$	$\mathcal{O}_{\lambda,\boldsymbol{p}}(SL_n(k))$	$A_n^{Q,\Gamma}(k)$	

all satisfy the Nullstellensatz over k.

Proof. Each of the algebras listed contains a sequence of subalgebras satisfying the hypotheses of Theorem II.7.17 (Exercise II.7.E). □

II.7.19. Theorem. *Let A be a k-algebra with a $\mathbb{Z}_{\geq 0}$-filtration $(A_d)_{d \geq 0}$ such that $A_0 = k \cdot 1$ and all A_d are finite dimensional over k. Assume that $\mathrm{gr}\, A$ can be generated by homogeneous elements $y_1, \ldots, y_m$ satisfying relations $y_i y_j = q_{ij} y_j y_i$ for $i > j$, for some $q_{ij} \in k^\times$. Then A is noetherian and satisfies the Nullstellensatz over k.*

Proof. Obviously $\mathrm{gr}\, A$ is noetherian, since it is a homomorphic image of $\mathcal{O}_q(k^m)$, whence A is noetherian by Theorem I.12.13. That A satisfies the Nullstellensatz is the first part of [**156**, Theorem 3.8]. $\square$

II.7.20. Corollary. *$\mathcal{O}_q(G)$ satisfies the Nullstellensatz over k.*

Proof. As we saw in the proof of Theorem I.8.18, there are k-algebra generators $u_1, \ldots, u_m$ for A and scalars $q_{ij} \in k^\times$ and $\alpha_{ij}^{st}, \beta_{ij}^{st} \in k$ such that

$$u_i u_j = q_{ij} u_j u_i + \sum_{s=1}^{j-1} \sum_{t=1}^{m} (\alpha_{ij}^{st} u_s u_t + \beta_{ij}^{st} u_t u_s)$$

for $1 \leq j < i \leq m$. The proof of Proposition I.8.17 then shows that A has a nonnegative filtration $(A_d)_{d \geq 0}$ such that $\mathrm{gr}\, A$ is generated as a k-algebra by homogeneous elements $y_1, \ldots, y_m$ satisfying $y_i y_j = q_{ij} y_j y_i$ for all $i > j$. Observe also that $A_0 = k \cdot 1$ and that $\dim_k(A_d) \leq m^d$ for all d. Therefore Theorem II.7.19 applies. $\square$

NOTES

As noted in (II.7.9), the Dixmier-Moeglin equivalence grew out of Dixmier's and Moeglin's work on enveloping algebras [**49**, **161**]. The cardinality arguments used in Propositions II.7.12 and II.7.16 are due to Amitsur [**5**].

EXERCISES

Exercise II.7.A. Let $A = \mathcal{O}_q(k^2)$ where q is not a root of unity. Show that $A/A(xy-1)$ is a faithful irreducible A-module, whence A is a primitive ring. [Hint: The cosets of the elements $1, x, y, x^2, y^2, \ldots$ form a basis for this module.] $\square$

Exercise II.7.B. If A is a commutative affine k-algebra, show that the rational primes of A are precisely the maximal ideals. $\square$

Exercise II.7.C. Let $A = \mathcal{O}_q(k^2)$ where q is not a root of unity. Show that $Z(\mathrm{Fract}\, A) = k$. [Hint: First show that $Z(\mathcal{O}_q((k^\times)^2)) = k$, and then use the fact that $\mathcal{O}_q((k^\times)^2)$ is simple (Exercise II.1.B).] $\square$

Exercise II.7.D. Show that $\mathcal{O}_q(k^2)$ is a Jacobson ring. [Hint: If q is not a root of unity, use Exercise II.7.A; if it is a root of unity, note that $\mathcal{O}_q(k^2)$ is a finitely generated module over its centre and consult PI theory.] $\square$

Exercise II.7.E. Show that the algebras listed in Corollary II.7.18 satisfy the hypotheses of Theorem II.7.17. ☐

Exercise II.7.F. Let $A = \mathcal{O}_q(k^n)$ with q not a root of unity, and let P be a primitive ideal of A with finite codimension. The following steps lead to showing that A/P is a field, that $A/P = k$ if k is algebraically closed, and to identifying P in the latter case.

(a) For $j = 1, \ldots, n$, show that if $x_j \notin P$, then the coset $x_j + P$ is invertible in A/P. [Hint: Recall that x_j is a normal element of A.]

(b) Show that at most one $x_j \notin P$. [Hint: Recall from Exercise II.1.B that $\mathcal{O}_q((k^\times)^2)$ is a simple ring.]

(c) Show that A/P is a homomorphic image of a polynomial algebra $k[x_j]$, and conclude that A/P is indeed a field. [The latter conclusion also follows from the fact that all primes of A are completely prime (Corollary II.6.10).]

(d) Now assume that k is algebraically closed. Show that

$$P = \langle x_i \mid i \neq j \rangle + \langle x_j - \lambda \rangle$$

for some j and some $\lambda \in k$. Conclude that all finite dimensional irreducible A-modules have dimension 1. ☐

Exercise II.7.G. Let $A = \mathcal{O}_q(GL_n(k))$ with q not a root of unity, and let P be a primitive ideal of A with finite codimension. Analyze A/P as follows.

(a) The first major step is to show that $X_{ij} \in P$ for all $i > j$. Assuming not, list these elements in the order

$$X_{n1}, X_{n-1,1}, X_{n,2}, X_{n-2,1}, X_{n-1,2}, X_{n3}, \ldots, X_{21}, X_{32}, \ldots, X_{n,n-1},$$

and let X_{lm} be the first element in this list which is not in P. Show that there exist $\alpha_{ij} \in k^\times$ such that $X_{lm}X_{ij} \equiv \alpha_{ij}X_{ij}X_{lm} \pmod{P}$ for all i, j.

(b) Use part (a) to see that X_{lm} is invertible modulo P.

(c) Show that $X_{lj} \in P$ for all $j \neq m$. [Hint: If $f \in k[t]$ is the minimal polynomial for the coset $X_{lj} + P$, multiply the equation $f(X_{lj} + P) = 0$ by X_{lm} and use parts (a) and (b).] Similarly, show that $X_{im} \in P$ for all $i \neq l$.

(d) Show that $X_{ij} \in P$ for all $i < l$ and $j > m$. [Hint: Use the relation between X_{im} and X_{lj}.]

(e) Observe that we now have $X_{l-1,j} \in P$ for all j. However, the *quantum Laplace expansion* of D_q along the $(l-1)$-st row [**180**, Corollary 4.4.4] shows that D_q lies in the ideal generated by $X_{l-1,1}, \ldots, X_{l-1,n}$. Conclude from this contradiction that indeed $X_{ij} \in P$ for all $i > j$.

(f) Give a symmetric argument to show that $X_{ij} \in P$ for all $i < j$.

(g) Show that A/P is a homomorphic image of the Laurent polynomial algebra $k[X_{11}^{\pm 1}, \ldots, X_{nn}^{\pm 1}]$, and conclude that A/P is a field. [Again, this conclusion also follows from Corollary II.6.10.]

(h) Now assume that k is algebraically closed. Show that

$$P = \langle X_{ij} \mid i \neq j \rangle + \langle X_{ii} - \lambda_i \mid i = 1, \ldots, n \rangle$$

for some $\lambda_i \in k^\times$. Conclude that all finite dimensional irreducible A-modules have dimension 1. □

CHAPTER II.8

THE DIXMIER-MOEGLIN EQUIVALENCE

The main goal of this chapter is to establish sufficient conditions for the Dixmier-Moeglin equivalence in a context that can be applied to quantized coordinate rings. By Lemma II.7.15, we have the implications

$$\text{locally closed} \quad \Longrightarrow \quad \text{primitive} \quad \Longrightarrow \quad \text{rational}$$

for prime ideals of a noetherian k-algebra satisfying the Nullstellensatz. In the original context of enveloping algebras, closing the loop – i.e., proving that rational primes are locally closed – was the most difficult of the three implications. In the context of quantized coordinate rings, we proceed by taking advantage of the stratifications which we have not yet exploited relative to primitive ideals. This adds another link to the chain above, and provides a third criterion for primitivity.

More precisely, the main result of the chapter (Theorem II.8.4) shows that under the conditions of the Stratification Theorem, if H-spec A is finite and A satisfies the Nullstellensatz over k, then not only does A satisfy the Dixmier-Moeglin equivalence, but a prime ideal of A is primitive if and only if it is maximal within its H-stratum. This theorem applies to all our main examples when appropriate parameters are non-roots of unity (Corollary II.8.5).

The second goal of the chapter is to establish symmetry in the primitive spectrum of an algebra A as above, via the action of H. Observe that the H-strata of prim A are stable under H; thus, the most one could hope for would be that each H-stratum of prim A consist of a single H-orbit. We prove this (under hypotheses as above) when k is algebraically closed (Theorem II.8.14). One can also obtain this result from a very general theorem of Moeglin-Rentschler and Vonessen concerning rational actions of arbitrary affine algebraic groups (Theorem II.8.9), which we discuss briefly for the sake of general interest.

Maximality within H-strata

II.8.1. Example. Let $A = \mathcal{O}_q(k^2)$ with k algebraically closed and q not a root of unity, and let $H = (k^\times)^2$ act on A as in Example II.1.6(a). In Example II.2.3(a), we saw that there are four H-strata in spec A, as pictured in Diagram II.8.1 below.

$$
\begin{array}{ccc}
\boxed{\begin{array}{c} - - -\ \langle x, y - \beta \rangle\ - - - \\[2pt] (\beta \in k^\times) \\[20pt] \langle x \rangle \end{array}} &
\boxed{\begin{array}{c} \langle x, y \rangle \end{array}} &
\boxed{\begin{array}{c} - - -\ \langle x - \alpha, y \rangle\ - - - \\[2pt] (\alpha \in k^\times) \\[20pt] \langle y \rangle \end{array}} \\[30pt]
& \boxed{\langle 0 \rangle} &
\end{array}
$$

DIAGRAM II.8.1. H-STRATA IN $\operatorname{spec}\mathcal{O}_q(k^2)$

Now let us place the primitive ideals of A, which we determined in Example II.7.2, in this picture. Two of the H-strata here are singletons, and the single prime in each is primitive. The other two strata are in inclusion-preserving bijection with $\operatorname{spec} k[z^{\pm 1}]$, and the primitive ideals within these strata match up with the primitive ideals of $k[z^{\pm 1}]$, that is, the maximal ideals.

To summarize: The primitive ideals of A are precisely those prime ideals which are maximal in their H-strata.

The example above suggests 'maximality within H-stratum' as an additional criterion for primitivity. One relation of this condition with the other conditions we have been studying is given in the following easy lemma.

II.8.2. Lemma. *Let H be a group acting by automorphisms on a ring R. If H-spec R is finite, then any prime of R which is maximal within its H-stratum in spec R is locally closed.*

Proof. Let $J \in H$-spec R, and let P be a maximal element of $\operatorname{spec}_J R$. If J is maximal within H-spec R, then all primes containing J lie in $\operatorname{spec}_J R$. In this case, P is a maximal ideal of R and so is clearly locally closed.

Now assume that J is not maximal within H-spec R, let $K_1, \ldots, K_n$ be the H-primes properly containing J, and observe that the ideal $K = K_1 \cap \cdots \cap K_n$ properly contains J (because $K_1 K_2 \cdots K_n \not\subseteq J$). Since $(P : H) = J$, it follows that $K \not\subseteq P$. If Q is any prime properly containing P, then $Q \notin \operatorname{spec}_J R$ because P is maximal within that H-stratum. Since $Q \supseteq P \supseteq J$, the H-prime $(Q : H)$ must properly contain J, whence $(Q : H) = K_i \supseteq K$ for some i. Thus all primes properly containing P must contain K, and so the intersection of these primes is larger than P. Therefore P is locally closed in $\operatorname{spec} R$. $\square$

Combining Lemmas II.7.15 and II.8.2, we see that to establish the Dixmier-Moeglin equivalence for a noetherian k-algebra A satisfying the Nullstellensatz and with a group H acting as automorphisms on A such that H-spec A is finite, it suffices to find conditions which ensure that rational primes are maximal in their H-strata.

For application to quantized coordinate rings, the appropriate conditions are those of the Stratification Theorem – H should be a k-torus acting rationally on A, as in the following proposition. Results of this type have also been proved for rational actions of more general algebraic groups, as we discuss in the following section.

II.8.3. Proposition. *Let A be a noetherian k-algebra, with k infinite, let H be a k-torus acting rationally on A by k-algebra automorphisms, and let $J \in H$-spec A. Then every rational prime in $\mathrm{spec}_J A$ is maximal in $\mathrm{spec}_J A$. Moreover, if there exists a rational prime in $\mathrm{spec}_J A$, then the field $Z(\mathrm{Fract}\, A/J)^H$ is algebraic over k, that is, J is H-rational in the sense of (II.6.1).*

Proof. We may obviously assume that $J = 0$ and that there exists at least one rational prime $P \in \mathrm{spec}_0 A$. By parts (b) and (c) of the Stratification Theorem (II.2.13), the set $\mathcal{E}_0$ of regular H-eigenvectors is a denominator set in A, and $\mathrm{spec}_0 A$ is homeomorphic to $\mathrm{spec}\, A_0$ via localization and contraction, where $A_0 = A[\mathcal{E}_0^{-1}]$. Thus $PA_0 \in \mathrm{spec}\, A_0$ and to see that P is maximal in $\mathrm{spec}_0 A$ it suffices to show that PA_0 is a mxaimal ideal of A_0. Since $\mathrm{Fract}(A_0/PA_0) \cong \mathrm{Fract}(A/P)$ as k-algebras (cf. Exercise II.6.D), the rationality of P implies that $Z(\mathrm{Fract}\, A_0/PA_0)$ is algebraic over k.

Theorem II.2.13(e) says that $Z(A_0)$ is a Laurent polynomial ring over the field $K = Z(\mathrm{Fract}\, A)^H$. Set $Q = PA_0 \cap Z(A_0)$. Then K embeds in the domain $Z(A_0)/Q$, which in turn embeds in $Z(\mathrm{Fract}\, A_0/PA_0)$. Thus K and $Z(A_0)/Q$ are algebraic over k. In particular, $Z(A_0)/Q$ must be a field, i.e., Q is a maximal ideal of $Z(A_0)$. Therefore we conclude from Theorem II.2.13(d) that PA_0 is a maximal ideal of A_0, as desired. $\square$

II.8.4. Theorem. *Let A be a noetherian k-algebra, with k infinite, and let H be a k-torus acting rationally on A by k-algebra automorphisms. Assume that H-spcc A is finite, and that A satisfies the Nullstellensatz over k. Then A satisfies the Dixmier-Moeglin equivalence, and the primitive ideals of A are precisely the primes maximal in their H-strata:*

$$\mathrm{prim}\, A = \{\text{locally closed prime ideals}\}$$
$$= \{\text{rational prime ideals}\}$$
$$= \bigsqcup_{J \in H\text{-spec}\, A} \{\text{maximal elements of}\, \mathrm{spec}_J A\}.$$

Proof. Lemmas II.7.15 and II.8.2 and Proposition II.8.3. $\square$

II.8.5. Corollary. *Let A be one of the following algebras:*

    (a) $\mathcal{O}_q(k^n)$, $\mathcal{O}_q((k^\times)^n)$, $\mathcal{O}_q(k^n)$ or $\mathcal{O}_q((k^\times)^n)$. *No restrictions on q or (q_{ij}) are needed, but assume that k contains a non-root of unity.*

    (b) $\mathcal{O}_q(M_n(k))$, $\mathcal{O}_q(GL_n(k))$, *or* $\mathcal{O}_q(SL_n(k))$, *where q is not a root of unity.*

(c) $\mathcal{O}_{\lambda,\boldsymbol{p}}(M_n(k))$, $\mathcal{O}_{\lambda,\boldsymbol{p}}(GL_n(k))$, or $\mathcal{O}_{\lambda,\boldsymbol{p}}(SL_n(k))$, where λ is not a root of unity.

(d) $A_n^{Q,\Gamma}(k)$, where $q_1, \ldots, q_n$ are non-roots of unity.

(e) $\mathcal{O}_q(G)$, where q is not a root of unity.

Let H be the k-torus acting on A described in (II.1.14–18).

Then A satisfies the Dixmier-Moeglin equivalence, and the primitive ideals of A are precisely the primes maximal in their H-strata.

Proof. By Exercise II.2.G, the action of H on A is rational, and H-spec A is finite by Theorem II.5.14, Theorem II.5.17, or Corollary II.4.12. (Exercise II.1.J shows that finiteness of H-spec passes from $\mathcal{O}_q(k^n)$ and $\mathcal{O}_q(k^n)$ to $\mathcal{O}_q((k^\times)^n)$ and $\mathcal{O}_q((k^\times)^n)$.) Further, A satisfies the Nullstellensatz over k by Corollary II.7.18 or Corollary II.7.20. Therefore the result follows from Theorem II.8.4. $\square$

II.8.6. Example. Let $A = \mathcal{O}_q(SL_2(k))$ with k algebraically closed and q not a root of unity, and let $H = (k^\times)^2$ act on A as in Example II.1.6(b). In Example II.2.3(b), we saw that there are four H-strata in spec A, as pictured in Diagram II.8.6 below.

$$- - - \langle \bar{a} - \lambda, \bar{b}, \bar{c}, \bar{d} - \lambda^{-1} \rangle - - -$$
$$(\lambda \in k^\times)$$

$$\langle \bar{b}, \bar{c} \rangle$$

$$\langle \bar{b} \rangle \qquad - - - - - - \langle \bar{b} - \lambda \bar{c} \rangle - - - - - - \qquad \langle \bar{c} \rangle$$
$$(\lambda \in k^\times)$$

$$\langle 0 \rangle$$

DIAGRAM II.8.6. H-STRATA IN spec $\mathcal{O}_q(SL_2(k))$

Corollary II.8.5 allows us to immediately read off the primitive ideals of A from this picture – all prime ideals of A are primitive except for $\langle \bar{b}, \bar{c} \rangle$ and $\langle 0 \rangle$.

II.8.7. Example. Now let $A = \mathcal{O}_q(GL_2(k))$ with k algebraically closed and q not a root of unity, and let $H = (k^\times)^4$ act on A as in Example II.1.6(d). By Exercise II.2.M, there are only four H-primes in A, namely $\langle 0 \rangle$, $\langle b \rangle$, $\langle c \rangle$, $\langle b, c \rangle$, and so there are again just four H-strata in spec A. They can be pictured as in Diagram II.8.7 below (Exercise II.8.B), where two groups of primes have been left unlabelled.

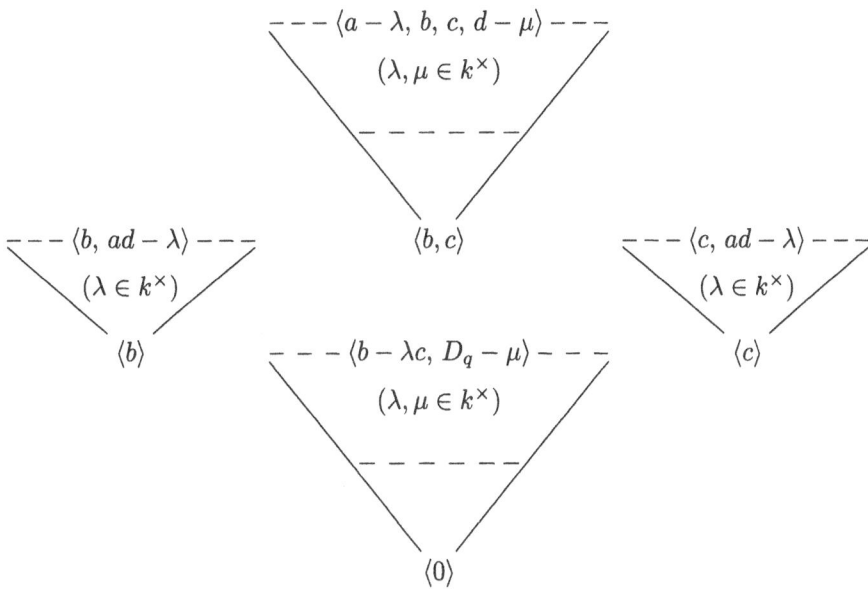

DIAGRAM II.8.7. H-STRATA IN $\operatorname{spec} \mathcal{O}_q(GL_2(k))$

Thus, by Corollary II.8.5, the primitive ideals of A are as follows:

$$\langle a - \lambda, b, c, d - \mu \rangle \qquad\qquad (\lambda, \mu \in k^\times)$$
$$\langle b, ad - \lambda \rangle \qquad\qquad (\lambda \in k^\times)$$
$$\langle c, ad - \lambda \rangle \qquad\qquad (\lambda \in k^\times)$$
$$\langle b - \lambda c, D_q - \mu \rangle \qquad\qquad (\lambda, \mu \in k^\times).$$

THE MOEGLIN-RENTSCHLER-VONESSEN TRANSITIVITY THEOREM

Although it will not be needed for our study of quantum algebras, we detour for a bit to show that a version of Theorem II.8.4 can be proved with the k-torus H replaced by an arbitrary algebraic group, at least when k is algebraically closed. This general result is based on a deep theorem which states that H acts transitively on certain sets of rational primes, i.e., each of these sets consists of a single H-orbit.

II.8.8. Definitions. Given a noetherian k-algebra A, let $\operatorname{Rat} A$ denote the collection of all rational prime ideals of A, equipped with the relative Zariski topology. If H is a group acting by k-algebra automorphisms on A, then there is an H-stratification

$$\operatorname{Rat} A = \bigsqcup_{J \in H\text{-spec } A} \operatorname{Rat}_J A,$$

where $\mathrm{Rat}_J A = (\mathrm{Rat}\, A) \cap (\mathrm{spec}_J A)$. Note that H permutes each of the H-strata $\mathrm{Rat}_J A$; of course, some of these sets may be empty. The theorem below gives sufficient conditions under which H acts transitively on the H-strata of $\mathrm{Rat}\, A$; this is equivalent to saying that the H-orbits within $\mathrm{Rat}\, A$ coincide with the nonempty H-strata.

II.8.9. Transitivity Theorem. *Let A be a noetherian k-algebra, with k algebraically closed, and let H be an affine algebraic group over k, acting rationally on A by k-algebra automorphisms. Then H acts transitively on each H-stratum in $\mathrm{Rat}\, A$.*

Proof. See [**162**, Théorème 2.12(ii)] for the characteristic zero case and [**208**, Theorem 2.2] for the general case. The hypotheses of these results do not require A to be noetherian, only that suitable factor rings are Goldie. $\square$

The Transitivity Theorem is much more accessible in the case when H is a k-torus, as we shall see in the following section (Proposition II.8.13).

II.8.10. Theorem. *Let A be a noetherian k-algebra, with k algebraically closed, and let H be an affine algebraic group over k, acting rationally on A by k-algebra automorphisms. Assume that H-spec A is finite, and that A satisfies the Nullstellensatz over k. Then A satisfies the Dixmier-Moeglin equivalence, and the primitive ideals of A are precisely the primes maximal in their H-strata.*

Proof. By Lemmas II.7.15 and II.8.2, we have

$$\text{maximal in } H\text{-stratum} \quad \Longrightarrow \quad \text{locally closed}$$
$$\Longrightarrow \quad \text{primitive} \quad \Longrightarrow \quad \text{rational}$$

in spec A. Hence, it suffices to show that any rational prime ideal P in A is maximal within its H-stratum in spec A. (This has been proved by Vonessen in the same generality as the Transitivity Theorem (i.e., without assuming the Nullstellensatz) [**208**, Theorem 2.3], but we prefer to apply the Transitivity Theorem directly, as follows.)

Let $J = (P : H)$, so that $P \in \mathrm{spec}_J A$, and choose a prime ideal $Q \supsetneq P$ maximal within $\mathrm{spec}_J A$. The implications given above show that Q must be rational. Now $P, Q \in \mathrm{Rat}_J A$, and so Theorem II.8.9 says that there exists $h \in H$ such that $h(Q) = P$. Since $h(Q) = P \subseteq Q$, Exercise II.2.H shows that $h(Q) = Q$. Thus $P = Q$, and therefore P is maximal within $\mathrm{spec}_J A$, as desired. $\square$

Under suitable conditions, Theorem II.8.10 extends to algebras over non-algebraically closed fields, as follows (we do not define all the terms here).

II.8.11. Theorem. *Let A be a noetherian k-algebra, with k infinite. Assume that $A \otimes \overline{k}$ is noetherian, and that $A \otimes \overline{k}$ satisfies the Nullstellensatz over $\overline{k}$. Let H be the group of k-rational points of a k-affine algebraic group $\overline{H}$, suppose that H acts* *rationally on A by k-algebra automorphisms, and assume that H-spec A is finite. Finally, assume that either k is perfect or $\overline{H}$ is reductive.*

Then A satisfies the Dixmier-Moeglin equivalence, and the primitive ideals of A are precisely the prime ideals maximal within their H-strata.

Proof. [**80**, Theorem 2.12] □

H-ORBITS OF PRIMITIVE IDEALS

II.8.12. Examples. Let k be algebraically closed, and let $q \in k^\times$ be a non-root of unity.

(a) Let $A = \mathcal{O}_q(k^2)$ and let $H = (k^\times)^2$ act on A as in Example II.1.6(a). As observed in that example, there are just six H-orbits in spec A. Recall from Example II.7.2 that all primes of A except $\langle x \rangle$ and $\langle y \rangle$ are primitive. Hence, there are four H-orbits in prim A:

$$\{\langle x, \, y - \beta \rangle \mid \beta \in k^\times\} \qquad \{\langle x, y \rangle\} \qquad \{\langle x - \alpha, \, y \rangle \mid \alpha \in k^\times\}$$

$$\{\langle 0 \rangle\}$$

There are also exactly four H-primes in A (Example II.1.11), and thus exactly four H-strata in either spec A (cf. Example II.8.1) or prim A. The H-orbits in prim A given above coincide with the H-strata of prim A. In other words, H acts transitively on each H-stratum in prim A.

(b) Now let $A = \mathcal{O}_q(SL_2(k))$ and let $H = (k^\times)^2$ act on A as in Example II.1.6(b). There are again six H-orbits in spec A and four H-strata (cf. Example II.8.6), and four H-orbits in prim A:

$$\{\langle \overline{a} - \lambda, \, \overline{b}, \, \overline{c}, \, \overline{d} - \lambda^{-1} \rangle \mid \lambda \in k^\times\}$$

$$\{\langle \overline{b} \rangle\} \qquad\qquad \{\langle \overline{b} - \lambda \overline{c} \rangle \mid \lambda \in k^\times\} \qquad\qquad \{\langle \overline{c} \rangle\}$$

We again observe that these H-orbits coincide with the H-strata in prim A.

II.8.13. Proposition. *Let A be a noetherian k-algebra, with k algebraically closed, let H be a k-torus acting rationally on A by k-algebra automorphisms, and let $J \in H$-spec A. Then H acts transitively on the set of rational primes within $\mathrm{spec}_J A$.*

Proof. We may assume that $J = 0$ and that there exist at least two rational primes P_1, P_2 in $\mathrm{spec}_0 A$. By the Stratification Theorem (II.2.13) and Proposition II.8.3, the set $\mathcal{E}_0$ of regular H-eigenvectors is a denominator set in A and P_1, P_2

induce maximal ideals in the localization $A_0 = A[\mathcal{E}_0^{-1}]$. By Theorem II.2.13, the ideals $Q_i = P_i A_0 \cap Z(A_0)$ are maximal ideals of $Z(A_0)$, and $Z(A_0)$ is a Laurent polynomial ring of the form $K[z_1^{\pm 1}, \ldots, z_n^{\pm 1}]$ where $K = (\operatorname{Fract} A)^H$ and the z_i are H-eigenvectors with linearly independent H-eigenvalues g_i. Since k is algebraically closed, $K = k$ by Proposition II.8.3, and so each

$$Q_i = \langle z_1 - \alpha_{i1}, \ldots, z_n - \alpha_{in} \rangle$$

for some $\alpha_{ij} \in k^\times$.

Since $g_1, \ldots, g_n$ are linearly independent elements of $X(H)$, Exercise II.8.C shows that there exists $h \in H$ such that $g_j(h) = \alpha_{1j}\alpha_{2j}^{-1}$ for all j. Then

$$h(z_j - \alpha_{1j}) = g_j(h)z_j - \alpha_{1j} = \alpha_{1j}\alpha_{2j}^{-1}(z_j - \alpha_{2j})$$

for all j, whence $h(Q_1) = Q_2$. Now

$$h(P_1)A_0 \cap Z(A_0) = h(Q_1) = Q_2 = P_2 A_0 \cap Z(A_0),$$

and therefore we conclude from Theorem II.2.13 that $h(P_1) = P_2$. $\square$

II.8.14. Theorem. *Let A be a noetherian k-algebra, with k algebraically closed, and let H be a k-torus acting rationally on A by k-algebra automorphisms. Assume that H-spec A is finite, and that A satisfies the Nullstellensatz over k.*

Then the H-orbits within prim A coincide with the H-strata of prim A. In particular, there are only finitely many H-orbits in prim A.

Proof. Theorem II.8.4 and Proposition II.8.13. $\square$

The final conclusion of the theorem above is sometimes recorded by saying that prim A is *H-finite*.

II.8.15. Corollary. *Let A be one of the algebras*

$$\mathcal{O}_q(k^n), \qquad \mathcal{O}_q((k^\times)^n), \qquad \mathcal{O}_q(M_n(k)), \qquad \mathcal{O}_q(GL_n(k)),$$
$$\mathcal{O}_q(SL_n(k)), \qquad \mathcal{O}_q(k^n), \qquad \mathcal{O}_q((k^\times)^n), \qquad \mathcal{O}_{\lambda,\boldsymbol{p}}(M_n(k)),$$
$$\mathcal{O}_{\lambda,\boldsymbol{p}}(GL_n(k)), \qquad \mathcal{O}_{\lambda,\boldsymbol{p}}(SL_n(k)), \qquad A_n^{Q,\Gamma}(k), \qquad \mathcal{O}_q(G),$$

with non-root of unity restrictions as in Corollary II.8.5. Assume that k is algebraically closed, and let H be the k-torus acting on A described in (II.1.14–18).

Then the H-orbits within prim A are precisely the H-strata $\operatorname{prim}_J A$ for J in H-spec A.

Proof. As in the proof of Corollary II.8.5, A and H satisfy the hypotheses of Theorem II.8.14. $\square$

NOTES

Theorem II.8.4 and most of the applications to quantized coordinate rings (Corollary II.8.5(a)–(d)) were proved by Goodearl and Letzter [80]. The Dixmier-Moeglin equivalence for $\mathcal{O}_q(G)_{\mathbb{C}}$ and the fact that the H-orbits in spec $\mathcal{O}_q(G)$ coincide with the strata prim$_w \mathcal{O}_q(G)$ were proved by Hodges-Levasseur for the case $G = SL_n(\mathbb{C})$ [94, 95] and by Joseph in general [115, 116].

EXERCISES

Exercise II.8.A. Let A be a noetherian k-algebra, with k infinite, and let H be a k-torus acting rationally on A by k-algebra automorphisms.

(a) Show that any locally closed prime of A is maximal in its H-stratum.

(b) Show that an H-prime J of A is H-rational if and only if there exists a rational prime in spec$_J A$. □

Exercise II.8.B. Verify the details of Example II.8.7. □

Exercise II.8.C. Let H be a k-torus, where k is an algebraically closed field, let $g_1, \ldots, g_n$ be linearly independent elements of $X(H)$, and let $\beta_1, \ldots, \beta_n \in k^{\times}$. Show that there exists $h \in H$ such that $g_i(h) = \beta_i$ for all i. [Hint: Identify H with $\mathrm{Hom}(X(H), k^{\times})$.] □

CHAPTER II.9

CATENARITY

The term 'catenarity' refers to a chain condition on prime ideals which reflects the behaviour of chains of irreducible subvarieties of affine algebraic varieties. Since this property holds for the coordinate ring of any affine algebraic variety, we look for it in quantized coordinate rings as well. In this chapter, we discuss the main results to date.

CHAINS OF PRIME IDEALS

II.9.1. Definitions. Recall that the *length* of a chain

$$P_0 \subset P_1 \subset \cdots \subset P_r \tag{C}$$

of prime ideals in a ring R is the number of gaps, namely r. We will sometimes need to say that (C) runs *from P_0 to P_r*. The chain (C) is *saturated* provided no prime can be fitted into any of the gaps, i.e., there do not exist any index i and prime P such that $P_{i-1} \subset P \subset P_i$.

The *height* of a prime ideal Q of R is the supremum of the lengths of all chains of primes contained in Q, that is, chains (C) with $P_r \subseteq Q$. Note that it suffices to consider such chains with $P_r = Q$ and P_0 a minimal prime of R.

We say that R is *catenary* (or that it satisfies the *saturated chain condition for prime ideals*) if for every pair $P \subset Q$ of distinct, comparable primes of R, all saturated chains of primes from P to Q have the same length. Note that the given condition passes from 'looser' to 'tighter' pairs of primes: if $P \subseteq P' \subset Q' \subseteq Q$ are primes of R, and if all saturated chains of primes from P to Q have the same length, then the same holds for chains from P' to Q'. Consequently, to check whether R is catenary, it suffices to consider saturated chains from any minimal prime to any maximal ideal.

II.9.2. The minimal situation for failure of catenarity can be illustrated as follows:

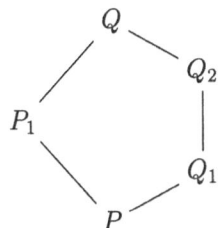

Here $P \subset P_1 \subset Q$ and $P \subset Q_1 \subset Q_2 \subset Q$ are supposed to be saturated chains of primes. There is a famous example of Nagata [169] showing that this can occur even in a commutative noetherian domain. Catenarity typically fails for the enveloping algebra of a semisimple Lie algebra; in the cases of $U(\mathfrak{sl}_4(\mathbb{C}))$ and $U(\mathfrak{sl}_5(\mathbb{C}))$, this can be derived from the information displayed in [17, p. 39]. For some algebras in which the lack of catenarity is easier to establish, see [12].

A classical result is that any commutative affine algebra over a field is catenary (e.g., [154, §14.H, Corollary 3]); this was extended to affine PI algebras by Schelter [193]. It is a famous result of Gabber that the enveloping algebra of any finite dimensional solvable Lie algebra over an algebraically closed field of characteristic zero is catenary [57], and his methods underlie the results below.

Easy examples of catenary rings include any ring with classical Krull dimension 0 or 1, since in such a ring a chain of primes of length 1 is already saturated. It is also easy to see that our standard examples $\mathcal{O}_q(k^2)$ and $\mathcal{O}_q(SL_2(k))$ are catenary, as follows. (The case when q is not a root of unity is the one of interest here, but these algebras are also catenary when q is a root of unity, since then they are PI algebras and Schelter's theorem applies.)

II.9.3. Examples. Let k be algebraically closed and $q \in k^\times$ a non-root of unity.

(a) First let $A = \mathcal{O}_q(k^2)$, and recall the picture of spec A given in Diagram II.1.2. It is clear from this picture that any saturated chain from $\langle 0 \rangle$ to a maximal ideal of A has one of the following forms, where $\alpha, \beta \in k$:

$$\langle 0 \rangle \subset \langle x \rangle \subset \langle x, y - \beta \rangle \qquad \langle 0 \rangle \subset \langle y \rangle \subset \langle x - \alpha, y \rangle.$$

Since these chains all have length 2, we conclude that A is catenary.

(b) Now let $A = \mathcal{O}_q(SL_2(k))$, and recall spec A from Diagram II.1.3. In this case, any saturated chain from $\langle 0 \rangle$ to a maximal ideal of A has the form

$$\langle 0 \rangle \subset \langle \beta \bar{b} - \gamma \bar{c} \rangle \subset \langle \bar{b}, \bar{c} \rangle \subset \langle \bar{a} - \lambda, \bar{b}, \bar{c}, \bar{d} - \lambda^{-1} \rangle$$

for some $\beta, \gamma, \lambda \in k$ with $\lambda \neq 0$ and β, γ not both zero. Again we see that A is catenary.

(c) A third easy example is $A_1^q(k)$ (Exercise II.9.A).

GABBER'S METHOD

Gabber's original method made use of some special features of the structure of solvable enveloping algebras. The homological aspects of his method were abstracted by Björk [15], and an axiomatic version of the full line was given by Goodearl and Lenagan [72]. It involves an interplay between homological conditions and growth properties, controlled by normal elements. Normal elements we have met earlier (Definition II.4.1), and the homological conditions are discussed in Appendix I.15. The required growth properties are given in terms of Gel'fand-Kirillov dimension, denoted GK.dim, for which we refer the reader to [127].

II.9.4. Definition. We say that the prime spectrum of a ring R satisfies *normal separation* (or that spec R is *normally separated*) if for every pair $P \subset Q$ of distinct, comparable primes in R, the ideal Q/P of R/P contains a nonzero normal element of R/P, that is, there exists an element of $Q \setminus P$ which is normal modulo P.

For instance, all the generators x_i in $\mathcal{O}_q(k^n)$ are normal elements. It is not hard to check that spec $\mathcal{O}_q(k^2)$ and spec $\mathcal{O}_q(SL_2(k))$ have normal separation when k is algebraically closed and q is not a root of unity (Exercise II.9.B).

II.9.5. Catenarity Theorem. *Let A be an affine, noetherian, Auslander-Gorenstein, Cohen-Macaulay k-algebra, with $\mathrm{GK.dim}(A) < \infty$. If spec A has normal separation, then A is catenary, and*

$$\mathrm{ht}(Q/P) = \mathrm{GK.dim}(A/P) - \mathrm{GK.dim}(A/Q)$$

for all primes $P \subset Q$ in A. In particular, if A is a prime ring, then

$$\mathrm{ht}(Q) + \mathrm{GK.dim}(A/Q) = \mathrm{GK.dim}(A)$$

for all $Q \in$ spec A.

Proof. [**72**, Theorem 1.6] □

The remainder of this chapter is devoted to applications of the above theorem. Notice that it is a generalisation of the catenarity result for commutative affine k-algebras mentioned in (II.9.2), since any such algebra is a factor of a polynomial k-algebra $k[x_1, \ldots, x_t]$ which satisfies the hypotheses of the theorem.

II.9.6. Definition. The final conclusion of Theorem II.9.5, namely that

$$\mathrm{ht}(Q) + \mathrm{GK.dim}(A/Q) = \mathrm{GK.dim}(A)$$

for all primes Q of A, is known as *Tauvel's height formula*. It was first established by Tauvel for enveloping algebras of finite dimensional solvable Lie algebras [**204**].

We now turn to the problem of verifying the various hypotheses of the Catenarity Theorem in our standard examples.

GK-DIMENSION AND HOMOLOGICAL CONDITIONS

II.9.7. Lemma. *Let $A = k[x_1][x_2; \tau_2, \delta_2] \cdots [x_n; \tau_n, \delta_n]$ be an n-fold iterated skew polynomial algebra over k. Then $\mathrm{gl.dim}(A) \leq n$. If $\tau_i(x_j) \in k^\times x_j$ for $1 \leq j < i \leq n$, then $\mathrm{GK.dim}(A) = n$.*

Proof. Set $A_0 = k$ and $A_i = k[x_1][x_2; \tau_2, \delta_2] \cdots [x_i; \tau_i, \delta_i]$ for $i = 1, \ldots, n$. Since $\mathrm{gl.dim}(A_i) \leq \mathrm{gl.dim}(A_{i-1}) + 1$ for all $i > 0$ [**158**, Theorem 7.5.3], it is clear that $\mathrm{gl.dim}(A) \leq n$.

Now assume that $\tau_i(x_j) \in k^\times x_j$ for $1 \leq j < i \leq n$. Then the set $V_{i-1} = k + kx_1 + \cdots + kx_{i-1}$ is a finite dimensional generating subspace for A_{i-1} such that $\tau_i(V_{i-1}) = V_{i-1}$. By [**97**, Lemma 2.2], $\mathrm{GK.dim}(A_i) = \mathrm{GK.dim}(A_{i-1}) + 1$. Therefore $\mathrm{GK.dim}(A) = n$. □

II.9.8. Lemma. *Let A be an affine noetherian k-algebra and $x \in A$ a regular normal element. If A has a finite dimensional generating subspace V such that $1 \in V$ and $xV = Vx$, then* $\mathrm{GK.dim}(A[x^{-1}]) = \mathrm{GK.dim}(A)$.

Proof. [**133**, Theorem 2]. □

II.9.9. Proposition. *The algebras*

$$\mathcal{O}_q(k^n), \qquad \mathcal{O}_q((k^\times)^n), \qquad \mathcal{O}_q(M_n(k)), \qquad \mathcal{O}_q(GL_n(k)),$$
$$\mathcal{O}_q(SL_n(k)), \qquad \mathcal{O}_q(k^n), \qquad \mathcal{O}_q((k^\times)^n), \qquad \mathcal{O}_{\lambda,\mathbf{p}}(M_n(k)),$$
$$\mathcal{O}_{\lambda,\mathbf{p}}(GL_n(k)), \qquad \mathcal{O}_{\lambda,\mathbf{p}}(SL_n(k)), \qquad A_n^{Q,\Gamma}(k), \qquad \mathcal{O}_q(G)$$

have finite GK-dimension and finite global dimension.

Proof. The cases $\mathcal{O}_q(k^n)$, $\mathcal{O}_q(M_n(k))$, $\mathcal{O}_q(k^n)$, $\mathcal{O}_{\lambda,\mathbf{p}}(M_n(k))$, and $A_n^{Q,\Gamma}(k)$ follow directly from Lemma II.9.7. Then, Lemma II.9.8 and [**158**, Corollary 7.4.3] cover the localizations $\mathcal{O}_q((k^\times)^n)$, $\mathcal{O}_q(GL_n(k))$, $\mathcal{O}_q((k^\times)^n)$, and $\mathcal{O}_{\lambda,\mathbf{p}}(GL_n(k))$.

Finiteness of GK-dimension for $\mathcal{O}_q(SL_n(k))$ and $\mathcal{O}_{\lambda,\mathbf{p}}(SL_n(k))$ follows from the fact that $\mathcal{O}_q(M_n(k))$ and $\mathcal{O}_{\lambda,\mathbf{p}}(M_n(k))$ have finite GK-dimension. Recall from Lemma I.2.9 and Exercise I.2.K that

$$\mathcal{O}_q(GL_n(k)) \cong \mathcal{O}_q(SL_n(k))[z^{\pm 1}]$$
$$\mathcal{O}_{\lambda,\mathbf{p}}(GL_n(k)) \cong \mathcal{O}_{\lambda,\mathbf{p}}(SL_n(k))[z^{\pm 1}].$$

Since global dimension increases in Laurent polynomial extensions [**158**, Theorem 7.5.3], we conclude that $\mathcal{O}_q(SL_n(k))$ and $\mathcal{O}_{\lambda,\mathbf{p}}(SL_n(k))$ have finite global dimension.

That $\mathcal{O}_q(G)$ has finite global dimension was proved in [**21**, Proposition 2.7] for the case that $\mathrm{char}\, k = 0$ and q is transcendental over $\mathbb{Q}$; the general case is similar. By the proofs of Theorem I.8.18 and Proposition I.8.17, $\mathcal{O}_q(G)$ has a nonnegative filtration $(A_d)_{d \geq 0}$ such that $\mathrm{gr}\, \mathcal{O}_q(G)$ is a homomorphic image of some quantum affine space $\mathcal{O}_q(k^m)$. We noted in the proof of Corollary II.7.20 that $\dim_k(A_d) < \infty$ for all d. It now follows from [**127**, Proposition 6.6] that

$$\mathrm{GK.dim}\big(\mathcal{O}_q(G)\big) = \mathrm{GK.dim}\big(\mathrm{gr}\, \mathcal{O}_q(G)\big) \leq \mathrm{GK.dim}\big(\mathcal{O}_q(k^m)\big) = m,$$

where the final equality holds by Lemma II.9.7. □

The actual GK-dimensions of the algebras above (except $\mathcal{O}_q(G)$) can be readily calculated (Exercise II.9.E). Concerning their global dimensions, see Exercise II.9.F. We now address the homological conditions, expanding on the discussion in Appendix I.15.

II.9.10. Lemma. *Let* $A = k[x_1][x_2; \tau_2, \delta_2] \cdots [x_n; \tau_n, \delta_n]$ *be an iterated skew polynomial algebra over* k. *Then* A *is Auslander-regular. Now suppose that for* $1 \le j < i \le n$, *we have* $\tau_i(x_j) \in k^\times x_j$ *and* $\delta_i(x_j) \in \sum_{s,t<i} kx_s x_t$. *Then* A *is also Cohen-Macaulay.*

Proof. Set $A_0 = k$ and $A_i = k[x_1][x_2; \tau_2, \delta_2] \cdots [x_i; \tau_i, \delta_i]$ for $i = 1, \ldots, n$. Auslander-regularity passes from A_{i-1} to A_i by Theorem I.15.3, and therefore A is Auslander-regular.

Now assume that $\tau_i(x_j) \in k^\times x_j$ and $\delta_i(x_j) \in \sum_{s,t<i} kx_s x_t$ for $1 \le j < i \le n$. It follows that each A_{i-1} can be made into a connected graded k-algebra with each x_j homogeneous of degree 1 (Exercise II.9.G), and that each component of this grading is stable under τ_i. Consequently, the Cohen-Macaulay property passes from A_{i-1} to A_i by Lemma I.15.4, and therefore A is Cohen-Macaulay. $\square$

II.9.11. Lemma. *Let* A *be a noetherian, Auslander-regular, Cohen-Macaulay* k-*algebra.*

(a) *If* $z \in A$ *is a central regular element, then* $A/\langle z \rangle$ *is Auslander-Gorenstein and Cohen-Macaulay.*

(b) *Assume that* $A = \bigoplus_{i=0}^\infty A_i$ *is a connected graded* k-*algebra. If* $c \in A$ *is a regular normal element such that* $cA_i = A_ic$ *for all* i, *then* $A[c^{-1}]$ *is Auslander-regular and Cohen-Macaulay.*

Proof. (a) Lemma I.15.4.

(b) Since c is regular and normal, there is a k-algebra automorphism σ of A such that $ca = \sigma(a)c$ for all $a \in A$, and the hypothesis on c implies that $\sigma(A_i) = A_i$ for all i. Hence, the skew polynomial algebra $B = A[y; \sigma^{-1}]$ is Auslander-regular and Cohen-Macaulay by Theorem I.15.3 and Lemma I.15.4. Now $cy - 1 = yc - 1$ is a central regular element in B, and so part (a) implies that $B/\langle cy - 1 \rangle$ is Auslander-Gorenstein and Cohen-Macaulay. Since $B/\langle cy - 1 \rangle \cong A[c^{-1}]$, the latter algebra is Auslander-Gorenstein and Cohen-Macaulay. Moreover, gl.dim$(A[c^{-1}]) \le$ gl.dim$(A) < \infty$ by [**158**, Corollary 7.4.3], and therefore $A[c^{-1}]$ is Auslander-regular. $\square$

Part (a) of the lemma cannot be improved to yield Auslander-regularity for $A/\langle z \rangle$ – for example, take $A = k[x]$ and $z = x^2$.

II.9.12. Proposition. *The algebras*

$$\mathcal{O}_q(k^n), \qquad \mathcal{O}_q((k^\times)^n), \qquad \mathcal{O}_q(M_n(k)), \qquad \mathcal{O}_q(GL_n(k)),$$
$$\mathcal{O}_q(SL_n(k)), \qquad \mathcal{O}_q(k^n), \qquad \mathcal{O}_q((k^\times)^n), \qquad \mathcal{O}_{\lambda,p}(M_n(k)),$$
$$\mathcal{O}_{\lambda,p}(GL_n(k)), \qquad \mathcal{O}_{\lambda,p}(SL_n(k)), \qquad A_n^{Q,\Gamma}(k)$$

are Auslander-regular and Cohen-Macaulay.

Proof. The cases $\mathcal{O}_q(k^n)$, $\mathcal{O}_q(M_n(k))$, $\mathcal{O}_q(k^n)$, and $\mathcal{O}_{\lambda,\mathbf{p}}(M_n(k))$ follow directly from Lemma II.9.10. Then, Lemma II.9.11(a) implies that the factors $\mathcal{O}_q(SL_n(k))$ and $\mathcal{O}_{\lambda,\mathbf{p}}(SL_n(k))$ are Auslander-Gorenstein and Cohen-Macaulay, while it follows from Lemma II.9.11(b) that the localizations $\mathcal{O}_q((k^\times)^n)$, $\mathcal{O}_q(GL_n(k))$, $\mathcal{O}_q((k^\times)^n)$, and $\mathcal{O}_{\lambda,\mathbf{p}}(GL_n(k))$ are Auslander-regular and Cohen-Macaulay. Since $\mathcal{O}_q(SL_n(k))$ and $\mathcal{O}_{\lambda,\mathbf{p}}(SL_n(k))$ have finite global dimension (Proposition II.9.9), they are actually Auslander-regular. We leave the case of $A_n^{Q,\Gamma}(k)$ to Exercise II.9.H. □

POLYNORMALITY

Recall the concepts of a *polynormal sequence* and a *polynormal ideal* from Definitions II.4.1. For example, the ideal $\langle a, b \rangle$ in $\mathcal{O}_q(M_2(k))$ cannot be generated by a set of normal elements, but it is generated by the polynormal sequence b, a.

II.9.13. Proposition. *All ideals in the algebras*

$$\mathcal{O}_q(k^n), \qquad \mathcal{O}_q((k^\times)^n), \qquad \mathcal{O}_q(k^n), \qquad \mathcal{O}_q((k^\times)^n)$$

are polynormal, and consequently the prime spectra of these algebras are normally separated.

Proof. First let $A = \mathcal{O}_q(k^n)$. To see that all ideals of A are polynormal, it suffices to show that for any ideals $I \supset J$ in A, there exists an element $u \in I \setminus J$ which is normal modulo J.

Define the *length* of an element $a \in A$ to be the number of distinct monomials $x^m = x_1^{m_1} x_2^{m_2} \cdots x_n^{m_n}$ appearing in a with nonzero coefficients. Choose an element $u \in I \setminus J$ of minimum length, say length d. Then $u = \alpha_1 x^{s_1} + \alpha_2 x^{s_2} + \cdots + \alpha_d x^{s_d}$ for some nonzero scalars α_p and some distinct n-tuples s_p of nonnegative integers. Fix $t \in \{1, \ldots, n\}$. Each $x^{s_p} x_t = \beta_{pt} x_t x^{s_p}$ where $\beta_{pt} = \prod_{i=1}^n q_{it}^{s_{pi}} \in k^\times$. Hence,

$$u x_t - \beta_{1t} x_t u = \alpha_2 (\beta_{2t} - \beta_{1t}) x_t x^{s_2} + \cdots + \alpha_d (\beta_{dt} - \beta_{1t}) x_t x^{s_d}.$$

This is an element of I with length less than d, and so it lies in J by minimality of d. Thus $u x_t \equiv \beta_{1t} x_t u \pmod{J}$ for all $t = 1, \ldots, n$, and so $u + J$ is normal in A/J, as desired. This completes the proof that all ideals of A are polynormal.

Now let $B = \mathcal{O}_q((k^\times)^n)$. If $I \supset J$ are ideals of B, then $I \cap A \supset J \cap A$ is a proper inclusion of ideals in A. The polynormality just proved for A provides an element $u \in (I \cap A) \setminus (J \cap A)$ which is normal modulo $J \cap A$; in fact, the proof shows that there exist $\beta_1, \ldots, \beta_n \in k^\times$ such that $u x_i - \beta_i x_i u \in J \cap A$ for all i. Hence, $u \in I \setminus J$ and $u x_i^{\pm 1} - \beta_i^{\pm 1} x_i^{\pm 1} u \in J$ for all i, whence $u + J$ is normal in B/J. As above, it follows that all ideals of B are polynormal.

Finally, let $P \subset Q$ be distinct, comparable primes in either A or B. There exists a polynormal sequence of generators for Q, say $u_1, \ldots, u_m$. If i is the least index such that $u_i \notin P$, then u_i is an element of $Q \setminus P$ which is normal modulo P. Therefore spec A and spec B have normal separation. $\square$

Stronger properties than polynormality can be proved for ideals in the quantum tori $\mathcal{O}_q((k^\times)^n)$ and $\mathcal{O}_q((k^\times)^n)$ – see Exercise II.9.I.

II.9.14. Theorem. *The algebras*

$$\mathcal{O}_q(k^n), \qquad \mathcal{O}_q((k^\times)^n), \qquad \mathcal{O}_q(k^n), \qquad \mathcal{O}_q((k^\times)^n)$$

are catenary, and they satisfy Tauvel's height formula.

Proof. These are affine noetherian k-algebras, they have finite GK-dimension by Proposition II.9.9, they are Auslander-regular and Cohen-Macaulay by Proposition II.9.12, and their prime spectra are normally separated by Proposition II.9.13. Therefore the desired conclusions follow from the Catenarity Theorem (II.9.5). $\square$

NORMAL SEPARATION

Quantum affine spaces are the only quantum algebras in which normal separation can be verified as easily as in Proposition II.9.13. This condition has, for example, also been verified in the algebras $A_n^{Q,\Gamma}(k)$ when the q_i are non-roots of unity [72, Theorem 3.12], and hence $A_n^{Q,\Gamma}(k)$ is catenary in that case [72, Theorem 3.13]. See Exercises II.9.A and II.9.J for discussions of these algebras when $n = 1, 2$.

In the situation of the Stratification Theorem, the task of checking normal separation in spec A can be reduced to looking at pairs of comparable H-primes, as shown in the following theorem.

II.9.15. Theorem. *Let A be a noetherian k-algebra, with k infinite, let H be a k-torus acting rationally on A by k-algebra automorphisms, and assume that the following condition holds:*

(*) *For every pair of distinct comparable H-primes $J \subset K$ in A, the factor K/J contains a nonzero normal H-eigenvector.*

Let $J \in H$-spec A, and let $\mathcal{N}_J$ (respectively, $\mathcal{E}_J$) denote the set of all nonzero normal (respectively, regular) H-eigenvectors in A/J.

   (a) *The sets $\mathcal{N}_J$ and $\mathcal{E}_J$ are denominator sets in A/J, the algebras $A_J^\nu = (A/J)[\mathcal{N}_J^{-1}]$ and $A_J = (A/J)[\mathcal{E}_J^{-1}]$ are H-simple, and $A_J^\nu \subseteq A_J$.*

   (b) *spec$_J$ A is homeomorphic to spec A_J^ν via localization and contraction, and spec A_J^ν is homeomorphic to spec $Z(A_J^\nu)$ via contraction and extension.*

   (c) *$Z(A_J^\nu) = Z(A_J)$, and this algebra is a Laurent polynomial ring over the field $Z(A_J^\nu)^H = Z(\operatorname{Fract} A/J)^H$.*

   (d) *spec A has normal separation.*

We refer to condition (*) above by saying that H-spec A *has normal H-separation.*

Proof. In proving parts (a),(b),(c), we may assume that $J = 0$. Theorem II.2.13 shows that $\mathcal{E}_0$ is a denominator set, A_0 is H-simple, and $Z(A_0)$ is a Laurent polynomial ring over $Z(A_0)^H = Z(\text{Fract } A)^H$.

(a) Since $\mathcal{N}_0$ is a multiplicatively closed set consisting of normal elements, it is a denominator set (Exercise II.9.K). Having factored out J, which is a prime ideal (Proposition II.2.9), A is a prime ring. Hence, all nonzero normal elements in A are regular. In particular, $\mathcal{N}_0 \subseteq \mathcal{E}_0$, and so $A_0^\nu \subseteq A_0$.

If I is a nonzero H-ideal of A_0^ν, then $I \cap A$ is a nonzero H-ideal of A. There are primes $Q_1, \ldots, Q_t$ minimal over $I \cap A$ such that $Q_1 Q_2 \cdots Q_t \subseteq I \cap A$, and so the ideals $K_j = (Q_j : H)$ are H-primes containing $I \cap A$ such that $K_1 K_2 \cdots K_t \subseteq I \cap A$. Since $K_j \supseteq I \cap A \neq 0$, condition (*) implies that each K_j contains an element of $\mathcal{N}_0$. Thus, there is an element of $\mathcal{N}_0$ lying in $K_1 K_2 \cdots K_t \subseteq I$, whence $I = A_0^\nu$. Therefore A_0^ν is H-simple.

(b) Condition (*) implies that $\text{spec}_0 A$ equals the set of those primes of A which are disjoint from $\mathcal{N}_0$. Hence, $\text{spec}_0 A \approx \text{spec } A_0^\nu$ via localization and contraction. Since A_0^ν is H-simple, Corollary II.3.9 shows that $\text{spec } A_0^\nu \approx \text{spec } Z(A_0^\nu)$ via contraction and extension. (As in Exercise II.3.A, the induced H-action on A_0^ν is rational.)

(c) It is clear that $Z(A_0^\nu) \subseteq Z(A_0)$. By Theorem II.2.13, we have

$$Z(A_0) = F[z_1^{\pm 1}, \ldots, z_m^{\pm 1}]$$

where $F = Z(\text{Fract } A)^H$ and the z_i are H-eigenvectors. Corollary II.3.9 shows that $Z(\text{Fract } A)^H = Z(A_0^\nu)^H$, and so $F \subseteq Z(A_0^\nu)$.

For $i = 1, \ldots, m$, set $C_i = \{c \in A_0^\nu \mid cz_i \in A_0^\nu\}$, and observe that C_i is an H-ideal of A_0^ν, nonzero because $z_i \in \text{Fract } A = \text{Fract } A_0^\nu$. Since A_0^ν is H-simple, $C_i = A_0^\nu$, and thus $z_i \in A_0^\nu$. Similarly, $z_i^{-1} \in A_0^\nu$ for all i, and therefore we have $Z(A_0) = Z(A_0^\nu)$.

(d) Let $P \subset Q$ be distinct comparable primes of A. If $(P : H) \neq (Q : H)$, then condition (*) yields an H-eigenvector $c \in (Q : H) \setminus (P : H)$ which is normal modulo $(P : H)$. Hence, $c \in Q$ and c is normal modulo P. Further, since c is an H-eigenvector not lying in $(P : H)$, it cannot lie in P. This verifies normal separation in the present case.

Now suppose that $(P : H) = (Q : H)$; there is no loss of generality in assuming that $(P : H) = (Q : H) = 0$. Then P and Q lie in $\text{spec}_0 A$, and part (b) implies that $PA_0^\nu \cap Z(A_0^\nu) \subset QA_0^\nu \cap Z(A_0^\nu)$ are distinct comparable primes of $Z(A_0^\nu)$. In particular, there must exist an element

$$z \in \left(QA_0^\nu \cap Z(A_0^\nu)\right) \setminus PA_0^\nu.$$

Write $z = cd^{-1}$ for some $c \in Q \setminus P$ and some $d \in \mathcal{N}_0$. Since z is central in A_0^ν while d is normal in A, we see that

$$cA = zdA = zAd = Azd = Ac,$$

and so c is normal in A. (Note that it would not suffice to have z just normal in A_0^ν.) Thus c is normal modulo P, and normal separation holds in this case too. $\quad\square$

II.9.16. Examples. Theorem II.9.15 makes it very easy to check normal separation in those of our examples for which we have explicit descriptions of the H-primes.

(a) For instance, suppose that $A = \mathcal{O}_q(k^n)$ with k infinite, and let $H = (k^\times)^n$ act on A as in (II.1.14). It is clear from Exercise II.2.K that each H-prime of A is generated by normal H-eigenvectors. Consequently, H-spec A has normal H-separation, and the theorem then implies that spec A has normal separation.

(b) Now let A be one of the algebras $\mathcal{O}_q(SL_2(k))$, $\mathcal{O}_q(GL_2(k))$, or $\mathcal{O}_q(M_2(k))$, with q not a root of unity, and let H be either $(k^\times)^2$ or $(k^\times)^4$, acting on A as in Examples II.1.6. Descriptions of the H-primes of A are given in Example II.1.13 and Exercises II.2.M, II.2.O, respectively. It is easily checked that each of these H-primes is generated by a polynormal sequence of H-eigenvectors (Exercise II.9.L), and normal H-separation in H-spec A follows. Therefore spec A has normal separation in these three cases too.

The remaining hypotheses of the Catenarity Theorem are all in place – A is affine and noetherian, it has finite GK-dimension by Proposition II.9.9, and it is Auslander-regular and Cohen-Macaulay by Proposition II.9.12. Therefore A is catenary, and it satisfies Tauvel's height formula.

II.9.17. Theorem. *If q is not a root of unity, the prime spectra of $\mathcal{O}_q(M_n(k))$, $\mathcal{O}_q(SL_n(k))$, and $\mathcal{O}_q(GL_n(k))$ have normal separation.*

Proof. First let $A = \mathcal{O}_q(M_n(k))$, and let $H = (k^\times)^{2n}$ act on A as in (II.1.15). By [**33**, Théorème 6.2.1], spec A has normal H-separation, and so it follows from Theorem II.9.15 that spec A is normally separated [**33**, Théorème 6.2.2]. It follows immediately that spec $\mathcal{O}_q(SL_n(k))$ is normally separated. Finally, since $\mathcal{O}_q(GL_n(k))$ is a homomorphic image of $\mathcal{O}_q(SL_{n+1}(k))$ (Exercise II.9.M), we conclude that spec $\mathcal{O}_q(GL_n(k))$ is normally separated. $\quad\square$

II.9.18. Corollary. *Let A be one of the algebras $\mathcal{O}_q(M_n(k))$, $\mathcal{O}_q(SL_n(k))$, or $\mathcal{O}_q(GL_n(k))$, where q is not a root of unity. Then A is catenary, and it satisfies Tauvel's height formula.*

Proof. Propositions II.9.9, II.9.12 and Theorems II.9.17, II.9.5. $\quad\square$

Twisting techniques, as in [**8**], can be used to extend Theorem II.9.17 to the algebras $\mathcal{O}_{\lambda,p}(M_n(k))$, $\mathcal{O}_{\lambda,p}(SL_n(k))$ and $\mathcal{O}_{\lambda,p}(GL_n(k))$ when λ is not a root of unity, and thus to verify Corollary II.9.18 in these cases.

We conclude the chapter with two other normal separation results. In the general case of $\mathcal{O}_q(G)$, the Auslander-regular and Cohen-Macaulay conditions have not been proved, and it remains open whether $\mathcal{O}_q(G)$ is always catenary.

II.9.19. Theorem. *If $A = \mathcal{O}_q(G)$, then spec A has normal separation.*

Proof. Let H denote the torus $\mathrm{Hom}(L, k^\times)$, acting on A as in (II.1.18). This action is rational by Exercise II.2.G. Now let $J \subset K$ be distinct, comparable H-primes in A. By Corollary II.4.12, $J = K_w$ for some $w \in W \times W$ and $\mathrm{spec}_J A = \mathrm{spec}_w A$. Then $J_w \subseteq J \subset K$ but $K \notin \mathrm{spec}_w A$, and so $(K/J_w) \cap \mathcal{E}_w$ must be nonempty. Note that $J \in \mathrm{spec}_w A$ and so $(J/J_w) \cap \mathcal{E}_w$ is empty. Since $\mathcal{E}_w$ consists of normal H-eigenvectors in A/J_w, none of which lie in J/J_w, it follows that K/J contains a nonzero normal H-eigenvector. This shows that H-spec A has normal H-separation. Therefore spec A has normal separation by Theorem II.9.15. $\square$

Because of the isomorphism $\mathcal{O}_q(SL_n(k)) \cong \mathcal{O}_q(SL_n(\mathbb{C}))_k$, Theorem II.9.19 can be used to show that $\mathrm{spec}\,\mathcal{O}_q(SL_n(k))$ has normal separation, provided $\mathrm{char}(k) \neq 2$ and q is not a root of unity. For some time, this was the only known means to prove normal separation in $\mathrm{spec}\,\mathcal{O}_q(SL_n(k))$.

The following result of Letzter [**134**] shows how normal separation can be obtained in certain Hopf algebras via calculations with twisted adjoint actions. This can be used, in particular, to give another proof of Theorem II.9.19 – see [**134**, Proposition 2.4]. For the concept of *lying over* used in the proof, see [**83**, p. 168].

II.9.20. Theorem. *Let H be a noetherian Hopf algebra over k such that every finite dimensional irreducible H-module is 1-dimensional. Let K be a noetherian subalgebra of H (not necessarily a sub-Hopf algebra) such that H is a finitely generated K-module on each side. Assume that for each $a \in K$, there exists a k-algebra automorphism τ of H such that $(\mathrm{ad}_\tau H)(a)$ is finite dimensional.*
Then spec H has normal separation.

Proof. Let $P \subset Q$ be distinct comparable primes of H, and let $B = Q/P$, viewed as an H-H-bimodule. If we had $P \cap K = Q \cap K$, then P and Q would both lie over any prime of K minimal over $P \cap K$. However, this would contradict the property INC for the extension $K \subseteq H$ (see [**83**, Theorem 10.6]). Hence, the inclusion $P \cap K \subset Q \cap K$ is proper, and so there exists an element $a \in (Q \cap K) \setminus P$. Let $b = a + P$ be the corresponding nonzero coset in B.

By assumption, there exists a k-algebra automorphism τ of H such that $(\mathrm{ad}_\tau H)(a)$ is finite dimensional. Note that P and Q are $(\mathrm{ad}_\tau H)$-stable, whence B becomes an H-module via ad_τ using the same formula as in H:

$$(\mathrm{ad}_\tau h)(x) = \sum \tau(h_1) x S(h_2)$$

for $h \in H$ and $x \in B$. Then $(\mathrm{ad}_\tau H)(b)$ is a nonzero finite dimensional $(\mathrm{ad}_\tau H)$-submodule of B.

Let C be an irreducible $(\mathrm{ad}_\tau H)$-submodule of $(\mathrm{ad}_\tau H)(b)$. Because of the assumption that all finite dimensional irreducible H-modules are 1-dimensional,

$C = kc$ for some nonzero element c. This 1-dimensional H-module must correspond to a character $\lambda \in X(H)$, that is, $(\mathrm{ad}_r h)(c) = \lambda(h)c$ for all $h \in H$. Recall from Definition I.9.25 that the left winding automorphism σ of H corresponding to λ is given by the formula $\sigma(h) = \sum \lambda(h_1)h_2$. We now compute that

$$\tau(h)c = \sum \tau(h_1\varepsilon(h_2))c = \sum \tau(h_1)c\varepsilon(h_2) = \sum \tau(h_1)cS(h_2)h_3$$
$$= \sum ((\mathrm{ad}_r h_1)(c))h_2 = \sum \lambda(h_1)ch_2 = c\sigma(h)$$

for all $h \in H$. It follows that c is a normal element of H/P, and therefore normal separation is proved. $\quad\square$

NOTES

Catenarity also holds for enveloping algebras of solvable Lie algebras over non-algebraically-closed fields of characteristic zero; this follows from Gabber's theorem by the methods of [214]. Lenagan has extended Gabber's theorem to complex solvable Lie superalgebras [132], and this may again be carried over to other fields of characteristic zero by the methods of [214]. Letzter and Lorenz have proved that the group algebra of any polycyclic-by-finite group (over an arbitrary field) is catenary [136].

As noted earlier, the Catenarity Theorem (II.9.5) is an axiomatization of Gabber's method [57] by Goodearl and Lenagan [72], making use of Björk's abstraction of Gabber's homological ideas [15]. The applications to $\mathcal{O}_q(k^n)$, $\mathcal{O}_q(GL_n(\mathbb{C}))$, and $\mathcal{O}_q(SL_n(\mathbb{C}))$ were given by Goodearl and Lenagan [72], who also proved that $U_q^+(\mathfrak{g})$ is catenary when q is not a root of unity. This machinery was applied by Oh to show that coordinate rings of quantum symplectic and Euclidean spaces are catenary [175], and by Cauchon to show that $\mathcal{O}_q(M_n(k))$ is catenary [33]. That normal separation in $\operatorname{spec} \mathcal{O}_q(G)$ follows from the results of Hodges-Levasseur [94, 95] and Joseph [115, 116] was observed by Brown and Goodearl [19].

EXERCISES

Exercise II.9.A. Let $A = A_1^q(k)$ where q is not a root of unity. Show that $z = xy - yx$ is a normal element of A, and that every nonzero prime of A contains z. [Hint: Show that $A[z^{\pm 1}]$ is a simple ring.] Then show that A is catenary. $\quad\square$

Exercise II.9.B. (a) Let A be either $\mathcal{O}_q(k^2)$ or $\mathcal{O}_q(SL_2(k))$, with k algebraically closed and q not a root of unity. Use the description of $\operatorname{spec} A$ given in Examples II.1.2 and II.1.3 to show that $\operatorname{spec} A$ has normal separation.

(b) Now let $A = A_1^q(k)$ with q not a root of unity. Show that $\operatorname{spec} A$ has normal separation. $\quad\square$

Exercise II.9.C. Let R be a noetherian ring.

(a) If M is a nonzero finitely generated R-module with p.dim$(M) = d < \infty$, show that $\operatorname{Ext}^d_R(M, R) \neq 0$, and conclude that $j(M) \leq d$.

(b) Show that if R is Auslander-regular, then gl.dim(R) equals the supremum of the grades of all nonzero finitely generated left R-modules. $\square$

Exercise II.9.D. Let A be an affine, noetherian, Cohen-Macaulay k-algebra with finite GK-dimension and finite global dimension. If d is the minimum GK-dimension for nonzero finitely generated A-modules, show that

$$\mathrm{gl.dim}(A) \geq \mathrm{GK.dim}(A) - d.$$

Show that equality holds in case A is also Auslander-regular. $\square$

Exercise II.9.E. Pin down the GK-dimensions of the algebras mentioned in Proposition II.9.9 (except for $\mathcal{O}_q(G)$). $\square$

Exercise II.9.F. Use Propositions II.9.9 and II.9.12 to show that gl.dim$(A) =$ GK.dim(A) when A is any of the algebras

$\mathcal{O}_q(k^n)$,	$\mathcal{O}_q(M_n(k))$,	$\mathcal{O}_q(GL_n(k))$,
$\mathcal{O}_q(SL_n(k))$,	$\mathcal{O}_q(k^n)$,	$\mathcal{O}_{\lambda,\boldsymbol{p}}(M_n(k))$,
$\mathcal{O}_{\lambda,\boldsymbol{p}}(GL_n(k))$,	$\mathcal{O}_{\lambda,\boldsymbol{p}}(SL_n(k))$,	$A_n^{Q,\Gamma}(k)$.

[Hint: Exercise II.9.D.] On the other hand, for $A = \mathcal{O}_q((k^\times)^2)$ with q not a root of unity, show that gl.dim$(A) = 1$ while GK.dim$(A) = 2$. [Hint: Since A is simple, it has no nonzero finite dimensional modules.] $\square$

Exercise II.9.G. Let $A = k[x_1][x_2; \tau_2, \delta_2] \cdots [x_n; \tau_n, \delta_n]$ be an iterated skew polynomial algebra over k, such that $\tau_i(x_j) \in k^\times x_j$ and $\delta_i(x_j) \in \sum_{s,t<i} k x_s x_t$ for $1 \leq j < i \leq n$. Show that there is a grading on A such that A becomes a connected graded k-algebra with each x_j homogeneous of degree 1. $\square$

Exercise II.9.H. Let $A = A_n^{Q,\Gamma}(k)$; this algebra is Auslander-regular by Lemma II.9.10. Show that A is Cohen-Macaulay as follows. Define a k-algebra B with generators $z, x_1, y_1, \ldots, x_n, y_n$ satisfying the first four of the relations given for $A_n^{Q,\Gamma}(k)$ in (I.2.6) as well as

$$zx_j = x_j z \qquad\qquad\qquad\qquad zy_j = y_j z$$
$$x_j y_j = z^2 + q_j y_j x_j + \sum_{l<j}(q_l - 1)y_l x_l$$

for all j. Show that B can be expressed as an iterated skew polynomial algebra satisfying the hypotheses of Lemma II.9.10; thus B is Auslander-regular and Cohen-Macaulay. Then observe that $B/\langle z - 1 \rangle \cong A$ and apply Lemma II.9.11. $\square$

Exercise II.9.I. The concept of a *polycentral* ideal is defined by replacing 'normal' by 'central' throughout Definition II.4.1. Modify the proof of Proposition II.9.13 to show that all ideals of $\mathcal{O}_q((k^\times)^n)$ and $\mathcal{O}_q((k^\times)^n)$ are polycentral. (In fact, as long as k is infinite, all ideals of these algebras are centrally generated – this follows from Corollary II.3.9, since under the standard actions of $H = (k^\times)^n$, these algebras are H-simple.) □

Exercise II.9.J. Let $A = A_2^{Q,\Gamma}(k)$ where q_1 and q_2 are non-roots of unity. Show that the elements $z_j = x_j y_j - y_j x_j$ are normal in A. Now let $H = (k^\times)^2$ act on A as in (II.1.17). Show that the H-primes of A are

$$\langle 0 \rangle \qquad \langle z_1 \rangle \qquad \langle z_1, x_2 \rangle \qquad \langle z_1, y_2 \rangle \qquad \langle z_1, x_2, y_2 \rangle \qquad \langle z_2 \rangle.$$

Then use Theorem II.9.15 to show that spec A is normally separated, and conclude from the Catenarity Theorem that A is catenary. □

Exercise II.9.K. If X is a multiplicatively closed set generated by normal elements in a ring R, show that X is an Ore set. Consequently, if R is noetherian, X is a denominator set. □

Exercise II.9.L. Let A be one of the algebras $\mathcal{O}_q(SL_2(k))$, $\mathcal{O}_q(GL_2(k))$, or $\mathcal{O}_q(M_2(k))$, with q not a root of unity, and let H be either $(k^\times)^2$ or $(k^\times)^4$, acting on A as in Examples II.1.6. Show that each H-prime of A is generated by a polynormal sequence of H-eigenvectors, and conclude that H-spec A satisfies normal H-separation. □

Exercise II.9.M. Show that $\mathcal{O}_q(GL_n(k))$ is isomorphic to

$$\mathcal{O}_q(SL_{n+1}(k))/\langle \overline{X}_{n+1,1}, \ldots, \overline{X}_{n+1,n}, \overline{X}_{1,n+1}, \ldots, \overline{X}_{n,n+1} \rangle \quad \square$$

CHAPTER II.10

PROBLEMS AND CONJECTURES

To conclude Part II, we highlight a selection of open problems and conjectures related to the material presented so far. Further problems and conjectures will be discussed at the end of Part III.

DEFINITION OF QUANTUM GROUPS

Perhaps the most fundamental problem in the subject of quantum groups is the lack of a precise definition – all that we have is a set of examples which, by general consensus, constitutes this area of study. In particular, this means that much of the investigation of any given quantum group has to be done via direct calculation with the generators and relations of that particular algebra, and that a lot of this work has to be redone with each new quantum group that appears on the scene. Thus, we have

II.10.1. Problem. *Find suitable axioms to define "quantum groups", "quantized enveloping algebras", and "quantized coordinate rings" in the algebraic setting.*

In other words, find a set of axioms designed to cover all the examples of "algebraic quantum groups" that we have been discussing, while at the same time being strong enough to prove theorems. For the sake of contrast (and perhaps inspiration or competition), we note that the analogue of Problem II.10.1 in the context of compact groups has a solution going back to a 1987 paper of Woronowicz [**212**] (see also [**34**, §13.3], [**125**, §1.4]). Given such a set of axioms, a long-term goal then arises, namely to find all algebras satisfying these axioms, and to discover which are the "simplest" or most basic ones, in some sense. At the present stage of the subject, it is most natural to look for axioms covering quantized enveloping algebras of semisimple Lie algebras and quantized coordinate rings of semisimple algebraic groups. The wider view deserves study as well, but at present there are few examples to work with – for instance, aside from a few particular cases, no schemes to construct quantized enveloping algebras of solvable Lie algebras have been proposed.

More specific versions of these problems can be raised for quantized enveloping algebras of a particular Lie algebra, or for quantized coordinate rings of a particular algebraic group, as follows.

II.10.2. Problem. *Given a finite dimensional Lie algebra $\mathfrak{g}$, find axioms for Hopf algebras to qualify as quantized enveloping algebras of this particular $\mathfrak{g}$.*

For instance, what properties should be demanded of a Hopf algebra in order that we should consider it a quantized enveloping algebra of $\mathfrak{sl}_3(\mathbb{C})$? Once we have settled on the precise properties, how many possible examples are there?

II.10.3. Problem. *Given an affine algebraic group G, find axioms for Hopf algebras to qualify as quantized coordinate rings of this particular G.*

This direction has been investigated by Ohn [**176**], who took as his basic criterion that a Hopf algebra H qualifies as a quantized coordinate ring of G provided the category of finite dimensional H-comodules is similar to the category of finite dimensional rational G-modules (in terms of dimensions of irreducible objects and multiplicities of irreducible summands of tensor products). In the case when $G = SL_3(\mathbb{C})$, he determined nearly all the Hopf algebras satisfying the given conditions.

Another very interesting possibility arises from formal deformation theory. Gerstenhaber, Giaquinto and Schack proved that the $\mathbb{C}$-form of $\mathcal{O}_q(G)$ has a *preferred presentation*, meaning that the classical coordinate ring $\mathcal{O}(G)$ can be equipped with a new multiplication such that it remains a Hopf algebra with respect to its original comultiplication and becomes isomorphic (as a Hopf algebra) to $\mathcal{O}_q(G)$ (see [**63**], and also [**64**]). So far, this result is just an existence theorem – the proof indicates how to construct the desired new multiplication, but it has not been calculated explicitly. Continuing along this line, we pose the following version of Problem II.10.3: Given a complex affine algebraic group G, find all $\mathbb{C}$-algebra multiplications on $\mathcal{O}(G)$ such that $\mathcal{O}(G)$, together with this new multiplication and its original comultiplication, becomes a Hopf algebra.

THE QUANTUM GEL'FAND-KIRILLOV CONJECTURE

This conjecture was briefly discussed in (I.2.11–14). We restate it here:

II.10.4. Conjecture. *Let A be a k-form of either a quantized enveloping algebra or a quantized coordinate ring. Then the quotient division ring* Fract A *is isomorphic to* Fract $\mathcal{O}_q(K^t)$ *for some quantum affine space $\mathcal{O}_q(K^t)$, where K is a purely transcendental field extension of k.*

This conjecture has been verified in several cases, for instance (as recorded in Theorem I.2.14), for $A = \mathcal{O}_q(M_n(k))$ or $\mathcal{O}_{\lambda,\boldsymbol{p}}(M_n(k))$. Since Fract $\mathcal{O}_q(M_n(k)) =$ Fract $\mathcal{O}_q(GL_n(k))$ and Fract $\mathcal{O}_{\lambda,\boldsymbol{p}}(M_n(k)) =$ Fract $\mathcal{O}_{\lambda,\boldsymbol{p}}(GL_n(k))$, the cases of $\mathcal{O}_q(GL_n(k))$ and $\mathcal{O}_{\lambda,\boldsymbol{p}}(GL_n(k))$ follow immediately. These cases were done by Cliff [**36**, Propositions 3–5], Panov [**177**, Theorem 3.8], and Mosin-Panov [**164**, Theorem 1.24]. Further, the conjecture has been verified for $U_q^+(\mathfrak{sl}_n(k))$ with q not a root of unity, for $U_q^+(\mathfrak{g})$ with q transcendental over $\mathbb{Q}$, for $\mathcal{O}_q(G)$, $U_q^+(\mathfrak{g})$, and

$\check{U}_q^{\geq 0}(\mathfrak{g})$ with $q \in \mathbb{C}^\times$ not a root of unity, and for $\check{U}_q(\mathfrak{sl}_n(\mathbb{C}))$ with q transcendental over $\mathbb{C}$, by Alev-Dumas [**3**, Théorème 2.15], Iohara-Malikov [**103**, Theorem 3.5], Joseph [**117**], Caldero [**28**, Theorems 3.3, 3.1], and Fauquant-Millet [**56**, Corollaire 2.1]. It was also established by Musson for generic quantum symplectic and Euclidean spaces [**167**, Theorems 1.3, 2.4]. Analogous results have been obtained for four-dimensional Sklyanin algebras (Corwin-Gel'fand-Goodman [**38**]), and for the enveloping algebras of the Lie superalgebras $\mathfrak{osp}(1, 2r)$ (Musson [**168**]). Panov has recently proved that the conclusion of the conjecture holds for a class of iterated skew polynomial algebras which includes $\mathcal{O}_{\lambda,p}(M_n(k))$, $A_n^{Q,\Gamma}(k)$, and $U_q^+(\mathfrak{g})$ [**178**, Main Theorem].

One can also ask whether the conjecture holds for prime factors, although then it seems appropriate to drop the requirement that the field extensions appearing be purely transcendental. Thus, we ask: If A is a k-form of either a quantized enveloping algebra or a quantized coordinate ring, then given any prime ideal P of A, must Fract A/P be isomorphic to Fract $\mathcal{O}_q(K^t)$ for some t and some field extension K of k? This was first proved to hold when A is either $\mathcal{O}_q(M_n(k))$ with q not a root of unity, or $A_n^{Q,\Gamma}(k)$ with $\langle q_i, \gamma_{ij} \rangle$ torsionfree, by Cauchon [**31**, Théorèmes II.2.1, III.3.2.1]. He has recently incorporated these results into a theorem which holds for a class of iterated skew polynomial algebras including those in [**178**], thus covering generic cases of $\mathcal{O}_{\lambda,p}(M_n(k))$ and $U_q^+(\mathfrak{g})$, as well as quantum symplectic and Euclidean spaces [**32**, Théorème 6.1.1]. Moreover, the result holds for $A = \mathcal{O}_q(G)_\mathbb{C}$ when q is transcendental over $\mathbb{Q}$ [**32**, Théorème 6.4.2].

H-PRIMES IN $\mathcal{O}_q(SL_n(k))$

Assume that q is not a root of unity. Recall that there are several natural choices for a k-torus acting rationally on $\mathcal{O}_q(SL_n(k))$ by k-algebra automorphisms. The largest possibility, call it $\widehat{H}$, is the torus

$$\widehat{H} = \{(\alpha_1, \ldots, \alpha_n, \beta_1, \ldots, \beta_n) \in (k^\times)^{2n} \mid \alpha_1 \alpha_2 \cdots \alpha_n \beta_1 \beta_2 \cdots \beta_n = 1\},$$

acting on $\mathcal{O}_q(SL_n(k))$ as in (II.1.16). This is the choice needed in order to apply the skew polynomial results of Chapters II.5 and II.6. On the other hand, we also have the torus

$$H = \mathrm{Hom}(L, k^\times) \cong (k^\times)^{n-1},$$

where L is the weight lattice of $\mathfrak{sl}_n(\mathbb{C})$, acting on $\mathcal{O}_q(SL_n(k))$ via right winding automorphisms as in (II.1.18). This action corresponds to that of the subgroup

$$H_r = \{(1, \ldots, 1, \beta_1, \ldots, \beta_n) \in (k^\times)^{2n} \mid \beta_1 \beta_2 \cdots \beta_n = 1\} \subset \widehat{H}.$$

Alternatively, we could let H act on $\mathcal{O}_q(SL_n(k))$ via *left* winding automorphisms, corresponding to the action of the subgroup

$$H_l = \{(\alpha_1, \ldots, \alpha_n, 1, \ldots, 1) \in (k^\times)^{2n} \mid \alpha_1 \alpha_2 \cdots \alpha_n = 1\} \subset \widehat{H}.$$

Finally, since the above two actions of H commute with each other, we obtain an action of $H \times H$ on $\mathcal{O}_q(SL_n(k))$, corresponding to that of the subgroup

$$\{(\alpha_1, \ldots, \alpha_n, \beta_1, \ldots, \beta_n) \in (k^\times)^{2n} \mid \alpha_1\alpha_2 \cdots \alpha_n = \beta_1\beta_2 \cdots \beta_n = 1\} \subset \widehat{H}.$$

As discussed in (II.1.18), the $\widehat{H}$-primes and H_r-primes in $\mathcal{O}_q(SL_n(k))$ coincide; for similar reasons, these also coincide with the H_l-primes and $(H \times H)$-primes. At present, we need finiteness results to prove this – for instance, once it is known that H_r-spec $\mathcal{O}_q(SL_n(k))$ is finite, it follows from Proposition II.2.9 that every H_r-prime of $\mathcal{O}_q(SL_n(k))$ is $\widehat{H}$-prime, whence the H_r-primes coincide with the $\widehat{H}$-primes.

II.10.5. Problem. *Find a direct method to prove that the $\widehat{H}$-prime, H_r-prime, H_l-prime, and $(H \times H)$-prime ideals in $\mathcal{O}_q(SL_n(k))$ all coincide.*

FINITENESS OF H-spec

II.10.6. Problem. *Find conditions on a noetherian k-algebra A and a k-torus H acting rationally on A by k-algebra automorphisms, which imply that H-spec A is finite.*

What is desired here is a theorem which can cover all the quantized coordinate rings we have discussed, without needing either iterated skew polynomial presentations or long calculations.

COMPLETE PRIMENESS IN $\mathcal{O}_q(G)$ AND $\mathcal{O}_{q,p}(G)$

Recall from Corollary II.6.10 that in many generic quantized coordinate rings, all prime ideals are completely prime. In particular, this holds for $\mathcal{O}_q(M_n(k))$, $\mathcal{O}_q(GL_n(k))$, and $\mathcal{O}_q(SL_n(k))$ when q is not a root of unity, and for $\mathcal{O}_{\lambda,p}(M_n(k))$, $\mathcal{O}_{\lambda,p}(GL_n(k))$, and $\mathcal{O}_{\lambda,p}(SL_n(k))$ when the group $\langle \lambda, p_{ij} \rangle$ is torsionfree. Joseph has proved that all prime ideals of $\mathcal{O}_q(G)$ are completely prime when $q \in \mathbb{C}^\times$ is a non-root of unity [**115**, Theorem 11.4]. This conclusion is based on the full weight of the calculations in [**115**], and so we mention the problem of finding a shorter proof, and one adaptable to other base fields. For an alternative proof in case $q \in \mathbb{C}^\times$ is transcendental over $\mathbb{Q}$, see [**32**, Théorème 6.4.2].

Hodges, Levasseur, and Toro have investigated multiparameter quantized coordinate rings $\mathcal{O}_{q,p}(G)$, where $q \in \mathbb{C}^\times$ is a non-root of unity, $p : L \times L \to \mathbb{C}^\times$ is an alternating bicharacter, and the multiplication in $\mathcal{O}_{q,p}(G)$ is obtained from that in $\mathcal{O}_q(G)$ by a "twisting" involving p [**96**]. They proved that much of the structure theory of $\mathcal{O}_q(G)$ carries over to $\mathcal{O}_{q,p}(G)$. For instance, the H-primes in $\mathcal{O}_{q,p}(G)$ (where again $H = \text{Hom}(L, \mathbb{C}^\times)$ acts via right winding automorphisms) are parametrized by the double Weyl group. Since these H-primes are homogeneous with respect to the $L \times L$ grading on $\mathcal{O}_{q,p}(G)$, general twisting arguments can be used to show that all the H-primes in $\mathcal{O}_{q,p}(G)$ are completely prime. However, the case of arbitrary primes remains open, and we make the

II.10.7. Conjecture. *All prime ideals in $\mathcal{O}_{q,p}(G)$ are completely prime.*

NORMAL SEPARATION

Recall from Definition II.9.4 that the prime spectrum of a ring R has *normal separation* if for every pair $P \subset Q$ of distinct, comparable primes in R, the ideal Q/P of R/P contains a nonzero normal element of R/P. This condition is useful in several directions, and is a key ingredient in the Catenarity Theorem (II.9.5). It has been verified for many quantum algebras, such as $\mathcal{O}_q(k^n)$ (Theorem II.9.14), $A_n^{Q,\Gamma}(k)$ when the q_i are non-roots of unity [**72**, Theorem 3.12], $\mathcal{O}_q(M_n(k))$ when q is not a root of unity [**33**, Théorème 6.2.1], $\mathcal{O}_q(G)$ with q not a root of unity (Theorem II.9.19), and $U_q^+(\mathfrak{g})$ with $q \in \mathbb{C}^\times$ not a root of unity [**27**, Corollaire 3.2]. These cases were handled with different methods, and – except for the first – required substantial calculations. A common, smoother method would be desirable:

II.10.8. Problem. *Find easily verifiable conditions, satisfied by quantized coordinate rings (and related rings), which imply normal separation.*

Theorem II.9.15 suggests one possible approach to this problem. Suppose that A is a noetherian k-algebra, with k infinite, and that H is a k-torus acting rationally on A by k-algebra automorphisms. Then spec A has normal separation if the following condition holds:

(*) For every pair of distinct comparable H-primes $J \subset K$ in A, the factor K/J contains a nonzero normal H-eigenvector.

Thus, for present purposes, it would be enough to find easily verifiable conditions, satisfied by quantized coordinate rings, which imply condition (*).

QUANTUM MATRICES

In many respects, $\mathcal{O}_q(M_n(k))$ has proved one of the most difficult quantized coordinate rings to deal with, in spite of its seemingly nice structure as an iterated skew polynomial ring. While $\mathcal{O}_{\lambda,p}(M_n(k))$ appears more complicated, in practice one can treat it either by modifying proofs for $\mathcal{O}_q(M_n(k))$, or by transferring results from $\mathcal{O}_q(M_n(k))$ using the twisting methods of [**8**].

Normal separation has long been conjectured for these algebras. As we have seen, $\mathcal{O}_q(M_2(k))$ has normal separation (Examples II.9.16), and Cauchon has proved it for $\mathcal{O}_q(M_n(k))$ [**33**, Théorème 6.2.1]. These cases were proved via condition (*) of Theorem II.9.15.

Rather than remain satisfied with normal separation in quantum matrices, we ask for the following stronger and more precise property:

II.10.9. Conjecture. *Let $H = (k^\times)^{2n}$ act on $\mathcal{O}_q(M_n(k))$ and $\mathcal{O}_{\lambda,p}(M_n(k))$ as in (II.1.15). Then every H-prime of these algebras is generated by a polynormal sequence of quantum minors.*

By the same arguments developed in [74], it would suffice to verify this conjecture for the case of $\mathcal{O}_q(M_n(k))$. It does hold for $\mathcal{O}_q(M_2(k))$ (this follows from Exercise II.2.O), and Goodearl and Lenagan have proved it for $\mathcal{O}_q(M_3(k))$ [75].

It is likely that if Conjecture II.10.9 is established, that will be done (or accompanied) by a complete description of the H-prime ideals in $\mathcal{O}_q(M_n(k))$. That would provide the first step towards describing the full prime spectrum via the Stratification Theorem.

CATENARITY

II.10.10. Conjecture. *Quantized coordinate rings are catenary.*

The only general methods available are PI theory for root-of-unity cases, and Gabber's method (as axiomatized in Theorem II.9.5) for generic cases. Concerning the latter, it is not hard to show that quantized coordinate rings are affine noetherian algebras with finite GK-dimension; the tricky hypotheses are the homological conditions and normal separation. For instance, $\mathcal{O}_q(G)$ is known to have normal separation (Theorem II.9.19), but for $G \neq SL_n(k)$, it is not known whether $\mathcal{O}_q(G)$ is Auslander-regular or Cohen-Macaulay.

TOPOLOGICAL STRUCTURE OF PRIMITIVE SPECTRA

Much of the work on prime and primitive spectra of quantized coordinate rings that we have discussed is concerned with simply finding these spectra as sets, that is, describing and/or parametrizing the prime or primitive ideals in these algebras. A much deeper problem is to describe these spectra as topological spaces. Piecewise descriptions are given by the Stratification Theorem, but these are not sufficient to pin down the Zariski topologies.

II.10.11. Problem. *Describe the prime and primitive spectra of quantized coordinate rings as topological spaces.*

It is easier to think about this problem for the case of primitive spectra, since they correspond to affine varieties in the classical situation – if V is an affine algebraic variety over an algebraically closed field k, then $\operatorname{prim} \mathcal{O}(V) = \max \mathcal{O}(V)$ is Zariski-homeomorphic to V. Consider the case of $\mathcal{O}_q(G)$ with $q \in \mathbb{C}^\times$ a non-root of unity. As discussed in Theorem II.4.16, Hodges and Levasseur (for the case $G = SL_n(\mathbb{C})$) and Joseph (for general G) found bijections between $\operatorname{prim} \mathcal{O}_q(G)$ and the set of symplectic leaves of G, bijections which are equivariant for the natural actions of a maximal torus H of G. Rearranging this result slightly, we obtain an H-equivariant surjection

$$\pi : G \twoheadrightarrow \operatorname{prim} \mathcal{O}_q(G)$$

such that the fibres of π are the symplectic leaves in G.

II.10.12. Conjecture. π *is Zariski-continuous; in fact, the Zariski topology on* prim $\mathcal{O}_q(G)$ *should be the quotient topology induced from G via π.*

The second part of the conjecture asserts that the Zariski-closed subsets of prim $\mathcal{O}_q(G)$ are precisely those sets X such that $\pi^{-1}(X)$ is Zariski-closed in G. If this holds, then the Zariski topology on prim $\mathcal{O}_q(G)$ is completely determined by the map π and the Zariski topology on G. The conjecture is easily checked for the case $G = SL_2(\mathbb{C})$, but is open otherwise. It can also be widened by varying the field – for instance, if k is algebraically closed, one conjectures that prim $\mathcal{O}_q(SL_n(k))$ is a topological quotient of the variety $SL_n(k)$.

Similar conjectures can be raised for other quantized coordinate rings. The case of quantum affine spaces has been established, modulo a small technical assumption, by Goodearl and Letzter, as follows. Let $A = \mathcal{O}_q(k^n)$ with k algebraically closed, and assume that -1 is not in the subgroup $\langle q_{ij} \rangle \subseteq k^\times$. Then prim A is a topological quotient of k^n, and this quotient structure is equivariant for the standard actions of $(k^\times)^n$ on A and on k^n; similarly, spec A is a $(k^\times)^n$-equivariant topological quotient of spec $\mathcal{O}(k^n)$ [**81**, Theorem 4.11]. This theorem also extends to certain cocycle twists of commutative affine algebras, thus covering a class of quantized coordinate rings called "quantum toric varieties" in [**102**] (see [**81**, Theorem 6.3] and [**71**, Theorem 4.5]).

PART III. QUANTIZED ALGEBRAS AT ROOTS OF UNITY

CHAPTER III.1

FINITE DIMENSIONAL MODULES
FOR AFFINE PI ALGEBRAS

The purpose of this chapter is to develop some basic theory to help analyse the finite dimensional modules over quantized enveloping algebras and quantized function algebras in the cases where the parameter q is a non-trivial root of unity. Our point of departure will be those fundamental ideas of PI theory which are summarised in Appendix I.13.

III.1.1. Kaplansky's Theorem (I.13.3) imposes strong restrictions on the primitive factors of a PI ring. Our first aim is to show that for the PI rings with which we'll be concerned, there are "many" such primitive factors, and indeed "many" prime factors. Although most of the results we are interested in here are valid under much weaker conditions, in order to give self-contained proofs we'll assume throughout that the following hypotheses, sufficient for the applications we have in mind, are valid – namely, that

> **(H)** A is an affine algebra over the field k, and A is a finitely generated module over a central subalgebra Z_0.

For example, if $q \in k^\times$ is an ℓth root of unity, then both $\mathcal{O}_q(k^n)$ and $\mathcal{O}_q((k^\times)^n)$ satisfy hypotheses **(H)** – take $Z_0 = k[x_1^\ell, \ldots, x_n^\ell]$ in the first case, and $Z_0 = k[x_1^{\pm\ell}, \ldots, x_n^{\pm\ell}]$ in the second.

The Artin-Tate Lemma (I.13.4), the Hilbert Basis Theorem, and Example I.13.2(3) yield part 1 of the

Proposition. *Assume hypotheses* **(H)**, *with A generated by t elements as a Z_0-module.*

1. *$Z(A)$ and Z_0 are affine k-algebras, so they and A are noetherian PI rings.*
2. *Let $\mathfrak{p}$ be a prime ideal of Z_0. Then*

$$1 \leq \dim_{\mathrm{Fract}(Z_0/\mathfrak{p})}\big(\mathrm{Fract}(Z_0/\mathfrak{p}) \otimes (A/\mathfrak{p}A)\big) \leq t.$$

It follows that $\mathfrak{p}A$ is contained in at least 1 and at most t prime ideals P of A with $P \cap Z_0 = \mathfrak{p}$.

3. *Let $\mathfrak{m}$ be a maximal ideal of Z_0, and let m be the number of non-isomorphic irreducible A-modules V with $\mathrm{ann}_{Z_0}(V) = \mathfrak{m}$. Then $1 \leq m \leq t$.*

4. *Let V be an irreducible A-module. Then V is finite dimensional over k, and $\mathrm{ann}_{Z_0}(V)$ is a maximal ideal of Z_0. In particular, $A/\mathrm{ann}_A(V)$ and $Z_0/\mathrm{ann}_{Z_0}(V)$ are finite dimensional over k.*

5. *If P is a primitive ideal of A, then $P \cap Z_0$ is a maximal ideal of Z_0. Any maximal ideal $\mathfrak{m}$ of Z_0 is contained in at least 1 and at most t primitive (equivalently, maximal) ideals P of A with $P \cap A = \mathfrak{m}$.*

Proof. 2. For this part of the result, relax the affine hypotheses, assuming only that Z_0 is noetherian. Suppose first that $\mathfrak{p}$ were the Jacobson radical of Z_0. Then $\mathfrak{p}A \neq A$ by Nakayama's Lemma, and the desired conclusions are clear. Localizing at $\mathfrak{p}$ in Z_0 allows us to deduce the result under the present hypotheses.

3. This is just a restatement of a special case of 2, in view of 4.

4. Set $\overline{A} = A/\mathrm{ann}_A(V)$. Thus $\overline{A}$ is a central simple algebra by Kaplansky's Theorem (I.13.3), and $Z(\overline{A})$ is an affine k-algebra by the Artin-Tate Lemma (I.13.4). Therefore $Z(\overline{A})$ is a finite extension field of k. Since $Z_0/\mathrm{ann}_{Z_0}(V)$ embeds in $Z(\overline{A})$, we deduce that $\mathrm{ann}_{Z_0}(V)$ is a maximal ideal of Z_0 and that $Z_0/\mathrm{ann}_{Z_0}(V)$ is finite dimensional over k. Since V is a finitely generated Z_0-module, it follows that $\dim_k(V) < \infty$. It is then immediate that $A/\mathrm{ann}_A(V)$ is finite dimensional over k.

5. This follows from 3 and 4.   □

Notice one important consequence of Proposition III.1.1 – studying the irreducible A-modules is equivalent to studying the irreducible $(A/\mathfrak{m}A)$-modules, as $\mathfrak{m}$ ranges through the maximal ideals of Z_0. Each irreducible A-module will occur once and once only as a module over one of these finite dimensional k-algebras.

Recall (II.7.10) that a ring R is a *Jacobson ring* if every prime ideal of R is an intersection of primitive ideals. Hilbert's Nullstellensatz ensures that this is a property of commutative affine algebras, and the "Lying Over" result obtained as 2 of the above proposition allows us to extend this to the

Corollary. *Let A satisfy hypotheses* (**H**). *Then A is a Jacobson ring.*

Proof. We may assume without loss of generality that A is prime, and aim to show that its Jacobson radical $J(A)$ is zero. Since $Z(A)$ is Jacobson,

$$\bigcap \{\mathfrak{m} : \mathfrak{m} \in \mathrm{maxspec}(Z(A))\} = 0.$$

From Corollary I.13.3(1) and Proposition III.1.1(5) it follows that $J(A) = 0$, as claimed.   □

III.1.2. Regular primes. Suppose that A is a prime algebra over the algebraically closed field k and that A satisfies hypotheses (**H**). We know from Theorem I.13.5 that there is an upper bound on the k-dimensions of the irreducible

A-modules, namely PI-deg(A). To see that this upper bound is attained, we shall call a prime ideal P of a prime PI-ring R *regular* if the PI degree of R/P equals the PI degree of R. By Posner's Theorem (I.13.3), P is regular if and only if the minimal degree of R/P equals the minimal degree of R.

Examples. 1. If R is a full matrix ring $M_n(C)$ over a commutative domain C then all its prime ideals are regular since they have the form $M_n(\mathfrak{p})$ for the prime ideals $\mathfrak{p}$ of C.

2. Let

$$R = \begin{pmatrix} k[X] & \langle X \rangle \\ k[X] & k[X] \end{pmatrix} \subseteq M_2(k[X]).$$

Then the prime ideals

$$M = \begin{pmatrix} \langle X \rangle & \langle X \rangle \\ k[X] & k[X] \end{pmatrix} \quad \text{and} \quad N = \begin{pmatrix} k[X] & \langle X \rangle \\ k[X] & \langle X \rangle \end{pmatrix}$$

of R are not regular, since in both cases the factor ring is isomorphic to k. All the other prime ideals of R are regular.

3. Let $R = \mathcal{O}_q(k^2)$ where $q \in k^\times$ is a primitive ℓth root of unity for some $\ell > 1$. Then PI-deg(R) $= \ell$ by Example I.14.3(1). If P is any prime ideal of R containing either x or y, then R/P is commutative and so has PI degree 1. Hence, these prime ideals are not regular. However, all other prime ideals of R are regular. This follows from the fact that all prime ideals in the localization $\mathcal{O}_q((k^\times)^2)$ are regular, as we'll see at the end of (III.1.4).

The pattern evident in Example 2 – that "most" maximal ideals are regular – is typical, as we now show. In part 2, "closed" refers to the Zariski topology; see (II.1.1).

Lemma. *Let A be a prime k-algebra satisfying hypotheses* (**H**).
   1. *A prime ideal P of A is regular (respectively, non-regular) if and only if P is an intersection of regular (respectively, non-regular) maximal ideals of A.*
   2. *The set of non-regular prime (respectively, non-regular maximal) ideals of A is a proper closed subset of* spec A *(respectively,* maxspec A*). In particular, there exist regular maximal ideals, and their intersection is 0.*

Proof. That any prime ideal P of A is an intersection of some family of maximal ideals is immediate from Corollary III.1.1 together with the fact that all primitive ideals of A are maximal. Hence, $P = P_1 \cap P_2$ where P_1 (respectively, P_2) is the intersection of the regular (respectively, non-regular) maximal ideals containing P. Since P is prime, $P = P_1$ or $P = P_2$. Thus we have shown than any prime ideal of A is either an intersection of a family of regular maximal ideals or an intersection of a family of non-regular maximal ideals. Note that if P is not regular, then no prime ideal $Q \supseteq P$ can be regular, since

$$
\begin{aligned}
\text{minimal degree of } A/Q \;\;&\leq\;\; \text{minimal degree of } A/P \\
&<\;\; \text{minimal degree of } A.
\end{aligned}
$$

Thus in this case, P must be an intersection of non-regular maximal ideals.

    1. We have just seen that any non-regular prime ideal of A is an intersection of non-regular maximal ideals. The converse follows from Example I.13.2(5), since if P is an intersection of non-regular maximal ideals $\mathfrak{m}_i$, then A/P embeds in $\prod_i (A/\mathfrak{m}_i)$.

    Now if P is regular, then by what we have just proved, it cannot be an intersection of non-regular maximal ideals, and so by the remarks above, P must be an intersection of regular maximal ideals. On the other hand, if P is not regular then no maximal ideals containing P are regular, and hence P cannot be an intersection of regular maximal ideals.

    2. Let I be the intersection of the set of non-regular maximal ideals of A, and note from part 1 that I also equals the intersection of the set of non-regular prime ideals of A. Since 0 is a regular prime ideal, $I \neq 0$ by part 1. Applying Example I.13.2(5) again, we see that A/I satisfies a polynomial identity of degree less than the minimal degree of A. Hence, all prime ideals containing I are non-regular. Therefore

$$
\{P \in \operatorname{spec} A : P \text{ is not regular}\} = \{P \in \operatorname{spec} A : P \supseteq I\}
$$
$$
\{\mathfrak{m} \in \operatorname{maxspec} A : \mathfrak{m} \text{ is not regular}\} = \{\mathfrak{m} \in \operatorname{maxspec} : \mathfrak{m} \supseteq I\},
$$

and part 2 is proved. $\quad\square$

III.1.3. Azumaya algebras. To understand better the regular primes, we recall that a ring R is called an *Azumaya algebra over its centre* C if

    (1) R is a finitely generated projective C-module; and

    (2) The ring homomorphism

$$
\begin{aligned}
\Theta : R \otimes_C R^{\mathrm{op}} &\longrightarrow \operatorname{End}_C(R) \\
a \otimes b &\longmapsto (x \mapsto axb)
\end{aligned}
$$

      is an isomorphism.

Examples. 1. If C is a commutative ring and $n \geq 1$ then $M_n(C)$ is an Azumaya algebra over C.

    2. A central simple algebra R with centre C is an Azumaya algebra over C. (It's standard and easy to prove that $R \otimes_C R^{\mathrm{op}}$ is simple. Thus the map Θ in the definition is injective, and comparing C-dimensions gives surjectivity.)

3. The Azumaya property is local: R is Azumaya over C if and only if $R_{\mathfrak{m}}$ is Azumaya over $C_{\mathfrak{m}}$ for all maximal ideals $\mathfrak{m}$ of C. (The proof is routine.)

The relationship between the Azumaya property and regular primes is revealed in the following subsection, where we also see a different method of checking whether this property holds.

III.1.4. Suppose that R is a prime Azumaya algebra over C and let $\mathfrak{m}$ be a maximal ideal of C. Then $\dim_{C/\mathfrak{m}}(R/\mathfrak{m}R)$ equals the minimum number of generators of $R_{\mathfrak{m}}$ as a $C_{\mathfrak{m}}$-module, by Nakayama's lemma. By 1 of the definition of Azumaya, this equals the rank of the free $C_{\mathfrak{m}}$-module $R_{\mathfrak{m}}$, and hence equals $\dim_{\mathrm{Fract}(C)}(R \otimes_C \mathrm{Fract}(C))$. Thus $n^2 := \dim_{C/\mathfrak{m}}(R/\mathfrak{m}R)$ is constant as $\mathfrak{m}$ varies (and is necessarily a square since $R \otimes_C \mathrm{Fract}(C)$ is a central simple algebra with centre $\mathrm{Fract}(C)$). We say that the Azumaya algebra R has *rank* n^2 over C.

The key result in the theory of Azumaya algebras is the

Artin-Procesi Theorem. *Let R be a prime ring with centre C, and n a positive integer. Then the following are equivalent:*
1. *R is an Azumaya algebra of rank n^2 over C.*
2. *R is a PI ring of PI degree n whose prime ideals are all regular.*
3. *R is a PI ring of PI degree n whose maximal ideals are all regular.*

Proof. [**158**, 13.7.14], [**191**, Theorem 6.1.35] □

Example. Let $q \in k^\times$ be a primitive ℓth root of unity, where $\ell > 1$. With the aid of the Artin-Procesi Theorem, it is clear from Example III.1.2(3) that $\mathcal{O}_q(k^2)$ is not an Azumaya algebra; similarly, $\mathcal{O}_q(k^n)$ is not Azumaya for any $n \geq 2$. On the other hand, quantum tori $\mathcal{O}_q((k^\times)^n)$ are Azumaya, at least when k is algebraically closed [**44**, Proposition 7.2], Exercise III.1.B. To prove this, we let the group $H = (k^\times)^n$ act on the algebra $A = \mathcal{O}_q((k^\times)^n)$ by k-algebra automorphisms as in (II.1.14). It is known that the induced action on maxspec A is transitive (e.g., [**79**, 1.14(ii)]): Given any $M, M' \in \mathrm{maxspec}\, A$, there exists $h \in H$ such that $h(M) = M'$. Since A has at least one regular maximal ideal (Lemma III.1.2(2)), all its maximal ideals must be regular, and therefore the Artin-Procesi Theorem implies that A is Azumaya.

III.1.5. For later use we need to show that an analogue of a basic fact in algebraic geometry, that a finite morphism is closed [**89**, Exercise II.3.5(b)], holds in the present context. This is the following

Lemma. *Let R be a noetherian subring of the centre of a ring T, and suppose that T is a finitely generated R-module. Then the contraction maps*

$$\pi : \mathrm{spec}\, T \longrightarrow \mathrm{spec}\, R \qquad\qquad \pi' : \mathrm{maxspec}\, T \longrightarrow \mathrm{maxspec}\, R$$
$$P \longmapsto P \cap R \qquad\qquad\qquad\qquad M \longmapsto M \cap R$$

are closed – that is, π and π' map closed sets to closed sets (in the Zariski topologies).

Proof. Note that T is also a noetherian ring. Let $\mathcal{V}$ be a closed subset of spec T, so that $\mathcal{V} = \{P \in \operatorname{spec} T : I \subseteq P\}$ for some ideal I of T. After replacing I by $\bigcap \{P : P \in \mathcal{V}\}$, we may assume that I is semiprime. Set

$$\mathcal{W} = \{\mathfrak{p} \in \operatorname{spec} R : I \cap R \subseteq \mathfrak{p}\}.$$

We claim that $\pi(\mathcal{V}) = \mathcal{W}$. It's obvious that $\pi(\mathcal{V}) \subseteq \mathcal{W}$. In proving the converse there is no harm in assuming that I is prime, since we may consider in turn each of the primes minimal over I. Thus $I \cap R \in \operatorname{spec} R$. Let $\mathfrak{p} \in \mathcal{W}$. Since "Going Up" holds for the inclusion $R \subseteq T$ [**158**, Corollary 10.2.10(ii)], there exists a prime ideal P of T with $I \subseteq P$ and $P \cap R = \mathfrak{p}$. That is, $P \in \mathcal{V}$ and $\pi(P) = \mathfrak{p}$, so that $\mathfrak{p} \in \pi(\mathcal{V})$. This shows that $\pi(\mathcal{V}) = \mathcal{W}$, as claimed, and proves that $\pi(\mathcal{V})$ is closed.

If M is a maximal ideal of T and $\mathfrak{m} = M \cap R$, then $R/\mathfrak{m}$ is a domain which embeds in the field $Z(T/M)$. Since T is a finitely generated R-module, $Z(T/M)$ is a finitely generated $(R/\mathfrak{m})$-module, and it follows that $R/\mathfrak{m}$ is a field, that is, $\mathfrak{m}$ is a maximal ideal of R. Thus contraction does give a map $\pi' : \operatorname{maxspec} T \to \operatorname{maxspec} R$.

Let $\mathcal{V}' = \{M \in \operatorname{maxspec} T : I \subseteq M\}$ be a closed subset of maxspec T, where I is a semiprime ideal of T. Then $\pi'(\mathcal{V}')$ is contained in the closed set $\mathcal{W}' = \{\mathfrak{m} \in \operatorname{maxspec} R : I \cap R \subseteq \mathfrak{m}\}$. The proof above shows that if $\mathfrak{m} \in \mathcal{W}'$, then $\mathfrak{m} = P \cap R$ for some $P \in \operatorname{spec} T$ such that $P \supseteq I$. Then T/P is a finitely generated module over the field $R/\mathfrak{m}$, whence T/P is simple and so $P \in \mathcal{V}'$. This shows that $\mathfrak{m} \in \pi'(\mathcal{V}')$. Thus $\pi'(\mathcal{V}') = \mathcal{W}'$ is closed in maxspec R. $\square$

III.1.6. Characterisation of regular maximal ideals.

We are now in a position to characterise the regular maximal ideals of affine PI algebras over algebraically closed fields. The result is the following.

Theorem. *Suppose that k is algebraically closed, that the k-algebra A satisfies hypotheses (**H**) of (III.1.1), and that A is prime. Let n be the PI degree of A. Let Z be the centre of A, let M be a maximal ideal of A, and denote the maximal ideal $M \cap Z$ of Z by $\mathfrak{m}$. Then the following are equivalent:*

1. *M is a regular maximal ideal of A.*
2. *$A_{\mathfrak{m}}$ is Azumaya over $Z_{\mathfrak{m}}$.*
3. *$M = \mathfrak{m}A$.*
4. *The unique irreducible left (A/M)-module has the maximum possible k-dimension amongst irreducible A-modules, namely n.*
5. *$A/M \cong M_n(k)$.*

Proof. Observe that the centre of $A_{\mathfrak{m}}$ is just $Z_{\mathfrak{m}}$.

That n is the maximum k-dimension of irreducible A-modules follows from Lemma III.1.2(2) and Theorem I.13.5. The latter theorem also gives the equivalence of 1, 4 and 5.

($2 \implies 3$): Suppose that $A_{\mathfrak{m}}$ is Azumaya over $Z_{\mathfrak{m}}$. Then $M_{\mathfrak{m}} = \mathfrak{m}A_{\mathfrak{m}}$ by [**158**, Proposition 13.7.9]. Hence $M/\mathfrak{m}A$ is torsion with respect to the image of the multiplicatively closed set $Z \setminus \mathfrak{m}$ in $A/\mathfrak{m}A$. But these elements are units in $A/\mathfrak{m}A$, so $M = \mathfrak{m}A$ as claimed.

($3 \implies 5$): Suppose that $M = \mathfrak{m}A$. Since, by Nakayama's Lemma, the minimum size of a set of generators for the $Z_{\mathfrak{m}}$-module $A_{\mathfrak{m}}$ is $\dim_k(A/\mathfrak{m}A)$, we deduce that
$$\dim_k(A/\mathfrak{m}A) \geq \dim_{\mathrm{Fract}(Z)}(A \otimes_Z \mathrm{Fract}(Z)) = n^2.$$

Hence $A/M \cong M_t(k)$ for some $t \geq n$. By the maximality of n noted above, $t = n$.

($5 \implies 2$): From 5, we know that M is a regular maximal ideal. By [**158**, Proposition 13.7.5], all prime ideals of $A_{\mathfrak{m}}$ are regular. Therefore $A_{\mathfrak{m}}$ is Azumaya over its centre by the Artin-Procesi Theorem. $\square$

Corollary. *Let k, A, n, Z be as in the theorem, and assume that A is a free Z-module. Then $\dim_k(A/\mathfrak{m}A) = n^2$ for all maximal ideals $\mathfrak{m}$ of Z, and A is a free Z-module of rank n^2.*

Proof. Let r be the rank of A as a free Z-module. For any maximal ideal $\mathfrak{m}$ of Z, we have $A/\mathfrak{m}A$ free of rank r over $Z/\mathfrak{m} = k$, and thus $\dim_k(A/\mathfrak{m}A) = r$. Since A has at least one regular maximal ideal M by Lemma III.1.2(2), we conclude from the theorem that $r = n^2$. $\square$

III.1.7. The Azumaya locus. Suppose that A is a prime k-algebra satisfying hypotheses (**H**). Let Z be the centre of A. Motivated by Theorem III.1.6 and Lemma III.1.2(2) we define the *Azumaya locus of A over Z* to be the set

$$\mathcal{A}_A := \{\mathfrak{m} \in \mathrm{maxspec}\, Z : A_{\mathfrak{m}} \text{ is Azumaya over } Z_{\mathfrak{m}}\}.$$

Thus if k is algebraically closed, $\mathcal{A}_A$ is nonempty by Lemma III.1.2(2) and Theorem III.1.6. Moreover $\mathcal{A}_A$ is open in maxspec Z. For $\mathcal{A}_A$ is the complement of the image under the contraction map of the set of non-regular maximal ideals of A, so our claim follows from Lemma III.1.2(2) and Lemma III.1.5. Summarising, we have proved the

Theorem. *Suppose that k is algebraically closed, that the k-algebra A satisfies hypotheses (**H**) of (III.1.1), and that A is prime. Then $\mathcal{A}_A$ is a nonempty open (and hence dense) subset of maxspec Z.* $\square$

Examples. 1. If A is Azumaya, then of course $\mathcal{A}_A = \mathrm{maxspec}\, Z$.

2. In Example III.1.2(2), $Z(R)$ can be identified with $k[X]$, and $\mathcal{A}_R = (\mathrm{maxspec}\, k[X]) \setminus \{\langle X \rangle\}$.

3. Let $A = U(\mathfrak{sl}_2(k))$ and assume that k is algebraically closed of characteristic 2. Then, using the notation of (I.9.11)(a), the centre Z of A is given by

$$Z = k[e^2, f^2, h],$$

Exercise III.1.A. The Poincaré-Birkhoff-Witt Theorem says that the monomials $e^i f^j h^l$ form a k-basis for A, and so A is a free Z-module of rank 4. It follows from Corollary III.1.6 that PI-deg$(A) = 2$, and so a maximal ideal M of A is regular if and only if A/M is noncommutative, which occurs precisely when $h \notin M$. Therefore, by Theorem III.1.6,

$$\mathcal{A}_A = \{\mathfrak{m} \in \operatorname{maxspec} Z : h \notin \mathfrak{m}\}.$$

III.1.8. The singular locus. In this paragraph we make use of a few terms and notations introduced in Appendix I.15. If Z is a commutative noetherian ring, its *singular locus* is the set

$$\mathcal{S}_Z := \{\mathfrak{m} \in \operatorname{maxspec} Z : \operatorname{gl.dim}(Z_\mathfrak{m}) = \infty\}.$$

The name derives from the case when Z is an affine algebra over an algebraically closed field, since then Z is the coordinate ring of an affine algebraic variety, and the maximal ideals in $\mathcal{S}_Z$ correspond to the singular points of the variety.

Lemma. *Suppose that A is a prime k-algebra satisfying hypotheses* (**H**)*. Let $Z = Z(A)$. If* gl.dim$(A) < \infty$*, then $\mathcal{A}_A$ and $\mathcal{S}_Z$ are disjoint.*

Proof. Let $\mathfrak{m} \in \mathcal{A}_A$, and note that $A_\mathfrak{m}/\mathfrak{m}A_\mathfrak{m} \neq 0$. Since $A_\mathfrak{m}$ is assumed to be Azumaya over $Z_\mathfrak{m}$, it is a projective $Z_\mathfrak{m}$-module. Hence, any $A_\mathfrak{m}$-module projective resolution of $A_\mathfrak{m}/\mathfrak{m}A_\mathfrak{m}$ is also a $Z_\mathfrak{m}$-module projective resolution; thus, p.dim$_{Z_\mathfrak{m}}(A_\mathfrak{m}/\mathfrak{m}A_\mathfrak{m})$ is finite. Since $Z_\mathfrak{m}/\mathfrak{m}Z_\mathfrak{m} = k$, it is a direct summand of $A_\mathfrak{m}/\mathfrak{m}A_\mathfrak{m}$ as $Z_\mathfrak{m}$-modules, and so p.dim$_{Z_\mathfrak{m}}(Z_\mathfrak{m}/\mathfrak{m}Z_\mathfrak{m})$ is finite. It follows that $Z_\mathfrak{m}$ has finite global dimension [**190**, Corollary 9.55], and therefore $\mathfrak{m} \notin \mathcal{S}_Z$. $\square$

In the setting of the lemma, the Azumaya locus $\mathcal{A}_A$ is contained in the complement of the singular locus $\mathcal{S}_Z$. This inclusion is, in general, proper. For instance, in Example III.1.7(3), Z is a polynomial ring in three indeterminates, whence gl.dim$(Z) = 3$, and so $\mathcal{S}_Z$ is empty. However, $\mathcal{A}_A$ is a proper subset of maxspec Z – as noted in the example, $\mathcal{A}_A = \{\mathfrak{m} \in \operatorname{maxspec} Z : h \notin \mathfrak{m}\}$.

Under appropriate conditions, the gap can be closed. We state the following theorem only in the setting of the current section, although it holds more generally. In Chapter III.8 we prove a variant of this theorem, sufficient for application to the Hopf algebras with which we are concerned.

Theorem. *Suppose that k is algebraically closed and A is a prime k-algebra satisfying hypotheses (H) of (III.1.1), with centre Z. Assume also that A is Auslander-regular and Macaulay, and that $A_{\mathfrak{p}}$ is Azumaya over $Z_{\mathfrak{p}}$ for all height 1 prime ideals $\mathfrak{p}$ of Z. Then*

$$\mathcal{A}_A = (\mathrm{maxspec}\, Z) \setminus \mathcal{S}_Z.$$

Proof. [**21**, Theorem 3.8] □

NOTES

Most of the above PI ring theory can be found in [**191**]. The homological background for (III.1.8) is sketched in Appendix I.15, where further references are given. Theorem III.1.8 was proved in [**21**], based on earlier work of Le Bruyn [**131**] for the graded case.

EXERCISES

Exercise III.1.A. Check that the centre of $U(\mathfrak{sl}_2(k))$ over a field k of characteristic 2 is $k[e^2, f^2, h]$, as claimed in (III.1.7), Example 3.

Exercise III.1.B. Let q be a primitive ℓth root of 1 and let k be an algebraically closed field. Recall from (II.1.14) that $H = (k^{\times})^n$ acts as k-algebra automorphisms of $A = \mathcal{O}_q((k^{\times})^n)$. Show that H permutes the maximal ideal spectrum of A transitively. Deduce (as suggested in Example III.1.4) that A is an Azumaya algebra.

THE FINITE DIMENSIONAL
REPRESENTATIONS OF $U_\epsilon(\mathfrak{sl}_2(k))$

Throughout this chapter let ℓ be an odd positive integer, $\ell \neq 1$, and let ϵ be a primitive ℓth root of 1 in the field k. *We shall assume here and throughout most of Part III that k has characteristic 0.* In this chapter we work out the details of the irreducible $U_\epsilon(\mathfrak{sl}_2(k))$-modules. This has several purposes: it will help us to develop some intuition for the sorts of algebras which occur in the root of unity case, and to illustrate the theory summarised in Chapter III.1; the case of $\mathfrak{sl}_2(k)$ is used, just as in the classical theory, in many proofs valid for all quantized enveloping algebras at roots of unity, and so has to be developed separately; and some ways of thought which appear natural when discussing $U_\epsilon(\mathfrak{sl}_2(k))$ help lead us to some useful definitions which we shall introduce in Chapter III.4.

III.2.1. The k-algebra $U_\epsilon := U_\epsilon(\mathfrak{sl}_2(k))$ has generators and relations as in (I.3.1), but with q replaced by ϵ. Thus U_ϵ is the k-algebra with generators E, F, $K^{\pm 1}$ and relations

$$KEK^{-1} = \epsilon^2 E \qquad KFK^{-1} = \epsilon^{-2}F \qquad EF - FE = \frac{K - K^{-1}}{\epsilon - \epsilon^{-1}}. \quad (1)$$

As in the generic case, this algebra can be presented as an iterated skew polynomial ring $k[K^{\pm 1}][E; \tau_1][F; \tau_2, \delta_2]$. Also, we write $U_\epsilon^{\geq 0}$, $U_\epsilon^{\leq 0}$, and U_0 for the respective subalgebras of U_ϵ generated by E, $K^{\pm 1}$, by F, $K^{\pm 1}$, and by $K^{\pm 1}$.

The fundamental difference in the structure of U_ϵ as compared with $U_q(\mathfrak{sl}_2(k))$ with q generic lies in the sizes of their centres – for, as we've already noted in (I.4.5), Remark 2, the centre of $U_q(\mathfrak{sl}_2(k))$ is the polynomial algebra $k[C_q]$. On the other hand a straightforward calculation (Exercise III.2.A) shows that

$$Z_0 \quad := \quad k[E^\ell, F^\ell, K^{\pm \ell}]$$

is a central subalgebra of U_ϵ. By the PBW Theorem, Corollary I.3.1, Z_0 is a polynomial algebra in 3 indeterminates, and U_ϵ is a free Z_0-module of rank ℓ^3, with free basis

$$\{E^i K^j F^r : 0 \leq i, j, r \leq \ell - 1\}. \quad (2)$$

We're thus in the realm of algebras satisfying a polynomial identity, so that the theory recalled in Appendix I.13 and Chapter III.1 applies.

In particular, *every* irreducible U_ϵ-module V is finite dimensional and is killed by a maximal ideal of Z_0 by Proposition III.1.1(4). Indeed, writing Z for the centre of U_ϵ and assuming momentarily that k is algebraically closed, then Theorem I.13.5 yields

$$\dim_k V \le \left(\dim_{\mathrm{Fract}(Z)}\left(U_\epsilon \otimes_Z \mathrm{Fract}(Z)\right)\right)^{1/2} \le \ell^{3/2}.$$

By Lemma III.1.2(2) the first of these inequalities is best possible. As we shall see below, this implies that

$$\dim_{\mathrm{Fract}(Z)}\left(U_\epsilon \otimes_Z \mathrm{Fract}(Z)\right) = \ell^2;$$

in other words, ℓ is the maximal dimension of irreducible U_ϵ-modules when k is algebraically closed.

III.2.2. Baby Verma modules. Let

$$S := k\langle E, K^{\pm 1}, F^\ell \rangle \subseteq U_\epsilon.$$

Let $b, \lambda \in k$ with $\lambda \ne 0$. Since $SE = ES$ is an ideal of S and $S/SE \cong k[F^\ell, K^{\pm 1}]$ is commutative, there is a one-dimensional S-module $k(b, \lambda)$ given by

$$Ev = 0 \qquad\qquad Kv = \lambda v \qquad\qquad F^\ell v = bv$$

for $v \in k = k(b, \lambda)$. Define the *baby Verma module*

$$Z_b(\lambda) := U_\epsilon \otimes_S k(b, \lambda). \tag{3}$$

By the PBW theorem, Corollary I.3.1,

$$\dim_k(Z_b(\lambda)) = \ell.$$

We can also define $Z_b(\lambda)$ as a quotient of the Verma module $M(\lambda)$ – that is, in the notation of (I.4.2),

$$Z_b(\lambda) \cong M(\lambda)/U_\epsilon(m_\ell - bm_0), \tag{4}$$

(Exercise III.2.B). So from (I.4.2)(2), (3) and (4), $Z_b(\lambda)$ has a k-basis consisting of $\{\widehat{m}_0, \dots, \widehat{m}_{\ell-1}\}$ with

$$K\widehat{m}_i \;=\; \lambda\epsilon^{-2i}\widehat{m}_i, \tag{5}$$

$$F\widehat{m}_i \;=\; \begin{cases} \widehat{m}_{i+1}, & i < \ell - 1, \\ b\widehat{m}_0, & i = \ell - 1, \end{cases} \tag{6}$$

$$E\widehat{m}_i \;=\; \begin{cases} 0, & i = 0, \\ [i]_\epsilon \dfrac{\lambda\epsilon^{1-i} - \lambda^{-1}\epsilon^{i-1}}{\epsilon - \epsilon^{-1}}\widehat{m}_{i-1}, & i > 0. \end{cases} \tag{7}$$

Lemma. *Let $b, \lambda \in k$ with $\lambda \neq 0$.*

1. The distinct U_0-weightspaces of $Z_b(\lambda)$ are

$$Z_b(\lambda)_{\epsilon^{-2i}\lambda} = k\widehat{m}_i, \qquad (0 \leq i \leq \ell - 1).$$

2. $Z_b(\lambda)$ is annihilated by the maximal ideal $\langle E^\ell, \ F^\ell - b, \ K^\ell - \lambda^\ell \rangle$ of Z_0.

Proof. 1. This follows from (5), since ϵ is a primitive ℓth root of 1 with ℓ odd.

2. Consider the action of E^ℓ, F^ℓ and K^ℓ on $\widehat{m}_0$. □

Theorem. *Let $b, \lambda \in k$ with $\lambda \neq 0$. Consider the baby Verma module $Z_b(\lambda)$. There are four cases, which we label as follows:*

(A) $\lambda = \pm\epsilon^i$ for some i, $0 \leq i \leq \ell - 1$; $b = 0$.
(B) $\lambda = \pm\epsilon^i$ for some i, $0 \leq i \leq \ell - 1$; $b \neq 0$.
(C) $\lambda \neq \pm\epsilon^i$ for any i, $0 \leq i \leq \ell - 1$; $b \neq 0$.
(D) $\lambda \neq \pm\epsilon^i$ for any i, $0 \leq i \leq \ell - 1$; $b = 0$.

In cases (B), (C) and (D), $Z_b(\lambda)$ is an irreducible module. In case (A), $Z_0(\pm\epsilon^{\ell-1})$ is irreducible, but, for $0 \leq i < \ell - 1$, the elements $\widehat{m}_{i+1}, \ldots, \widehat{m}_{\ell-1}$ span the unique proper U_ϵ-submodule of $Z_0(\pm\epsilon^i)$.

Proof. Let $b, \lambda \in k$ with λ nonzero. Suppose that M is a nonzero submodule of $Z_b(\lambda)$. By 1 of the lemma, $Z_b(\lambda)$ and hence M is the sum of its U_0-weightspaces, and so M is spanned by those $\widehat{m}_j$ which it contains. Let $t \geq 0$ be minimal such that $\widehat{m}_t \in M$, so by (6)

$$M = \sum_{j=t}^{\ell-1} k\widehat{m}_j. \tag{8}$$

If $t = 0$ then of course $M = Z_b(\lambda)$ by (8).

Cases (B) and (C): Since $M \neq 0$, (8) shows that $\widehat{m}_{\ell-1} \in M$. Then $\widehat{m}_0 \in M$ by (6) since $b \neq 0$. Therefore $M = Z_b(\lambda)$.

Cases (A) and (D): By (7), $E\widehat{m}_t = 0$, so that

$$[t]_\epsilon(\lambda\epsilon^{1-t} - \lambda^{-1}\epsilon^{t-1}) \quad = \quad 0. \tag{9}$$

Suppose that $t \neq 0$. Then $[t]_\epsilon \neq 0$ by the hypothesis on ϵ, and so

$$\lambda^2 \quad = \quad \epsilon^{2(t-1)}.$$

In particular, $\lambda = \pm\epsilon^{(t-1)}$ and in case (D) we have a contradiction. Thus $t = 0$ in case (D). Finally, in case (A) the above calculation can be reversed to show that (9) holds with $t = i + 1$, so that $E\widehat{m}_{i+1} = 0$. It follows easily that $\sum_{j=i+1}^{\ell-1} k\widehat{m}_j$ is the unique proper submodule of $Z_b(\lambda)$ in this case, as claimed. □

Notation. For obvious reasons, in case (A) the unique irreducible quotient of $Z_0(\pm\epsilon^i)$ is denoted $L(i, \pm)$. It's easy to see that the formulae (I.4.3)(5), (6) and (7) giving the structure constants for $L(n, +)$ with respect to the basis $\{\overline{m}_0, \ldots, \overline{m}_n\}$ also serve for the $L(n, +)$ just defined here, with respect to the basis given by the cosets of $\{\widehat{m}_0, \ldots, \widehat{m}_n\}$ in $Z_0(\epsilon^n)/\sum_{j=n+1}^{\ell-1} k\widehat{m}_j$. And one can also confirm that there are non-split exact sequences

$$0 \longrightarrow L(\ell - n - 2, \pm) \longrightarrow Z_0(\pm\epsilon^n) \longrightarrow L(n, \pm) \longrightarrow 0$$

for $0 \leq n < \ell - 1$. (See Exercise III.2.D.)

III.2.3. The irreducible U_ϵ-modules. To see that we can use Theorem III.2.2 to describe *all* the irreducible U_ϵ-modules, we need to assume that k is algebraically closed. Let M be an irreducible U_ϵ-module and recall that by Proposition III.1.1(4) there is a maximal ideal $\mathfrak{m}$ of Z_0 with $\mathfrak{m}M = 0$. By our assumption on k and Hilbert's Nullstellensatz there are scalars b, c and α, with $\alpha \neq 0$, such that

$$\mathfrak{m} \quad = \quad \langle E^\ell - c, \ F^\ell - b, \ K^\ell - \alpha \rangle;$$

in other words, E^ℓ, F^ℓ and K^ℓ act on M by (respectively) multiplication by the scalars b, c and α.

There are 3 cases to discuss.

Case I. Suppose that
$$c = 0.$$

Then $M' := \{m \in M : Em = 0\}$ is a non-zero U_0-submodule of M, using relations (III.2.1)(1). Thus M' contains a U_0-eigenvector m, with eigenvalue λ. There is thus a U-module homomorphism

$$\theta : M(\lambda) \longrightarrow M$$
$$m_0 \longmapsto m$$

which is onto thanks to the simplicity of M. Then

$$\theta(m_\ell - bm_0) = \theta(F^\ell m_0 - bm_0) = F^\ell m - bm = 0.$$

By (III.2.2)(4), θ factors to give an epimorphism from $Z_b(\lambda)$ onto M. Hence M is one of the irreducible modules discussed in Theorem III.2.2.

Case II. Suppose that
$$b = 0, \quad c \neq 0.$$

Twist M by the automorphism ω of Lemma I.3.2. (See Exercise III.2.G.) Since $\omega(E) = F$ and $\omega(F) = E$, the twisted module $^\omega M$ belongs to case I. So $^\omega M$ is described by Theorem III.2.2, and belongs there to case (B) or to case (C), since $c \neq 0$. Thus

$$^\omega M \quad \cong \quad Z_c(\lambda)$$

for some $\lambda \in k^\times$, so that

$$M \quad = \quad {}^\omega({}^\omega M) \quad \cong \quad {}^\omega Z_c(\lambda).$$

It's easy to write down a basis for $^\omega Z_c(\lambda)$ for which the actions of E, F and K can be explicitly given (Exercise III.2.E).

Case III. Suppose that

$$b \neq 0 \neq c.$$

We'll show that

$$\dim_k M \quad = \quad \ell \tag{10}$$

and give a basis for M with an explicit action of the generators. The calculations in this case can also be done assuming only that $b \neq 0$; then if $c = 0$, parts of Case I are reproduced.

Let m_0 be a U_0-eigenvector in M; note that m_0 exists since k is algebraically closed. Let $Km_0 = \lambda m_0$, $\lambda \in k^\times$, and set

$$m_i \quad = \quad F^i m_0, \qquad 0 \leq i < \ell. \tag{11}$$

Thus

$$Fm_i \quad = \quad \begin{cases} m_{i+1} & i < \ell - 1 \\ bm_0 & i = \ell - 1, \end{cases} \tag{12}$$

and

$$Km_i \quad = \quad \epsilon^{-2i} \lambda m_i, \qquad 0 \leq i < \ell. \tag{13}$$

Since these eigenvalues are distinct, the subspace $M' := \sum_{i=0}^{\ell-1} km_i$ of M has

$$\dim_k M' \quad = \quad \ell. \tag{14}$$

We claim that

$$M' \quad \text{is a submodule of } M. \tag{15}$$

By construction, we only have to check that $Em_i \in M'$ for $0 \leq i < \ell$. Recall from (I.4.5) that

$$C_\epsilon \quad := \quad FE + \frac{K\epsilon + K^{-1}\epsilon^{-1}}{(\epsilon - \epsilon^{-1})^2} \quad \in \quad Z(U_\epsilon).$$

By Proposition III.1.1(4), C_ϵ acts on M by multiplication by a scalar. In particular, since

$$FEm_0 \quad = \quad C_\epsilon m_0 - \frac{K\epsilon + K^{-1}\epsilon^{-1}}{(\epsilon - \epsilon^{-1})^2} m_0,$$

(13) shows that there exists $d \in k$ with $FEm_0 = dm_0$. Hence

$$bEm_0 = F^\ell Em_0 = F^{\ell-1} dm_0 = dm_{\ell-1},$$

so that, setting $a := b^{-1} d$,

$$Em_0 = am_{\ell-1}. \tag{16}$$

Now, using Exercise I.4.D(iv), for all $i = 1, \dots, \ell - 1$,

$$Em_i = EF^i m_0 = F^i Em_0 + [i]_\epsilon F^{i-1}[K; 1 - i] m_0.$$

This implies, using (11), (12), (13) and (16), that

$$Em_i = \begin{cases} am_{\ell-1}, & i = 0, \\ \left(ab + \dfrac{(\epsilon^i - \epsilon^{-i})(\lambda \epsilon^{1-i} - \lambda^{-1}\epsilon^{i-1})}{(\epsilon - \epsilon^{-1})^2}\right) m_{i-1}, & i > 0. \end{cases} \tag{17}$$

This proves (15). Since M is irreducible, (10) is proved and we've found an explicit description of M.

Conversely, given scalars a, b and λ as above, (so that $b \neq 0$), (12), (13) and (17) can be used to define an irreducible U_ϵ-module, which is annihilated by $\mathfrak{m} = \langle E^\ell - c, \ F^\ell - b, \ K^\ell - \alpha \rangle$ provided λ is an ℓth root of α and a satisfies an equation of degree ℓ which can be written down using (16) (Exercise III.2.F).

III.2.4. The finite dimensional algebras $U_\epsilon/\mathfrak{m}U_\epsilon$. We continue in this paragraph to assume that k is algebraically closed. A very fruitful way of thinking about the representation theory and structure of the algebra U_ϵ (and of similar algebras, as we shall see), is to consider the bundle of finite dimensional algebras

$$U_\chi := U_\epsilon/\mathfrak{m}_\chi U_\epsilon$$

as $\mathfrak{m}_\chi$ ranges through maxspec Z_0. The ideals $\mathfrak{m}_\chi$ are parametrised by $k \times k \times k^\times$ – given $\chi = (c, b, \alpha) \in k \times k \times k^\times$, we set

$$\mathfrak{m}_\chi := \langle E^\ell - c, \ F^\ell - b, \ K^\ell - \alpha \rangle.$$

Clearly,

$$\dim_k(U_\chi) = \ell^3,$$

and in fact the cosets of the elements (III.2.1)(2) afford a basis of U_χ. It turns out, for reasons we'll examine later (III.6.5), that we may assume that $c = 0$. Then, by case I of (III.2.3), all irreducible U_χ-modules have the forms discussed in Theorem III.2.2. There are 3 cases:

Case I. Suppose that $b \in k$ is arbitrary and

$$\chi = (0, b, \alpha), \quad \alpha \neq \pm 1.$$

Here we are in Case (C) or (D) of Theorem III.2.2. Thus, for λ an ℓth root of α the U_χ-modules $Z_0(\lambda)$ are irreducible. Moreover, by (III.2.2)(5) each $Z_0(\lambda)$ has a basis of distinct U_0-eigenvectors, and by (III.2.2)(7) there is (up to a scalar multiple) a unique eigenvector killed by E. It follows at once that for distinct ℓth roots of α, say λ and μ, $Z_0(\lambda)$ is not isomorphic to $Z_0(\mu)$. Hence U_χ admits ℓ different ℓ-dimensional irreducible modules, and

$$U_\chi \cong M_\ell(k)^{\oplus \ell}.$$

Case II. Suppose that

$$\chi = (0, 0, \pm 1).$$

These are the cases covered by Case (A) of Theorem III.2.2. There are ℓ distinct irreducible U_χ-modules in these cases, of dimensions $1, 2, \ldots, \ell$. Thus

$$\dim_k(U_\chi / J(U_\chi)) = \frac{\ell(\ell+1)(2\ell+1)}{6}.$$

Note that $U_{(0,0,1)}$ and $U_{(0,0,-1)}$ are interchanged by the automorphism ω of U_ϵ defined in (I.3.2), and so are isomorphic algebras.

Case III. Suppose that

$$\chi = (0, b, \pm 1), \quad b \neq 0.$$

We're now in Case (B) of Theorem III.2.2. Therefore each irreducible U_χ-module has dimension ℓ, but in fact there are only $(\ell+1)/2$ isomorphism classes of irreducible modules. One can show, by methods we'll discuss in Chapters III.6 and III.9, that

$$U_\chi \cong M_\ell(k) \oplus \left(M_\ell(k[X]/\langle X^2 \rangle) \right)^{\oplus \frac{\ell-1}{2}}.$$

NOTES

Most of the above material is standard, although (III.2.4) may not have been written down in this form elsewhere. (III.2.3) and (III.2.4) are adapted from [107, Chapter 2].

Exercises

Exercise III.2.A. (Notation (III.2.1).) Show that Z_0 is contained in the centre of $U_\epsilon(\mathfrak{sl}_2(k))$. (Use Exercise I.4.D and the fact that $[\ell]_\epsilon = 0$.)

Exercise III.2.B. Confirm the presentation (III.2.2)(4) of $Z_b(\lambda)$.

Exercise III.2.C. Prove Lemma III.2.2(2).

Exercise III.2.D. In case (A) of Theorem III.2.2, show that there are non-split exact sequences of $U_\epsilon(\mathfrak{sl}_2(k))$-modules

$$0 \longrightarrow L(\ell - n - 2, \pm) \longrightarrow Z_0(\pm\epsilon^n) \longrightarrow L(n, \pm) \longrightarrow 0$$

for $0 \leq n < \ell - 1$.

Exercise III.2.E. Complete the details of (III.2.3) Case II.

Exercise III.2.F. Complete the details of (III.2.3) Case III. (One has to check that given a, b, λ with $b \neq 0$, an irreducible module can be defined using (III.2.3)(12), (13) and (17), which is a case III module provided a satisfies an appropriate condition.)

Exercise III.2.G. Let R be a ring, M a left R-module, and ω an automorphism of R. Define the *twist* $^\omega M$ of M by ω to be the left R-module which has the same structure of abelian group as M, with $r.m := \omega(r)m$ for $r \in R$ and $m \in {}^\omega M$, where on the right of this definition the action is the original R-module multiplication. Show that $^\omega M$ is a left R-module, and that if ω is an inner automorphism then $^\omega M \cong M$.

THE FINITE DIMENSIONAL
REPRESENTATIONS OF $\mathcal{O}_\epsilon(SL_2(k))$

In this chapter ϵ is again a primitive ℓth root of 1 in the field k of characteristic 0, with ℓ odd, $\ell \geq 3$. Our aims here are similar to those we had in the previous chapter, but this time we study the simplest case of a quantized function algebra at a root of unity, with a view to developing our intuition and ideas for the general case.

III.3.1. Recall from (I.1.9) that $\mathcal{O}_\epsilon(SL_2(k))$ is the Hopf algebra with generators a, b, c, d and relations

$$ab = \epsilon ba \qquad\qquad ac = \epsilon ca \qquad\qquad bc = cb$$
$$bd = \epsilon db \qquad\qquad cd = \epsilon dc \qquad\qquad ad - da = (\epsilon - \epsilon^{-1})bc,$$

together with the determinant relation $ad - \epsilon bc = 1$. The Hopf structure is given in (I.1.9).

Proposition. *Let Z_0 be the subalgebra of $\mathcal{O}_\epsilon(SL_2(k))$ generated by a^ℓ, b^ℓ, c^ℓ and d^ℓ.*

1. *Z_0 is in the centre of $\mathcal{O}_\epsilon(SL_2(k))$.*
2. *Z_0 is a sub-Hopf algebra of $\mathcal{O}_\epsilon(SL_2(k))$.*
3. *As a Hopf algebra, $Z_0 \cong \mathcal{O}(SL_2(k))$.*
4. *$\mathcal{O}_\epsilon(SL_2(k))$ is a projective Z_0-module of constant rank ℓ^3.*

Proof. 1. It's clear that b^ℓ and c^ℓ are in $Z(\mathcal{O}_\epsilon(SL_2(k)))$. Also, a^ℓ clearly commutes with b and c, and an easy induction to determine $[a^i, d]$ (Exercise III.3.A) shows that it commutes with d too. Similar calculations deal with d^ℓ.

2. That

$$\Delta(a^\ell) = a^\ell \otimes a^\ell + b^\ell \otimes c^\ell \tag{1}$$

follows from the q-binomial theorem (I.6.1), since $\Delta(a) = a \otimes a + b \otimes c$ and $x := a \otimes a$ and $y := b \otimes c$ satisfy $xy = \epsilon^2 yx$. The comultiplication of the other generators and the fact that $SZ_0 \subseteq Z_0$ follow by similar calculations.

3. This is also easy – we check that

$$a^\ell d^\ell - b^\ell c^\ell = 1$$

in Z_0. Thus the k-algebra map

$$\psi : \mathcal{O}(SL_2(k)) = k[x, y, z, t]/\langle xt - yz = 1\rangle \longrightarrow Z_0$$

sending x, y, z and t to a^ℓ, b^ℓ, c^ℓ and d^ℓ respectively is an epimorphism of algebras. That ψ is injective follows from a comparison of k-bases of the two algebras, using Example I.11.8. From (1) and its partner formulae for the other generators, and from the formulae for the antipode, there is no problem in confirming that ψ is an isomorphism of Hopf algebras.

4. This can be checked directly, but we prefer to deduce it as a special case of a general Hopf algebra result which we'll recall later – see Corollary III.4.7. □

III.3.2. Representation theory. Just as with $U_\epsilon(\mathfrak{sl}_2(k))$, $\mathcal{O}_\epsilon(SL_2(k))$ satisfies hypotheses **(H)** of (III.1.1). Hence Proposition III.1.1 applies – in particular every irreducible $\mathcal{O}_\epsilon(SL_2(k))$-module is finite dimensional over k and is annihilated by a maximal ideal of Z_0. We can study these irreducible representations by looking at the bundle of algebras

$$\mathcal{B}_\mathcal{O} \quad := \quad \{\mathcal{O}_\epsilon(SL_2(k))/\mathfrak{m}\mathcal{O}_\epsilon(SL_2(k)) : \mathfrak{m} \in \mathrm{maxspec}\, Z_0\}.$$

Assume for the rest of this chapter that

$$k \quad \text{is an algebraically closed field.}$$

Then each algebra in $\mathcal{B}_\mathcal{O}$ has dimension ℓ^3 by Proposition III.3.1(4) and the Nullstellensatz. There are 6 cases to consider.

Case I: $\mathfrak{m} = \langle b^\ell, c^\ell, a^\ell - \chi, d^\ell - \chi^{-1}\rangle$ for $\chi \in k^\times$. Set

$$\overline{\mathcal{O}} \quad := \quad \mathcal{O}_\epsilon(SL_2(k))/\mathfrak{m}\mathcal{O}_\epsilon(SL_2(k)).$$

So $\overline{b}$ and $\overline{c}$ are normal nilpotent elements in $\overline{\mathcal{O}}$, and hence

$$\overline{b}\overline{\mathcal{O}} + \overline{c}\overline{\mathcal{O}} \subseteq J(\overline{\mathcal{O}}).$$

But, since $\overline{d} = (\overline{a})^{-1}$,

$$\overline{\mathcal{O}}/(\overline{b}\overline{\mathcal{O}} + \overline{c}\overline{\mathcal{O}}) \cong k\langle\overline{a}\rangle = k[\hat{a} : \hat{a}^\ell = 1] \cong k^{\oplus \ell}.$$

Thus the algebras in this class (which are clearly all isomorphic) have exactly ℓ irreducibles, each of dimension 1.

Case II: $\mathfrak{m} = \langle b^\ell, c^\ell - \lambda, a^\ell - \chi, d^\ell - \chi^{-1}\rangle$ for $\chi, \lambda \in k^\times$. In this case $\overline{b}$ is normal and nilpotent, and

$$\overline{\mathcal{O}}/\overline{b}\overline{\mathcal{O}} \cong k\langle\hat{a}, \hat{c} : \hat{a}\hat{c} = \epsilon\hat{c}\hat{a}, \ \hat{a}^\ell = 1 = \hat{c}^\ell\rangle \cong M_\ell(k), \tag{2}$$

using Exercise III.3.C for the second isomorphism. So here $\overline{\mathcal{O}}$ has exactly one irreducible module, of dimension ℓ.

Case III: $\mathfrak{m} = \langle b^\ell - \lambda, \; c^\ell, \; a^\ell - \chi, \; d^\ell - \chi^{-1}\rangle$ for $\chi, \lambda \in k^\times$. Here $\overline{\mathcal{O}}$ has the same structure as in Case II, but with b and c interchanged.

Case IV: $\mathfrak{m} = \langle b^\ell - \lambda, \; c^\ell - \mu, \; a^\ell - \chi, \; d^\ell - \rho\rangle$ for $\lambda, \mu, \chi \in k^\times$ and $\rho \in k$, with $\chi\rho - \lambda\mu = 1$. Since $\overline{d} = (\overline{a})^{-1}(1 + \epsilon\overline{b}\overline{c})$,

$$\overline{\mathcal{O}} = k\langle \overline{a}, \overline{b}, \overline{c}\rangle = k\langle \overline{a}, \overline{c}\rangle[\overline{b}(\overline{c})^{-1}]$$

with $\overline{b}(\overline{c})^{-1}$ a central ℓth root of $\lambda\mu^{-1}$. Using Exercise III.3.C again, it follows easily that

$$\overline{\mathcal{O}} \cong (k^{\oplus\ell}) \otimes M_\ell(k) \cong M_\ell(k)^{\oplus\ell}.$$

Case V: $\mathfrak{m} = \langle b^\ell - \lambda, \; c^\ell - \mu, \; a^\ell - \chi, \; d^\ell - \rho\rangle$ for $\lambda, \mu, \rho \in k^\times$ and $\chi \in k$, with $\chi\rho - \lambda\mu = 1$. Here $\overline{\mathcal{O}}$ has the same structure as in Case IV, but with a and d interchanged.

Case VI: $\mathfrak{m} = \langle b^\ell - \lambda, \; c^\ell + \lambda^{-1}, \; a^\ell, \; d^\ell\rangle$ for $\lambda \in k^\times$. Observe that we may replace the relation $ad - da = (\epsilon - \epsilon^{-1})bc$ in (III.3.1) with

$$ad - \epsilon^2 da = 1 - \epsilon^2.$$

Since $\overline{c} = (\epsilon\overline{b})^{-1}(\overline{a}\overline{d} - 1)$,

$$\overline{\mathcal{O}} = k\langle \overline{a}, \overline{d}\rangle\langle \overline{b}\rangle \cong M_\ell(k)^{\oplus\ell},$$

using Exercise III.3.D.

III.3.3. Commentary on (III.3.2). From the above we see that the maximum dimension of irreducible $\mathcal{O}_\epsilon(SL_2(k))$-modules is ℓ. Thus from the definition of PI degree and from Lemma III.1.2(2),

$$\text{PI-deg}(\mathcal{O}_\epsilon(SL_2(k))) = \ell.$$

(In fact, we could already have said that the PI degree of $\mathcal{O}_\epsilon(SL_2(k))$ is *at most* ℓ, since PI-deg$(\mathcal{O}_\epsilon(M_2(k))) = \ell$ by (I.14.3(2)).) From (III.3.2) and Theorem III.1.6, the *regular* maximal ideals of $\mathcal{O}_\epsilon(SL_2(k))$ are precisely those lying over a maximal ideal of Z_0 from Classes II–VI. In other words, the complement of the Azumaya locus $\mathcal{A}_{\mathcal{O}_\epsilon(SL_2(k))}$ of $Z = Z(\mathcal{O}_\epsilon(SL_2(k)))$ is given by

$$\text{maxspec}(Z) \setminus \mathcal{A}_{\mathcal{O}_\epsilon(SL_2(k))} = \{\mathfrak{m} \in \text{maxspec}(Z) : b^\ell, c^\ell \in \mathfrak{m}\}. \qquad (3)$$

NOTES

The basic facts about the irreducible representations of $\mathcal{O}_\epsilon(SL_2(k))$ are worked out in [**43**].

EXERCISES

Exercise III.3.A. (Notation III.3.1.) Prove by induction a formula for $[a^i, d]$, where a and d are the generators of $\mathcal{O}_q(SL_2(k))$. Hence show that a^ℓ is central in $\mathcal{O}_\epsilon(SL_2(k))$ when ϵ is an ℓth root of 1. (Cf. Example I.14.3(2).)

Exercise III.3.B. Complete the details of the proof of Proposition III.3.1, parts (1), (2), (3).

Exercise III.3.C. Prove that

$$k\langle \hat{a}, \hat{c} : \hat{a}\hat{c} = \epsilon\hat{c}\hat{a}, \ \hat{a}^\ell = 1 = \hat{c}^\ell \rangle \cong M_\ell(k).$$

Exercise III.3.D. In Case VI of (III.3.2), prove that

$$k\langle \overline{a}, \overline{d} \rangle = k\langle \overline{a}, \overline{d} : \overline{a}\overline{d} - \epsilon^2 \overline{d}\overline{a} = 1 - \epsilon^2, \ \overline{a}^\ell = \overline{d}^\ell = 0 \rangle \cong M_\ell(k)$$

and that

$$\overline{\mathcal{O}} \cong k\langle \hat{a}, \hat{d} \rangle \langle \hat{b} : \hat{a}\hat{b} = \epsilon\hat{b}\hat{a}, \ \hat{b}\hat{d} = \epsilon\hat{d}\hat{b}, \ \hat{b}^\ell = 1 \rangle \cong M_\ell(k)^{\oplus \ell}.$$

CHAPTER III.4

BASIC PROPERTIES OF PI HOPF TRIPLES

Throughout this chapter k will be an algebraically closed field; positive characteristic will be allowed. In this chapter we'll forget about quantum groups, and consider a class of Hopf algebras satisfying a definition designed to capture some of the features we've already noted in the examples treated in Chapters III.2 and III.3, features which – as it turns out – occur more generally, not just for quantized enveloping algebras and quantized function algebras at roots of unity, but also, for instance, in enveloping algebras of modular Lie algebras.

III.4.1. Definition. A *PI Hopf triple* over k consists of a set of three k-algebras

$$Z_0 \subseteq Z \subseteq H \tag{1}$$

such that

$$H \text{ is a } k\text{-affine } k\text{-Hopf algebra,} \tag{2}$$
$$Z \text{ is the centre of } H, \tag{3}$$
$$Z_0 \text{ is a sub-Hopf algebra of } H, \tag{4}$$
$$H \text{ is a finitely generated } Z_0\text{-module,} \tag{5}$$

and

$$H \text{ is a prime ring.} \tag{6}$$

Much of our discussion applies without assuming (6), but this hypothesis holds in our main examples and simplifies the discussion at times, so it's convenient to assume it globally. Of course it implies that Z and Z_0 are domains.

Notice that if Z_1 and Z_2 are two subalgebras of H satisfying the conditions imposed on Z_0 in (1), (4) and (5) above, then the same is true for the subalgebra generated by $Z_1 \cup Z_2$. It follows (Exercise III.4.I) that an algebra H satisfying the conditions of the definition always has a unique maximal subalgebra with the properties required of Z_0 – we shall call this maximal central Hopf subalgebra the *Hopf centre* of H. Not every Hopf algebra which is a finite module over an affine central subalgebra is a finite module over a central Hopf subalgebra – see Exercise III.10.C.

Examples. 1. For the choices of k and ϵ stated at the starts of Chapters III.2 and III.3, $U_\epsilon(\mathfrak{sl}_2(k))$ and $\mathcal{O}_\epsilon(SL_2(k))$ yield PI Hopf triples with Z_0 as in those chapters.

2. For this class of examples, see Theorem I.13.2. Let $(\mathfrak{g}, [p])$ be a finite dimensional restricted Lie algebra over the field k of positive characteristic p. Then $H = U(\mathfrak{g})$, the enveloping algebra of $\mathfrak{g}$, forms a PI Hopf triple with Z_0 the subalgebra generated by the elements $\{x^p - x^{[p]} : x \in \mathfrak{g}\}$.

III.4.2. Lemma. *Let (III.4.1)(1) be a PI Hopf triple. Then Z_0 is an affine k-algebra. Thus H is a noetherian PI-algebra and Z is an affine algebra.*

Proof. The Artin-Tate Lemma (I.13.4) shows that Z_0 is an affine k-algebra. So Z_0 is noetherian by Hilbert's basis theorem. Hence H is a noetherian Z_0-module, and so *a fortiori* a noetherian ring. That H is a PI-ring follows from (3), (5) and (I.13.2), Example 3. Finally, Z is a finitely generated Z_0-module and therefore an affine k-algebra.   □

III.4.3. The restriction maps on spectra. Let (III.4.1)(1) be a PI Hopf triple. We define maps

$$\operatorname{spec} H \xrightarrow{\ \tau\ } \operatorname{spec} Z \xrightarrow{\ \mu\ } \operatorname{spec} Z_0$$

$$P \longmapsto P \cap Z$$

$$\mathfrak{p} \longmapsto \mathfrak{p} \cap Z_0$$

and write π for $\mu \circ \tau$. Here are the basic properties of this set-up, which don't depend on H being a Hopf algebra.

Lemma. *Keep the hypotheses and notation as above. Impose the Zariski topology on prime spectra.*

1. *τ, μ (and hence π) are continuous maps.*
2. *τ, μ (and hence π) are closed.*
3. *τ, μ (and hence π) are surjective and finite-to-one.*
4. *The restrictions τ', μ' and π' of these maps to the spectra of maximal ideals have images in the maximal ideal spectra, and 1, 2, 3 hold with τ', μ' and π' replacing τ, μ and π.*

Proof. 1. Clearly if I is an ideal of Z and $\mathfrak{V} = \{\mathfrak{p} \in \operatorname{spec} Z : I \subseteq \mathfrak{p}\}$, then $\tau^{-1}(\mathfrak{V}) = \{P \in \operatorname{spec} H : IH \subseteq P\}$. So τ is continuous. The other cases are similar.

2. This is a special case of Lemma III.1.5.

3. This is a special case of Proposition III.1.1(2).

4. This follows from Proposition III.1.1(5).   □

III.4.4. The bundle of algebras over $\mathrm{maxspec}\, Z_0$. Let (III.4.1)(1) be a PI Hopf triple. Recall that every irreducible H-module V has finite k-dimension and is killed by the maximal ideal $\mathrm{ann}_H(V) \cap Z_0 = \mathrm{ann}_{Z_0}(V) =: \mathfrak{m}$ of Z_0, by Proposition III.1.1(4). Thus we can study the irreducible H-modules by investigating the properties of the bundle

$$\mathfrak{B}_H \quad := \quad \{H/\mathfrak{m}H : \mathfrak{m} \in \mathrm{maxspec}\, Z_0\}$$

of finite dimensional k-algebras.

A particularly noteworthy member of $\mathfrak{B}_H$ is the algebra

$$\overline{H} \quad := \quad H/\mathfrak{m}_0 H$$

where $\mathfrak{m}_0$ is the augmentation ideal of Z_0; that is,

$$\mathfrak{m}_0 = \ker \varepsilon_H \cap Z_0 = \ker \varepsilon_{Z_0}.$$

Since by (4) $\mathfrak{m}_0$ is a Hopf ideal of Z_0, $\mathfrak{m}_0 H$ is a Hopf ideal of H, so that

$$\overline{H} \quad \text{is a finite dimensional Hopf } k\text{-algebra.}$$

Extending terminology from enveloping algebras in positive characteristic, $\overline{H}$ is called the *restricted Hopf algebra* (associated to the triple (III.4.1)(1)).

Example. When $H = U_\epsilon(\mathfrak{sl}_2(k))$, the algebra $\overline{H}$ is the Hopf k-algebra generated by $\overline{E}, \overline{F}, \overline{K}^{\pm 1}$ with relations as in (III.2.1), together with

$$\overline{E}^\ell = \overline{F}^\ell = 0 \qquad\qquad \overline{K}^\ell = \overline{K}^{-\ell} = 1.$$

It has dimension ℓ^3.

III.4.5. H is a projective Z_0-module. There are several ways to prove this. Here is what seems to us to be the simplest. We assume throughout this paragraph that (III.4.1)(1) is a PI Hopf triple. Let $\mathfrak{m}_0$ be the augmentation ideal of Z_0. Recall that if $\mathfrak{p}$ is a prime ideal of a commutative ring R, we can form the localization $R_\mathfrak{p}$ of R at $\mathfrak{p}$. In the simplest case where R is a domain, by definition

$$R_\mathfrak{p} \quad := \quad \{ac^{-1} : a \in R, \ c \in R \setminus \mathfrak{p}\} \subseteq \mathrm{Fract}(R).$$

Recall that

$$R_\mathfrak{p} \quad \text{is a flat } R\text{-module.} \tag{7}$$

Theorem. H is a finitely generated projective Z_0-module.

Proof. Let $r := \dim_k \overline{H}$. We first prove that

$$H_{\mathfrak{m}_0} := Z_{0\mathfrak{m}_0} \otimes_{Z_0} H \quad \text{is a free } Z_{0\mathfrak{m}_0}\text{-module of rank } r. \tag{8}$$

The map

$$\beta : H \otimes_{Z_0} H \longrightarrow H \otimes_k \overline{H}$$
$$x \otimes y \longmapsto \sum x y_1 \otimes \overline{y}_2 \tag{9}$$

is a well-defined homomorphism of left H-modules, where the H-module actions on these tensor products arise from multiplication in the lefthand factors. It is bijective, because the map

$$H \otimes_k \overline{H} \longrightarrow H \otimes_{Z_0} H$$
$$u \otimes \overline{v} \longmapsto \sum u S(v_1) \otimes v_2$$

is well-defined and is an inverse for β. Hence

$$H \otimes_{Z_0} H \quad \text{is free of rank } r \text{ as left } H\text{-module.} \tag{10}$$

Nakayama's lemma shows that there is an epimorphism of $Z_{0\mathfrak{m}_0}$-modules

$$f : Z_{0\mathfrak{m}_0}^{(r)} \longrightarrow H_{\mathfrak{m}_0}.$$

To prove (8) it suffices to show that f is injective. But, thanks to (10),

$$1 \otimes f : H_{\mathfrak{m}_0} \otimes_{Z_{0\mathfrak{m}_0}} Z_{0\mathfrak{m}_0}^{(r)} \longrightarrow H_{\mathfrak{m}_0} \otimes_{Z_{0\mathfrak{m}_0}} H_{\mathfrak{m}_0} = (H \otimes_{Z_0} H)_{\mathfrak{m}_0}$$

is a surjective left $H_{\mathfrak{m}_0}$-module homomorphism from $H_{\mathfrak{m}_0}^{(r)}$ onto $H_{\mathfrak{m}_0}^{(r)}$. Since $H_{\mathfrak{m}_0}$ is noetherian by Lemma III.4.2, $1 \otimes f$ must be injective. Hence f is also injective, proving (8).

Next we prove that, for every maximal ideal $\mathfrak{m}$ of Z_0,

$$H_{\mathfrak{m}} \quad \text{is a free } Z_{0\mathfrak{m}}\text{-module.} \tag{11}$$

Indeed (11) follows easily from (7) and Exercise III.4.A, using the winding automorphism of Z_0 corresponding to $\mathfrak{m}$ (I.9.25). But it's a standard fact in the homological algebra of commutative noetherian rings that a finitely generated module is projective if and only if it's locally free. So the proof of the theorem is complete.
□

Remarks. 1. The above theorem should be compared with the well known theorem of Nichols and Zoeller which states that a finite dimensional Hopf algebra is a free module over each of its Hopf subalgebras, [**163**, 3.1].

2. In general the theorem can't be strengthened to say that H is a free Z_0-module, even when H is commutative – see [**163**, Example 3.5.2]. We'll discuss this further later, in (III.10.2).

III.4.6. $Z_0 \subseteq H$ **is a Galois extension.** Recall from (I.9.18) the concept of a *right J-comodule algebra* A, where A is a k-algebra and J is a k-Hopf algebra. That is, we have a structure

$$\rho : A \longrightarrow A \otimes_k J$$

of right J-comodule on A, such that ρ is an algebra homomorphism. We define the *right coinvariants* $A^{\mathrm{coh}J}$ to be

$$A^{\mathrm{coh}J} \quad := \quad \{a \in A : \rho(a) = a \otimes 1\}.$$

Clearly $A^{\mathrm{coh}J}$ is a subalgebra of A. The name is justified by duality with the more familiar idea of *invariants*. Indeed A is a left J^*-module by Lemma I.9.16, and one can check (Exercise III.4.D) that

$$A^{\mathrm{coh}J} = A^{J^*} := \{a \in A : f.a = u^*(f)a = f(1)a \quad \text{for all } f \in J^*\},$$

where $u : k \longrightarrow J$ is the unit map.

Following Kreimer and Takeuchi [**128**], we say that the extension $A^{\mathrm{coh}J} \subseteq A$ is *right J-Galois* if the map

$$\begin{aligned}
\beta : A \otimes_{A^{\mathrm{coh}J}} A &\longrightarrow A \otimes_k J \\
a \otimes b &\longmapsto (a \otimes 1)\rho(b)
\end{aligned} \tag{12}$$

is bijective. The name is justified by the second of the examples below. Our interest will be in the third class of examples.

Examples. 1. Let $A = J$ be a finite dimensional Hopf algebra. Then $k \subseteq A$ is an A-Galois extension.

2. Classical Galois extensions. Let G be a finite group acting as k-automorphisms of a field E with $k \subseteq E$ and let $F = E^G$. So the dual $J = (kG)^*$ coacts on E. Now $F \subseteq E$ is classically Galois with Galois group $G \Longleftrightarrow G$ acts faithfully on $E \Longleftrightarrow [E : F] = |G|$. It's not hard to see that these equivalent conditions hold $\Longleftrightarrow F \subseteq E$ is J-Galois according to the above definition [**163**, 8.1.2].

3. Let H be a Hopf k-algebra, and let B be a central sub-Hopf algebra of H. Let B have augmentation ideal B^+. Thus $\overline{H} := H/B^+H$ is a quotient Hopf algebra of H with canonical epimorphism $\pi : H \longrightarrow \overline{H}$. So H is a right $\overline{H}$-comodule with

$$\rho = (1 \otimes \pi) \circ \Delta : H \longrightarrow H \otimes \overline{H}.$$

Lemma. *Keep the notation introduced above.*

    1. $B \subseteq H^{\mathrm{coh}\overline{H}}$.

    2. $\beta : H \otimes_B H \longrightarrow H \otimes_k \overline{H} : x \otimes y \mapsto (x \otimes 1)\rho(y)$ *is bijective.*

    3. *Suppose that H is a faithfully flat (left or right) B-module. Then $B = H^{\mathrm{coh}\overline{H}} = {}^{\mathrm{coh}\overline{H}}H$, and $B \subseteq H$ is a right (and left) $\overline{H}$-Galois extension.*

    4. *If $Z_0 \subseteq Z \subseteq H$ is a PI Hopf triple, then $Z_0 \subseteq H$ is $\overline{H}$-Galois.*

Proof. 1. Let $b \in B$. Then $\rho(b) = \sum b_1 \otimes \pi(b_2)$ and since $b_2 \in B$ we get $\rho(b) = \sum b_1 \otimes \varepsilon(b_2) = \sum b_1\varepsilon(b_2) \otimes 1 = b \otimes 1$. So $B \subseteq H^{\mathrm{coh}\overline{H}}$.

    2. This is the same as the proof of (III.4.5)(9): we check that the map

$$\mu : H \otimes_k \overline{H} \longrightarrow H \otimes_B H$$

$$x \otimes \overline{y} \longmapsto \sum xS(y_1) \otimes y_2$$

is well-defined and inverse to β.

    3. Assume that H is left faithfully flat over B. Consider the commutative diagram

$$
\begin{array}{ccccccc}
0 & \longrightarrow & B & \longrightarrow & H & \overset{\alpha}{\longrightarrow} & H \otimes_B H \\
& & & & \downarrow{\scriptstyle \mathrm{id}} & & \downarrow{\scriptstyle \beta} \\
0 & \longrightarrow & H^{\mathrm{coh}\overline{H}} & \longrightarrow & H & \overset{\gamma}{\longrightarrow} & H \otimes_k \overline{H}
\end{array}
\qquad (13)
$$

where $\alpha(h) = h \otimes_B 1 - 1 \otimes_B h$ and $\gamma(h) = (h \otimes 1) - \rho(h)$. There is an inclusion at the left hand column, by the first part of the lemma. By faithfully flat descent, the top row is exact (one checks that $((\ker \alpha)/B) \otimes_B H = 0$). Since β is injective, commutativity of the diagram forces $B = H^{\mathrm{coh}\overline{H}}$. Now part 2 of the lemma implies that the extension is $\overline{H}$-Galois.

    4. Note that every projective module over a commutative domain is locally free and hence faithfully flat. Thus part 4 is immediate from part 3, with $B = Z_0$, and Theorem III.4.5. □

Remark. With minor changes to the above proofs the same conclusion as in Lemma III.4.6(4) can be obtained under the weaker hypothesis that B is an affine commutative normal subalgebra of H over which H is a finitely generated module.

III.4.7. $Z_0 \subseteq H$ is a Frobenius extension. Let B be a subring of the ring A. Recall that $B \subseteq A$ is a *Frobenius extension* if

$$A \quad \text{is a finitely generated projective right } B\text{-module} \qquad (14)$$

and

$$\mathrm{Hom}_B(A_B, B_B) \quad \text{is isomorphic to } A \text{ as } (B, A)\text{-bimodules.} \qquad (15)$$

As we shall see in the next subsection, this definition is left-right symmetric, that is, $B \subseteq A$ is a Frobenius extension if and only if $B^{\mathrm{op}} \subseteq A^{\mathrm{op}}$ is a Frobenius extension.

Examples. 1. Suppose that $B = k$. Then a k-algebra A is a Frobenius extension of k if and only if A is a *Frobenius algebra*, [**199**, Chapter XIV.4].

2. Let $A = \oplus_{g \in G} A_g$ be a ring graded by the group G, and let T be a subgroup of finite index in G. Then A is a Frobenius extension of $B = \oplus_{t \in T} A_t$ if A is *strongly G-graded*, [**11**, Sec.1, Ex.(B)].

We'll prove in this paragraph that PI Hopf triples are Frobenius extensions, and then derive some consequences in (III.4.8). Recall that a *left integral* of a Hopf algebra K is a nonzero element $f \in K$ such that $gf = \varepsilon(g)f$ for all $g \in K$. If K is finite dimensional, then it has a left integral, which is moreover unique up to scalar multiples [**163**, Theorem 2.1.3].

Proposition. (Kreimer-Takeuchi) *Let J be a finite dimensional Hopf k-algebra, let A be a right J-comodule algebra, and suppose that $B = A^{\mathrm{coh}J} \subseteq A$ is a J-Galois extension. Then $B \subseteq A$ is a Frobenius extension.*

Sketch of the proof. [**128**, Theorem 1.7] Suppose that $B = A^{\mathrm{coh}J} \subseteq A$ is a J-Galois extension of k-algebras. Since J^* acts on A we can construct the smash product $A \# J^*$, [**163**, 4.1]. It's not hard to check that for $f \in J^*$, $a \in A$ and $b, b' \in B$,

$$f.(bab') \quad = \quad b(f.a)b',$$

so that J^* acts as (B, B)-bimodule endomorphisms of A. This gives a natural map

$$\pi : A \# J^* \longrightarrow \mathrm{End}_B(A_B).$$

One shows [**163**, Theorems 8.3.1, 8.3.3] that A is a finitely generated projective right B-module and that π is an isomorphism of algebras.

Next, use the (B, A)-bimodule structure on A to make the vector space $J^* \otimes A$ into a (B, A)-bimodule. One shows [**128**, Lemma 1.6] that there is a (B, A)-bimodule isomorphism $\tau : J^* \otimes A \to A \# J^*$ where $\tau(f \otimes x) = (1 \# f)(x \# 1)$ for $f \in J^*$ and $x \in A$. Then the composition

$$\omega = \pi\tau : J^* \otimes A \longrightarrow \mathrm{End}_B(A_B)$$

is a k-linear isomorphism, given by $\omega(f \otimes x)(a) = f(xa)$ for $f \in J^*$ and $x, a \in A$. Observe (since J^* acts on A as left B-module endomorphisms) that ω is a (B, A)-bimodule homomorphism. To complete the proof, we identify the sub-bimodule $\omega^{-1}\big(\mathrm{Hom}_B(A_B, B_B)\big)$ in $J^* \otimes A$.

Let I be the ideal of left integrals of J^* : that is,

$$I = \{f \in J^* : gf = \varepsilon(g)f \quad \text{for all } g \in J^*\}.$$

Since

$$B = \{b \in A : f(b) = \varepsilon(f)b \quad \text{for all } f \in J^*\},$$

we see that, for $f \in J^*$, f maps A into B if and only if $f \in I$. Now one deduces that

$$\omega^{-1}\left(\mathrm{Hom}_B(A_B, B_B)\right) = I \otimes A \subseteq J^* \otimes A.$$

Since $A \cong I \otimes A$ as (B, A)-bimodules (recall that $\dim_k(I) = 1$), the result follows. $\square$

From Lemma III.4.6(4) and the above proposition, we get the

Corollary. *Let* (III.4.1)(1) *be a PI Hopf triple. Then* $Z_0 \subseteq H$ *is a Frobenius extension.* $\square$

III.4.8 Frobenius extensions and bilinear forms. Let $B \subseteq A$ be rings. An *associative form from A to B* is a map

$$(\ ,\ ): A \times A \longrightarrow B,$$

additive in each argument, such that

$$
\begin{aligned}
(bx, y) &= b(x, y), \\
(x, yb) &= (x, y)b, \\
(xa, y) &= (x, ay)
\end{aligned}
$$

for all $x, y, a \in A$ and all $b \in B$. Suppose now that $B \subseteq A$ is a Frobenius extension, so there is an isomorphism

$$\theta: A \longrightarrow \mathrm{Hom}_B(A_B, B_B) \tag{16}$$

of (B, A)-bimodules. Then setting

$$(x, y) = \theta(x)(y) \tag{17}$$

for $x, y \in A$ clearly yields an associative form from A to B. Moreover, the bijectivity of θ ensures that $(\ ,\ )$ is *non-degenerate*; that is, for $x \in A$,

$$(x, A) = 0 \quad \Longrightarrow \quad x = 0$$

and

$$(A, x) = 0 \quad \Longrightarrow \quad x = 0.$$

Since A_B is finitely generated projective, there are elements $\{x_1, \ldots, x_n\}$ of A and $\{f_1, \ldots, f_n\}$ of $\mathrm{Hom}_B(A, B)$ which together form a dual basis for A_B. Set $y_i := \theta^{-1}(f_i) \in A$, $i = 1, \ldots, n$. By the properties of the dual basis and the non-degeneracy of the form $(\ ,\ )$, we have

$$A = \sum x_i B = \sum B y_i,$$

and indeed

$$a = \sum x_i(y_i, a) = \sum (a, x_i) y_i \tag{18}$$

for all $a \in A$. We say that subsets $\{x_1, \dots, x_n\}$ and $\{y_1, \dots, y_n\}$ satisfying (18) form *a dual projective pair* for $B \subseteq A$ relative to the form (,).

The existence of dual projective pairs characterises Frobenius extensions. The following result is due to Pareigis; the proof is routine (Exercise III.4.F), see e.g. [**11**, Theorem 1.1].

Lemma 1. *Let B be a subring of the ring A. Then A is a Frobenius extension of B if and only if there is an associative form from A to B relative to which a dual projective pair $\{x_1, \dots, x_n\}$, $\{y_1, \dots, y_n\}$ exists.* $\square$

One important immediate consequence of the lemma is that the definition of a Frobenius extension is left-right symmetric. Another important consequence is that Frobenius extensions are well-behaved under passage to quotients by ideals generated by elements of B; indeed the form in the factor is just the image of the form for $B \subseteq A$. First we note that the inclusion $B \subseteq A$ itself behaves well on passage to quotients. This is a consequence of

Lemma 2. *Let $B \subseteq A$ be a Frobenius extension with B commutative. Then B is a left and right B-module direct summand of A.*

Proof. The *trace ideal*

$$T \quad := \quad \operatorname{im}\{\tau : \operatorname{Hom}_B(A_B, B_B) \longrightarrow B : h \mapsto h(1)\}$$

equals B, by [**9**, Proposition A3]. Let $\tau(h) = 1$, so $h(1) = 1$ and $h(b) = b$ for all $b \in B$. Let $X = \ker h$, so the embedding of B in A splits the resulting exact sequence of right B-modules

$$0 \longrightarrow X \longrightarrow A \longrightarrow B \longrightarrow 0.$$

A similar argument applies on the left. $\square$

Corollary 1. *Let $B \subseteq A$ be a Frobenius extension with associated form (,) and dual projective pair $\{x_i\}, \{y_i\}$. Suppose that B is central in A and let I be an ideal of B. Then $\overline{B} := B/I \subseteq A/IA = \overline{A}$ is a Frobenius extension with form $\overline{(,)}$ and dual projective pair $\{\overline{x}_i\}, \{\overline{y}_i\}$.*

Proof. By Lemma 2, $\overline{B} \subseteq \overline{A}$. Since the conditions of Lemma 1 are clearly satisfied by $\overline{B} \subseteq \overline{A}$, the corollary follows. $\square$

Corollary 2. *Let* (III.4.1)(1) *be a PI Hopf triple. Then every algebra in* $\mathfrak{B}_H$ *is a Frobenius algebra.*

Proof. This follows from Corollary III.4.7 and Corollary 1 above.   □

It's not hard to write down the associative form for $Z_0 \subseteq H$ for the PI Hopf triple (III.4.1)(1). Retain the notation $\mathfrak{m}_0$ and $\overline{H}$ as in Corollary 2, and let $\gamma \in \overline{H}^*$ be a left integral (III.4.7) of $\overline{H}^*$. It follows from the proof of Proposition III.4.7 that the map θ of (III.4.8)(16) is given by

$$\theta(x) \quad = \quad \gamma(x-)$$

for $x \in H$. Hence (III.4.8)(17) shows that the form on H is defined by

$$(x, y) \quad = \quad \gamma(xy)$$

for all $x, y \in H$. From the definition of the $\overline{H}^*$-action on H, (Lemma I.9.16(a)), we deduce that

$$(x, y) \quad = \quad \sum \gamma(x_1 y_1) x_0 y_0,$$

where $\sum x_0 \otimes x_1$ and $\sum y_0 \otimes y_1$ are the images of x and y under the comodule structure map $H \longrightarrow H \otimes \overline{H}$ given in (III.4.6).

The Frobenius extension $B \subseteq A$ is *symmetric* if (,) is a symmetric form. If (III.4.1)(1) is a PI Hopf triple, then clearly $Z_0 \subseteq H$ is a symmetric extension if and only if all the algebras in $\mathfrak{B}_H$ are symmetric algebras with respect to the forms induced by (,). It's thus natural to ask:

Question. Which PI Hopf triples are symmetric?

We'll return to this question in Chapter III.10.

III.4.9. The Azumaya and unramified loci. Let (III.4.1)(1) be a PI Hopf triple. Recall the discussion of Azumaya algebras and the Azumaya locus from (III.1.3)–(III.1.7). From there, in particular from (III.1.7) and (III.1.6), we extract that the *Azumaya locus*

$$\begin{aligned} \mathcal{A}_H \quad :&= \quad \{\mathfrak{m} \in \mathrm{maxspec}\, Z : H_{\mathfrak{m}} \text{ is Azumaya over } Z_{\mathfrak{m}}\} \\ &= \quad \{\mathfrak{m} \in \mathrm{maxspec}\, Z : \mathfrak{m}H \in \mathrm{maxspec}\, H\} \end{aligned}$$

is a non-empty (and so dense) open subset of maxspec Z. Let

$$n = \left(\dim_{\mathrm{Fract}(Z)}(H \otimes_Z \mathrm{Fract}(Z))\right)^{1/2},$$

the PI-degree of H, (I.13.3). Then for all $\mathfrak{m} \in \mathcal{A}_H$,

$$H/\mathfrak{m}H \quad \cong \quad M_n(k), \tag{19}$$

by Theorem III.1.6. Put in another way: for every $\mathfrak{m} \in \mathcal{A}_H$ there is a unique irreducible H-module $V_\mathfrak{m}$ with $\mathfrak{m}V_\mathfrak{m} = 0$, and

$$\dim_k V_\mathfrak{m} = n,$$

the maximal k-dimension for irreducible H-modules.

We should think of $\mathcal{A}_H$ as the set of maximal ideals $\mathfrak{m}$ of Z for which the pairing $(\tau^{-1}(\mathfrak{m}), \mathfrak{m})$ has the simplest possible form, where $\tau : \operatorname{spec} H \to \operatorname{spec} Z$ is the restriction map (III.4.3). Continuing in this frame of mind, next we consider the most prevalent form of the pairing $(\mu^{-1}(\widehat{\mathfrak{m}}), \widehat{\mathfrak{m}})$ for a maximal ideal $\widehat{\mathfrak{m}}$ of Z_0. So consider a pair $S \subseteq R$ of k-affine commutative domains with R a finitely generated S-module. We shall say that a maximal ideal M of R is *unramified* (with respect to S) if $(M \cap S)R_M = M_M$. (Here, R_M denotes the localization of R at M, and $M_M = MR_M$.) The *unramified locus* $\mathcal{U}_S^R$ is then just the set of unramified maximal ideals of R. The basic facts we'll need are as follows. Here and elsewhere, by a phrase such as "for a generic maximal ideal" of the affine commutative algebra S we shall mean "for every maximal ideal in a dense open subset of maxspec S".

(1) The unramified locus $\mathcal{U}_S^R$ is an open subset of maxspec R, [**22**, Corollary 2.2].

(2) Suppose that $S \subseteq R$ is a *separable extension* (meaning that $\operatorname{Fract}(S) \subseteq \operatorname{Fract}(R)$ is a separable field extension). Then $\mathcal{U}_S^R$ is non-empty. Indeed, for a generic maximal ideal $\mathfrak{m}$ of S,

$$R/\mathfrak{m}R \quad \cong \quad k^{\oplus t}, \tag{20}$$

where $t = \dim_{\operatorname{Fract}(S)} \operatorname{Fract}(R)$, so that every maximal ideal of R lying over such a maximal ideal $\mathfrak{m}$ is unramified.

III.4.10. The fully Azumaya locus. We continue to work with a PI Hopf triple (III.4.1)(1), and keep the notation introduced in the above paragraphs. We define the *fully Azumaya locus* (of Z_0 with respect to H) to be

$$\mathcal{F}_H \quad := \quad \{\mathfrak{m} \in \operatorname{maxspec} Z_0 : \mu'^{-1}(\mathfrak{m}) \subseteq \mathcal{A}_H\}.$$

Since $\mathcal{A}_H$ is a non-empty open subset of maxspec Z and the morphism μ' is closed by Lemma III.4.3(4), it's an easy exercise (Exercise III.4.G) to prove the

Proposition. $\mathcal{F}_H$ *is a non-empty (and so dense) open subset of* maxspec Z_0. $\square$

Generalising (III.4.9)(19) we have the

Theorem. *Let* $\mathfrak{m} \in \mathcal{F}_H$ *and continue to set* $n = \operatorname{PI-deg}(H)$. *Then*

$$H/\mathfrak{m}H \quad \cong \quad M_n(Z/\mathfrak{m}Z).$$

Proof. We show first that

$$H/\mathfrak{m}H \quad \text{is a free } Z/\mathfrak{m}Z\text{-module of rank } n^2. \tag{21}$$

Let $\mathfrak{m}_1, \ldots, \mathfrak{m}_t$ be the maximal ideals of Z containing $\mathfrak{m}$. By our hypothesis on $\mathfrak{m}$ and Definition III.1.3(1), $H_{\mathfrak{m}_i}$ is a free $Z_{\mathfrak{m}_i}$-module, for $i = 1, \ldots, t$. Set $\mathfrak{n} := \cap_{i=1}^{t} \mathfrak{m}_i$, so that

$$\mathfrak{n}^r \quad \subseteq \quad \mathfrak{m}Z \quad \subseteq \quad \mathfrak{n} \tag{22}$$

for some $r \geq 1$. Standard homological algebra now implies that $H_{\mathfrak{n}}$ is a projective $Z_{\mathfrak{n}}$-module, and so, since $Z_{\mathfrak{n}}$ is a commutative semilocal domain, it follows from Exercise III.4.H that $H_{\mathfrak{n}}$ is a free $Z_{\mathfrak{n}}$-module, necessarily of rank n^2 since

$$\dim_{\mathrm{Fract}(Z)}(H \otimes_Z \mathrm{Fract}(Z)) \quad = \quad n^2$$

by definition of the PI-degree.

Hence $H_{\mathfrak{n}}/\mathfrak{m}H_{\mathfrak{n}}$ is a free $Z_{\mathfrak{n}}/\mathfrak{m}Z_{\mathfrak{n}}$-module of rank n^2. But (22) shows that the image in $Z/\mathfrak{m}Z$ of $\mathcal{C}_Z(\mathfrak{n})$, (the elements of Z which are regular *modulo* $\mathfrak{n}$), consists of units, so $H_{\mathfrak{n}}/\mathfrak{m}H_{\mathfrak{n}} = H/\mathfrak{m}H$ and $Z_{\mathfrak{n}}/\mathfrak{m}Z_{\mathfrak{n}} = Z/\mathfrak{m}Z$ and (21) is proved.

Let V_i be the unique irreducible $H/\mathfrak{m}_i H$-module and let P_i be its $H/\mathfrak{m}H$-projective cover, $1 \leq i \leq t$. Since $H/\mathfrak{m}_i H \cong M_n(k)$ for each i, as left $H/\mathfrak{m}H$-modules we have

$$H/\mathfrak{m}H \quad \cong \quad P_1^{(n)} \oplus P_2^{(n)} \oplus \ldots \oplus P_t^{(n)}. \tag{23}$$

Thus

$$H/\mathfrak{m}H \quad \cong \quad \mathrm{End}_{H/\mathfrak{m}H}(H/\mathfrak{m}H) \quad \cong \quad M_n\big(\mathrm{End}_{H/\mathfrak{m}H}(P_1 \oplus \ldots \oplus P_t)\big). \tag{24}$$

Now the map $z \mapsto (p \mapsto zp)$, for $z \in Z/\mathfrak{m}Z$ and $p \in P_1 \oplus \ldots \oplus P_t$, defines a ring homomorphism from $Z/\mathfrak{m}Z$ to $\mathrm{End}_{H/\mathfrak{m}H}(P_1 \oplus \ldots \oplus P_t)$. By (21) and (23) this map is injective, and comparing ranks in (21) and (24) shows that it is surjective. $\square$

Combining this with the discussion of the unramified locus in (III.4.9) we get the

Corollary. *Suppose that the PI Hopf triple (III.4.1)(1) is such that $Z_0 \subseteq Z$ is a separable extension. Then $\mathcal{W}_{Z_0}^Z := \{\mathfrak{m} \in \mathrm{maxspec}\, Z_0 : \mu'^{-1}(\mathfrak{m}) \subseteq \mathcal{U}_{Z_0}^Z\}$ is a non-empty open subset of $\mathrm{maxspec}\, Z_0$. For all $\mathfrak{m}$ in the dense open subset $\mathcal{F}_H \cap \mathcal{W}_{Z_0}^Z$ of $\mathrm{maxspec}\, Z_0$,*

$$H/\mathfrak{m}H \quad \cong \quad M_n(k)^{\oplus t}, \tag{25}$$

where $n = \mathrm{PI\text{-}deg}(H)$ and $t = \dim_{\mathrm{Fract}(Z_0)} \mathrm{Fract}(Z)$.

Proof. By our separability hypothesis and (III.4.9), the set $\mathcal{W}_{Z_0}^Z$ is non-empty and open for the same reason as $\mathcal{F}_H$ (Exercise III.4.G). Thus $\mathcal{F}_H \cap \mathcal{W}_{Z_0}^Z$ is non-empty and open, and so dense. The remainder of the corollary now follows from the above theorem and (III.4.9)(2). $\square$

III.4.11. Transitive actions by winding automorphisms. Notice that, by Theorem and Proposition III.4.10, at the generic point of $\mathcal{B}_H$ all the irreducible modules have the same dimension, namely the PI-degree of H. Case II for the algebra $U_\epsilon(\mathfrak{sl}_2(k))$ in (III.2.4) shows that this equidimensionality doesn't hold at every point of $\mathcal{B}_H$ in general; nevertheless it *is* true for $\mathcal{O}_\epsilon(SL_2(k))$, as can be read off from the six cases in (III.3.2). What is it which produces the differing behaviour of these two examples?

Suppose that (III.4.1)(1) is a PI Hopf triple. If, for all $\mathfrak{m} \in \text{maxspec } Z_0$, every irreducible $H/\mathfrak{m}H$-module had the same dimension, this would in particular be true for the restricted Hopf algebra $\overline{H}$, which would thus have all its irreducible modules of the same dimension as the trivial module – namely, one. In fact, as we'll see, this necessary condition is also sufficient. So let's assume in the rest of this paragraph that (III.4.1)(1) is a PI Hopf triple such that

$$\textit{every irreducible } \overline{H}\textit{-module is one-dimensional.} \tag{26}$$

Recall from (I.9.25) that the character group of one-dimensional representations $X(H)$ of H gives rise to two groups of automorphisms of H, the *left and right winding automorphisms of H*, denoted T^ℓ and T^r respectively. It's easy to see that under hypothesis (26) the irreducible $\overline{H}$-modules form a subgroup of $X(H)$; thus, since the map $\tau^\ell : X(H) \longrightarrow T^\ell : \lambda \longmapsto \tau^\ell_\lambda$ is a group antihomomorphism,

$$T^\ell_{\overline{H}} := \{\tau^\ell_\lambda : \lambda \in X(\overline{H})\}$$

is a finite subgroup of T^ℓ.

The result we're aiming for is the

Proposition. *Let* (III.4.1)(1) *be a PI Hopf triple. Assume* (26) *and let* $\mathfrak{m} \in$ maxspec Z_0. *Then (up to isomorphism) the irreducible $H/\mathfrak{m}H$-modules are permuted transitively by* $T^\ell_{\overline{H}}$.

Of course this implies the following corollary, which is already clear by inspection in (III.3.2) for $\mathcal{O}_\epsilon(SL_2(k))$.

Corollary. *With the above hypotheses and notation, the number of irreducible $H/\mathfrak{m}H$-modules is a divisor of the number of irreducible $\overline{H}$-modules.*

Proof of the proposition. Let V and W be irreducible $H/\mathfrak{m}H$-modules. Then W^* is an $H/(S^{-1}\mathfrak{m})H$-module, so $V \otimes W^*$ is an $H/\mathfrak{m}_0 H$-module – that is, an $\overline{H}$-module. Thus there exists $\lambda \in X(\overline{H})$ such that, for the corresponding irreducible $\overline{H}$-module k_λ, there is an embedding of $\overline{H}$-modules

$$k_\lambda \hookrightarrow V \otimes W^*.$$

Put another way, $\mathrm{Hom}_H(k_\lambda, V \otimes W^*)$ is nonzero. But by Exercises III.8.A and III.8.B(i),

$$0 \neq \mathrm{Hom}_H(k_\lambda, V \otimes W^*) \cong \mathrm{Hom}_H(k_\lambda, \mathrm{Hom}_k(W, V)) \cong \mathrm{Hom}_H(k_\lambda \otimes W, V). \quad (27)$$

Now $k_\lambda \otimes W = {}^{\tau_\lambda^\ell}W$, the twist of W by the automorphism τ_λ^ℓ. So this module and V are both irreducible, and hence (27) implies that they are isomorphic. $\quad \square$

Notes

The concept of a PI Hopf triple was introduced in [18]. Some of the ideas concerning Azumaya and unramified loci were introduced in [22] and [23]. Transitivity of the winding action on the irreducible modules annihilated by a fixed maximal ideal of Z_0 of $\mathcal{O}_\epsilon(G)$ was proved in [43]. An abstract Hopf algebraic approach to this result was later given in [134]; Letzter's argument was different to the one given here.

Exercises

Exercise III.4.A. Recall the definition in Exercise III.2.G of the *twist* of an R-module by an automorphism of the ring R. An obvious modification of this idea is used below. Suppose that M is a module over a commutative ring R, θ an automorphism of R, and $\mathfrak{m}, \mathfrak{n}$ maximal ideals of R such that $\theta(\mathfrak{m}) = \mathfrak{n}$. Prove that θ induces a ring isomorphism $\theta' : R_\mathfrak{m} \to R_\mathfrak{n}$. Let $M_\mathfrak{n}'$ be the $R_\mathfrak{m}$-module obtained from $M_\mathfrak{n} := R_\mathfrak{n} \otimes_R M$ via θ'. Prove that $M_\mathfrak{n}' \cong M_\mathfrak{m}$. Finally, conclude that $M_\mathfrak{m}$ is a free $R_\mathfrak{m}$-module if and only if $M_\mathfrak{n}$ is a free $R_\mathfrak{n}$-module.

Exercise III.4.B. Prove Lemma I.9.16(a): Right C-comodules are left C^*-modules.

Exercise III.4.C. Prove Lemma I.9.19.

Exercise III.4.D. Let A be a right J-comodule algebra. Show that $A^{\mathrm{coh}J} = A^{J^*}$. (See (III.4.6).)

Exercise III.4.E. Check Example III.4.7(2).

Exercise III.4.F. Prove Lemma III.4.8(1).

Exercise III.4.G. Prove Proposition III.4.10, and prove that the set $\mathcal{W}_{Z_0}^Z$ described in Corollary III.4.10 is a non-empty open subset of maxspec Z_0. [Hint: Nonzero ideals of Z have nonzero intersection with Z_0.]

Exercise III.4.H. Show that all projective modules over a commutative semilocal domain are free [92].

Exercise III.4.I. Prove that any Hopf algebra H contains a maximal central Hopf subalgebra Z_0, that is, Z_0 is a subalgebra of $Z(H)$ as well as a sub-Hopf algebra of H, and every sub-Hopf algebra of H which is contained in $Z(H)$ must be contained in Z_0.

POISSON STRUCTURES

In this chapter k is the complex field $\mathbb{C}$. The concepts of *Poisson bracket* and *quantization* introduced here are fundamental to much of our later work, and are also extremely important in many other contexts in representation theory.

III.5.1. Poisson algebras. A commutative k-algebra A is a *Poisson algebra* if it's endowed with a bilinear map

$$\{\ ,\ \} : A \times A \longrightarrow A,$$

called a *Poisson bracket*, such that A is a Lie algebra under $\{\ ,\ \}$, and the *Leibniz identity* holds: for all $a, b, c \in A$,

$$\{a, bc\} = \{a, b\}c + b\{a, c\}. \tag{1}$$

The second condition here can be restated as: for each $a \in A$, the map

$$X_a := \{a, -\} : A \longrightarrow A$$
$$b \longmapsto \{a, b\}$$

is a derivation, called the *Hamiltonian vector field* of a on A. The map

$$X : A \longrightarrow \mathrm{Der}_k\, A$$
$$a \longmapsto X_a$$

is a homomorphism of Lie algebras (thanks to the Jacobi identity for $\{\ ,\ \}$), where $\mathrm{Der}_k\, A$ denotes the Lie algebra of all k-linear derivations on A.

III.5.2. Symplectic leaves. From the Leibniz identity, we see that if $\mathfrak{m}$ is an ideal of A and $g \in \mathfrak{m}^2$, then

$$\{f, g\} \in \mathfrak{m} \tag{2}$$

for all $f \in A$. Thus, suppose $A = \mathcal{O}(Y)$, the coordinate ring of the affine algebraic variety Y. Let $T(Y)$ denote the tangent bundle of Y. Then the Poisson bracket $\{\ ,\ \}$ determines a *bivector* $u \in \Lambda^2 T(Y)$, via the following definition: if $P \in Y$, with defining maximal ideal $\mathfrak{m}_P$ in A, and $\alpha, \beta \in T^*(Y)_P = \mathfrak{m}_P/\mathfrak{m}_P^2$, the cotangent space at P, then

$$u_P(\alpha, \beta) \quad := \quad \{f, g\}(P), \tag{3}$$

where f and g are lifts of α and β to A. Note that u_P is a well-defined skew-symmetric form on $T^*(Y)_P$ in view of (2).

In general u_P is a degenerate form on $T^*(Y)_P$. Let

$$N(P) \quad := \quad \{\alpha \in T^*(Y)_P : u_P(\alpha, -) \equiv 0\},$$

and let

$$H(P) \quad := \quad N(P)^\perp \subseteq T(Y)_P.$$

Given any $d \in \mathrm{Der}_k A$, let us write d^P for the corresponding element of $T(Y)_P = (\mathfrak{m}_P/\mathfrak{m}_P^2)^*$, so that $d^P(\alpha) = d(f)(P)$ for $\alpha \in \mathfrak{m}_P/\mathfrak{m}_P^2$ and any lift f of α to A. Then one easily confirms (Exercise III.5.A) that

$$H(P) = k\text{-span}(\{X_f^P : f \in A\}) \subseteq T(Y)_P. \tag{4}$$

In order to define symplectic leaves, we shift our point of view to complex manifolds. Thus, assume now that Y is a *smooth* variety, so that it supports a natural manifold structure (e.g., [**166**, pp. 10-12]). The formula (3) extends the Poisson bracket to the algebra $C^\infty(Y)$ of complex-valued C^∞ functions on Y, since $T^*(Y)_P$ may be identified with $\overline{\mathfrak{m}}_P/\overline{\mathfrak{m}}_P^2$ where $\overline{\mathfrak{m}}_P$ is the ideal of functions in $C^\infty(Y)$ vanishing at P. Thus, Y now has the structure of a *Poisson manifold*.

A *symplectic leaf* L in Y is a connected (immersed) submanifold of Y which is maximal such that, for all $P \in L$, $H(P)$ is the tangent space to L at P. By the *Theorem of Frobenius*, (e.g. [**37**, Theorem 4.4.4]), Y decomposes as the disjoint union of symplectic leaves. The symplectic leaves in Y can also be defined as the equivalence classes for the relation of being connected by piecewise Hamiltonian curves, where a *Hamiltonian curve* on Y is a smooth map $\gamma : [0, 1] \to Y$ for which there is some $f \in C^\infty(Y)$ such that $\dot{\gamma}(t) = X_f^{\gamma(t)}$ for $0 < t < 1$, where $\dot{\gamma} = d\gamma/dt$.

When Y is connected and the form u is nondegenerate at every point, there is a unique leaf, and we say that Y is a *symplectic manifold*. In general, there is an induced structure of symplectic manifold on each leaf. The nondegeneracy condition ensures that a symplectic manifold always has even dimension.

The symplectic leaves of Y need not be subvarieties, although that does turn out to be the case in some of the standard examples.

III.5.3. Poisson groups. Suppose that A and B are two Poisson algebras. There is a canonical Poisson structure on $A \otimes B$ extending the two given ones, namely

$$\{a \otimes b, c \otimes d\} = \{a, c\} \otimes bd + ac \otimes \{b, d\}$$

for $a, c \in A$, $b, d \in B$. Suppose now that the Poisson algebra A is $\mathcal{O}(G)$, the coordinate ring of the algebraic group G. We say that G is a *Poisson algebraic group* if the Poisson and Hopf structures on A are compatible; that is, for all $a, b \in A$,

$$\Delta(\{a, b\}) = \{\Delta(a), \Delta(b)\},$$
$$S(\{a, b\}) = -\{S(a), S(b)\},$$

and

$$\varepsilon(\{a, b\}) = 0.$$

III.5.4. Quantization. As we shall understand it, quantization is a procedure whereby a noncommutative algebra is constructed whose noncommutativity encodes the Poisson structure of a (commutative) Poisson algebra. To describe this process, let R be a commutative k-algebra and let $h \in R$. Let A be an R-algebra with h not a zero divisor in A, and suppose that $\overline{A} := A/hA$ is a commutative k-algebra. (In practice, R will be $k[q]$ or $k[q^{\pm 1}]$ and h will be a linear polynomial in q.) For $a \in A$, write $\overline{a}$ for $a + hA \in \overline{A}$. We define a Poisson bracket on $\overline{A}$ as follows. Given $\alpha, \beta \in \overline{A}$, choose $a, b \in A$ with $\alpha = \overline{a}$, $\beta = \overline{b}$. By the commutativity of $\overline{A}$,

$$[a, b] \quad = \quad h\gamma(a, b)$$

for a unique element $\gamma(a, b)$ of A. Define

$$\{\alpha, \beta\} \quad = \quad \overline{\gamma(a, b)}; \tag{5}$$

it's easy to check that this is independent of our choice of lift of $\alpha, \beta \in \overline{A}$ to $a, b \in A$, and that this defines a Poisson bracket on $\overline{A}$, (Exercise III.5.C). We say that A is a *quantization* of the Poisson algebra $\overline{A}$.

Given a Poisson algebra $\overline{A}$, the *quantization problem* seeks an algebra A whose noncommutativity encodes the Poisson bracket on $\overline{A}$ as above. In the case when $\overline{A} = \mathcal{O}(G)$ is a Poisson group, then one seeks an R-Hopf algebra A which is a flat R-module with $\overline{A}$ a quotient Hopf algebra, and A a quantization of $\overline{A}$ in the above sense.

III.5.5. Examples. 1. Enveloping algebras. Let $\mathfrak{g}$ be a complex Lie algebra, and let A be the homogenisation of the enveloping algebra of $\mathfrak{g}$ with respect to the homogeneous element h. That is, A is the $k[h]$-algebra with generators a basis $\{x_1, \dots, x_n\}$ of $\mathfrak{g}$, and relations

$$x_i x_j - x_j x_i = h[x_i x_j].$$

In this case A is a quantization of $\overline{A} = \mathcal{O}(\mathfrak{g}^*)$ with the so-called Kirillov-Kostant Poisson structure on the latter. In particular, $\mathcal{O}(\mathfrak{g}^*) \cong k[x_1, \dots, x_n]$ and $\{x_i, x_j\} =$

$[x_i x_j]$. The symplectic leaves are the coadjoint orbits by a theorem of Kirillov. For details see e.g. [**35**, Example 1.3.16].

    2. Let $A = \mathcal{O}_q(SL_2(k))$ with generators a, b, c, d and relations as in (I.1.7), (I.1.9). With $R = k[q]$ and $h = q - 1$ in (III.5.4) we get a Poisson structure on $\overline{A} = \mathcal{O}(SL_2(k))$ with

$$\{a, b\} = ab \qquad \{a, c\} = ac \qquad \{b, c\} = 0$$
$$\{b, d\} = bd \qquad \{c, d\} = cd \qquad \{a, d\} = 2bc$$

(Exercise III.5.D). This is the so-called *standard Poisson bracket* on $\mathcal{O}(SL_2(k))$. Of course one can extend this to $\mathcal{O}(SL_n(k))$. The symplectic leaves of $SL_2(k)$ with respect to the Poisson structure above were calculated in [**94**, Theorem B.2.1]; there are leaves of dimensions 0 and 2, as follows (Exercise III.5.G):

0-dim. leaves : $\left\{ \begin{bmatrix} \alpha & 0 \\ 0 & \alpha^{-1} \end{bmatrix} \right\}$ $(\alpha \in k^{\times})$

2-dim. leaves : $\left\{ \begin{bmatrix} \alpha & 0 \\ \gamma & \alpha^{-1} \end{bmatrix} : \alpha, \gamma \in k^{\times} \right\}, \quad \left\{ \begin{bmatrix} \alpha & \beta \\ 0 & \alpha^{-1} \end{bmatrix} : \alpha, \beta \in k^{\times} \right\},$

$$\left\{ \begin{bmatrix} \alpha & \lambda\gamma \\ \gamma & \delta \end{bmatrix} : \alpha, \delta \in k, \ \gamma \in k^{\times}, \ \alpha\delta - \lambda\gamma^2 = 1 \right\} \qquad (\lambda \in k^{\times}).$$

Note that the 2-dimensional leaves are not Zariski-closed in $SL_2(k)$; they are, however, locally closed (or *quasi-affine*) subvarieties.

III.5.6. We can generalise somewhat the idea in (III.5.4). So let R, A and h be as before, but drop the assumption that $\overline{A} := A/hA$ is commutative. Consider now an element $\alpha \in Z(\overline{A})$, with $\alpha = \overline{a} = a + hA$ for some $a \in A$. Define a map $d_a : \overline{A} \longrightarrow \overline{A}$ by setting, for $\beta \in \overline{A}$,

$$d_a(\beta) \quad := \quad \overline{h^{-1}[a, b]},$$

where $b \in A$ with $\overline{b} = \beta$. We can easily check (Exercise III.5.E) that

Lemma. 1. d_a is a well-defined derivation of $\overline{A}$.

    2. If a_0 is a second element of A with $\overline{a_0} = \alpha$, then $d_a - d_{a_0}$ is an inner derivation of $\overline{A}$.

    3. Let ψ be an R-algebra automorphism of A. Then

$$d_{\psi(a)} \quad = \quad \psi \circ d_a \circ \psi^{-1}. \qquad \square$$

The lemma immediately implies the

Corollary. 1. $Z(\overline{A})$ *is a Poisson algebra with*

$$\{\alpha, \beta\} \quad := \quad d_a(b)$$

for $\alpha, \beta \in Z(\overline{A})$ *and* $a, b \in A$ *with* $\overline{a} = \alpha$, $\overline{b} = \beta$.

2. *The group of R-algebra automorphisms of A induces a group of algebra automorphisms of $\overline{A}$ restricting to Poisson algebra automorphisms of $Z(\overline{A})$.* $\square$

III.5.7. An application of the above ideas will be needed for the proof of Theorem III.6.2(3). Keep the notation of the previous paragraph, and assume that A is a Hopf R-algebra with bijective antipode, so that $\overline{A}$ is also a Hopf algebra, over R/hR. Let W be a sub-Hopf algebra of $\overline{A}$ with $W \subseteq Z(\overline{A})$. Let T be the smallest subalgebra of $Z(\overline{A})$ with $W \subseteq T$ and T closed under the Poisson bracket defined in Corollary III.5.6.

Proposition. *With the above notation, T is a sub-Hopf algebra of $\overline{A}$. The Hopf operations on T are compatible with the Poisson bracket.*

Proof. Let $T^\circ = \{t \in T : \Delta(t) \in T \otimes T\}$, so that T° is a subalgebra of T and $W \subseteq T^\circ$. We claim that

$$\{T^\circ, T^\circ\} \subseteq T^\circ. \tag{6}$$

For, let $\alpha, \beta \in T^\circ$, with $a, b \in A$ such that $\overline{a} = \alpha$, $\overline{b} = \beta$. Then

$$\Delta(\{\alpha, \beta\}) = \Delta(d_a(\beta)) = d_{\Delta(a)}(\Delta(\beta)) = \{\Delta(\alpha), \Delta(\beta)\}, \tag{7}$$

where the second equality in (7) is a simple calculation, (Exercise III.5.F). But $\{\Delta(\alpha), \Delta(\beta)\} \in T \otimes T$ by definition of T°, so $\{\alpha, \beta\} \in T^\circ$. This proves (6), and hence $T^\circ = T$. Moreover the antiautomorphism S of $\overline{A}$ preserves W and so maps T to itself. This proves the first statement in the proposition, and the second is clear from the definitions. $\square$

III.5.8. Isomorphisms along Hamiltonian flows. The most important mechanism by which Poisson structure impacts on the noncommutative algebras we're interested in is outlined in this paragraph. We continue with the hypotheses and notation introduced in (III.5.6). Thus we have a (noncommutative) prime algebra $\overline{A}$ whose centre $Z(\overline{A})$ admits a Poisson bracket $\{ \ , \ \}$ with the Hamiltonian vector fields $\{\alpha, -\}$ extending to derivations d_a of $\overline{A}$, for $\alpha \in Z(\overline{A})$.

Recall that the underlying field is $\mathbb{C}$. Suppose that $Z(\overline{A})$ contains a Poisson sub-Hopf algebra Z_0 which is an affine $\mathbb{C}$-algebra with $\overline{A}$ a finitely generated Z_0-module. Thus, as discussed in (III.4.4), there is a bundle of finite dimensional $\mathbb{C}$-algebras

$$\mathcal{B}_{\overline{A}} \quad := \quad \{\overline{A}/\mathfrak{m}\overline{A} : \mathfrak{m} \in \mathrm{maxspec}\, Z_0\}.$$

We assume further that

$$\overline{A} \quad \text{is a projective } Z_0\text{-module}$$

and

$$\text{maxspec } Z_0 \quad \text{is a smooth variety,}$$

the latter assumption being made so that symplectic leaves in maxspec Z_0 are defined. Then, by integrating the Hamiltonian vector fields d_a on $\overline{A}$, one proves:

Proposition. *Keep the above notation, with the stated hypotheses on $\overline{A}$ and on Z_0. Let $\mathfrak{m}$ and $\mathfrak{n}$ be in maxspec Z_0, and suppose that $\mathfrak{m}$ and $\mathfrak{n}$ belong to the same symplectic leaf of maxspec Z_0. Then*

$$\overline{A}/\mathfrak{m}\overline{A} \quad \cong \quad \overline{A}/\mathfrak{n}\overline{A}.$$

Proof. [**44**, Corollary 11.8], [**43**, Lemma 9.1] □

Notice that it is a consequence of Corollary III.5.6(2) that the group of automorphisms G of $\overline{A}$ induced by the R-automorphisms of A preserves the Poisson structure. Thus, suppose that G_0 is a subgroup of G with $g(z) \in Z_0$ for all $g \in G_0$, $z \in Z_0$, where Z_0 is the subalgebra of $Z(\overline{A})$ introduced above. Of course G_0 then has an induced action on maxspec Z_0, and our hypothesis on G ensures that G_0 will map each symplectic leaf in maxspec Z_0 to a symplectic leaf. We can therefore improve the proposition to:

Corollary. *Keep the notation and hypotheses on $\overline{A}$ and Z_0 already introduced in this paragraph. Let G_0 be as described above. Let $\mathfrak{m}$ and $\mathfrak{n}$ be in maxspec Z_0 and suppose that there exist symplectic leaves L and K, and $g \in G_0$, with*

$$\mathfrak{m} \in L, \quad \mathfrak{n} \in K, \quad \text{and} \quad g(L) = K.$$

Then

$$\overline{A}/\mathfrak{m}\overline{A} \quad \cong \quad \overline{A}/\mathfrak{n}\overline{A}. \quad □$$

III.5.9. The example of $\mathcal{O}_\epsilon(SL_2(k))$. Recall that the torus $k^\times$ acts as left and as right winding automorphisms of $\mathcal{O}_q(SL_2(k))$ and $\mathcal{O}_\epsilon(SL_2(k))$, according to the formulas given in (I.9.25). The resulting automorphisms together yield an action of $H = (k^\times)^2$ as defined in Example II.1.6(b) – for each $\lambda \in k^\times$ there is a left winding automorphism which multiplies the first row of the matrix of algebra generators of $\mathcal{O}_q(SL_2(k))$ by λ and the second row by λ^{-1}, while the right winding automorphisms have a similar effect on the columns. With $R = k[q]$, these automorphisms clearly induce R-algebra automorphisms as in (III.5.6), and hence by Corollary III.5.6(2) we obtain a 2-torus H of automorphisms of $\mathcal{O}_\epsilon(SL_2(k))$ restricting to Poisson automorphisms of the subalgebra $Z_0 \cong \mathcal{O}(SL_2(k))$ defined in (III.3.1), for the Poisson structure defined in Example III.5.5(2). Thus H (and indeed the one-parameter subgroups of right or left winding automorphisms) act on the set of symplectic leaves of maxspec $\mathcal{O}(SL_2(k))$. We can easily check that the orbits under these three actions are the same (Exercise III.5.H). Namely, there is a single orbit of 0-dimensional leaves

$$\left\{ \left\{ \begin{bmatrix} \alpha & 0 \\ 0 & \alpha^{-1} \end{bmatrix} \right\} : \alpha \in k^{\times} \right\};$$

the two leaves

$$\left\{ \begin{bmatrix} \alpha & 0 \\ \gamma & \alpha^{-1} \end{bmatrix} : \alpha, \gamma \in k^{\times} \right\}$$

and

$$\left\{ \begin{bmatrix} \alpha & \beta \\ 0 & \alpha^{-1} \end{bmatrix} : \alpha, \beta \in k^{\times} \right\}$$

are left invariant by H; and the remaining leaves form a single H-orbit.

Corollary III.5.8 now predicts that there will be (at most) four isomorphism classes of algebras $\mathcal{O}_\epsilon(SL_2(k))/\mathfrak{m}\mathcal{O}_\epsilon(SL_2(k))$ as $\mathfrak{m}$ ranges through the maximal ideals of Z_0, corresponding to the four orbits of symplectic leaves found above. A glance at the calculations we did in (III.3.2) confirms that this is indeed the case – in fact the two singleton orbits yield isomorphic algebras, so that there are only three distinct isomorphism classes.

We shall explain in (III.7.8) how this picture generalises to $\mathcal{O}_\epsilon(G)$ for an arbitrary semisimple group G.

NOTES

Poisson manifolds and symplectic leaves were introduced by Weinstein [210]; discussions of these concepts can be found, for instance, in [29, Part II] and [126, Chapter 1]. The symplectic leaves for the standard Poisson structure on a semisimple group G, and the orbits of these leaves under the action of the character group, were calculated in [94]. The results of (III.5.8) are taken from [44]. The generalisation of the construction of (III.5.4) described at the beginning of (III.5.6) is due to Hayashi [90].

EXERCISES

Exercise III.5.A. Prove (III.5.2(4)).

Exercise III.5.B. Prove that (III.5.3) defines a Poisson structure on the tensor product of the Poisson k-algebras A and B.

Exercise III.5.C. Check that definition (III.5.4)(5) is well-defined and yields a Poisson bracket on $\overline{A}$.

Exercise III.5.D. (i) Confirm the description of the standard Poisson bracket on $\mathcal{O}(SL_2(k))$ given in Example III.5.5(2).

(ii) Define an ideal I of a Poisson algebra A to be a *Poisson ideal* if $\{A, I\} \subseteq I$. Determine the prime Poisson ideals of $\mathcal{O}(SL_2(k))$. Compare the answer you get with the prime ideals of $\mathcal{O}_q(SL_2(k))$. Bearing in mind (II.8), one might suggest a plausible definition of "Poisson primitive ideal", and consider which Poisson primes of $\mathcal{O}(SL_2(k))$ are Poisson primitive.

(iii) As a more involved mini-project, one can determine the Poisson prime ideals of $\mathcal{O}^+(SL_3(\mathbb{C}))$. Which ones are Poisson primitive? Does a stratification of the Poisson prime and primitive spectra suggest itself?

Exercise III.5.E. Prove Lemma III.5.6.

Exercise III.5.F. Prove (III.5.7(7)).

Exercise III.5.G. Confirm the decomposition of maxspec $\mathcal{O}(SL_2(k))$ into symplectic leaves as in Example III.5.5(2), in the following steps. Let $\mathfrak{m}_P$ be a maximal ideal of $\mathcal{O}(SL_2(k))$.

(i) Suppose that $b, c \in \mathfrak{m}_P$. For $f \in \mathcal{O}(SL_2(k))$ show that $\{f, \mathfrak{m}_P\} \subseteq \mathfrak{m}_P$. Deduce that $H(P) = 0$ in this case (Notation III.5.2(4)). Thus, for each $\alpha \in k^\times$,

$$\left\{ \begin{bmatrix} \alpha & 0 \\ 0 & \alpha^{-1} \end{bmatrix} \right\}$$

is a 0-dimensional leaf.

(ii) Let $b \in \mathfrak{m}_P$, $c \notin \mathfrak{m}_P$. Show that for all $\mathfrak{m}$ in the leaf containing $\mathfrak{m}_P$, $b \in \mathfrak{m}$. Show that the bracket induces a structure of symplectic manifold on $(\mathcal{O}(SL_2(k))/\langle b \rangle)[c^{-1}]$, and deduce that

$$\left\{ \begin{bmatrix} \alpha & 0 \\ \gamma & \alpha^{-1} \end{bmatrix} : \alpha, \gamma \in k^\times \right\}$$

is the symplectic leaf containing P.

(iii) Note that the case $c \in \mathfrak{m}_P$, $b \notin \mathfrak{m}_P$ is symmetric to (ii).

(iv) Suppose finally that $\mathfrak{m}_P$ is not in cases (i), (ii) or (iii), so that for some $\gamma_0, \lambda \in k^\times$ and $\alpha_0, \delta_0 \in k$,

$$P = \begin{bmatrix} \alpha_0 & \lambda \gamma_0 \\ \gamma_0 & \delta_0 \end{bmatrix}.$$

Show that for every $\mathfrak{m}$ in the leaf containing $\mathfrak{m}_P$,

$$b - \lambda c \in \mathfrak{m}_P.$$

Deduce that

$$\left\{ \begin{bmatrix} \alpha & \lambda \gamma \\ \gamma & \delta \end{bmatrix} : \alpha, \delta \in k, \ \gamma \in k^\times, \ \alpha\delta - \lambda\gamma^2 = 1 \right\}$$

is the leaf containing $\mathfrak{m}_P$.

Exercise III.5.H. Check the statements regarding the H-orbits of symplectic leaves in $\mathcal{O}(SL_2(k))$ made in (III.5.9).

STRUCTURE OF $U_\epsilon(\mathfrak{g})$

Throughout this chapter we'll assume that $\mathfrak{g}$ is a finite dimensional semisimple complex Lie algebra and that ϵ is a primitive ℓth root of unity in the field k of characteristic 0, where

$$\ell \geq 3 \quad \textit{is odd, and prime to 3 if } \mathfrak{g} \textit{ contains a factor of type } G_2. \tag{1}$$

Our aim in this chapter is to examine how the structure of $U_\epsilon(\mathfrak{sl}_2(k))$ which we found in Chapter III.2 generalises to arbitrary semisimple $\mathfrak{g}$. We'll see (III.6.2) that $U_\epsilon(\mathfrak{g})$ is a PI Hopf triple in the language of Chapter III.4. Moreover the construction of (III.5.6) can be used to obtain a structure of Poisson group on the Hopf centre Z_0 of $U_\epsilon(\mathfrak{g})$, so that the representation theory of $U_\epsilon(\mathfrak{g})$ can be studied with the help of (III.5.8).

III.6.1. Construction of $U_\epsilon(\mathfrak{g})$. We start with the data associated with $\mathfrak{g}$ as assembled in (I.5), and form the quantized enveloping algebra $U_q(\mathfrak{g}; M)$ over the field $k' := k(q)$. Here, q is transcendental over k and M is any lattice between Q and P; note using (I.5.3) that M is necessarily W-invariant. Let $\mathcal{A} := k[q^{\pm 1}]$. Recall the braid group B_W from (I.6.7). Write, for $i = 1, \ldots, n$,

$$\overline{E}_i \quad := \quad (q_i - q_i^{-1})E_i \qquad\qquad \overline{F}_i \quad := \quad (q_i - q_i^{-1})F_i,$$

and let A be the smallest B_W-invariant $\mathcal{A}$-subalgebra of $U_q(\mathfrak{g}; M)$ containing $\{\overline{E}_i, \overline{F}_i, K_\lambda : 1 \leq i \leq n, \lambda \in M\}$. Finally, form the k-algebra

$$U_\epsilon(\mathfrak{g}; M) \quad := \quad A/\langle q - \epsilon \rangle.$$

As in the generic case (I.6.3), we write $U_\epsilon(\mathfrak{g}) = U_\epsilon(\mathfrak{g}; Q)$ and $\check{U}_\epsilon(\mathfrak{g}) = U_\epsilon(\mathfrak{g}; P)$.

Essentially from the definition and Chapter I.6 we get the

Theorem. (De Concini, Kac, Procesi) *Keep the above hypotheses and notation.*

1. $U_\epsilon(\mathfrak{g}; M)$ has the same generators and the same relations as $U_q(\mathfrak{g}; M)$, (namely as in (I.6.3)), but with q replaced by ϵ throughout.

2. $U_\epsilon(\mathfrak{g}; M)$ is a Hopf algebra with the same definitions for Δ, ε and S as $U_q(\mathfrak{g}; M)$, but with q replaced by ϵ.

3. $U_\epsilon(\mathfrak{g}; M)$ admits an action of B_W as in (I.6.7). Thus $U_\epsilon(\mathfrak{g}; M)$ contains root vectors E_{β_j} and F_{β_j}, $j = 1, \ldots, n$, and has as PBW basis over k the monomials

$$M_{r,k,\alpha} \quad = \quad F^r K_\alpha E^k,$$

where $r, k \in \mathbb{Z}_{\geq 0}^N$ and $\alpha \in M$, as in (I.6.11).

4. The analogues of the relations in Proposition I.6.10 with q replaced by ϵ hold for $U_\epsilon(\mathfrak{g}; M)$.

5. $U_\epsilon(\mathfrak{g}; M)$ is a $\mathbb{Z}_{\geq 0}^{(2N+1)}$-filtered algebra, with associated graded algebra $\mathrm{gr}(U)$ as in Proposition I.6.11, but with q replaced by ϵ and Q by M.

Proof. [40, 42] □

III.6.2. Central sub-Hopf algebra. For convenience we'll assume in the rest of this chapter that $M = P$, so we are considering $\check{U}_\epsilon(\mathfrak{g})$. We define Z_0 to be the smallest B_W-invariant subalgebra of $\check{U}_\epsilon(\mathfrak{g})$ containing

$$K_{\ell\alpha} = K_\alpha^\ell, \; E_i^\ell, \; F_i^\ell \tag{2}$$

for $\alpha \in P$ and $1 \leq i \leq n$.

Theorem. *Keep the above notation.*
1. Z_0 is in the centre of $\check{U}_\epsilon(\mathfrak{g})$.
2. Z_0 has a k-basis of monomials

$$M_{\ell r, \ell k, \ell\alpha} \quad = \quad F^{\ell r} K_{\ell\alpha} E^{\ell k},$$

where $r, k \in \mathbb{Z}_{\geq 0}^N$ and $\alpha \in P$. In particular, as an algebra Z_0 is a polynomial ring in $2N + n = \dim \mathfrak{g}$ generators over k, with n generators inverted.

3. Z_0 is a sub-Hopf algebra of $\check{U}_\epsilon(\mathfrak{g})$.
4. $\check{U}_\epsilon(\mathfrak{g})$ is a free Z_0-module of rank $\ell^{\dim \mathfrak{g}}$, with a k-basis of ordered monomials

$$F_{\beta_N}^{r_N} \ldots F_{\beta_1}^{r_1} K_\alpha E_{\beta_1}^{k_1} \ldots E_{\beta_N}^{k_N},$$

where $0 \leq r_i, k_i < \ell$ and α ranges through a set of coset representatives of $P/\ell P$.

Proof. 1. It's obvious from the conjugation relations that $K_{\ell\alpha} \in Z(\check{U}_\epsilon(\mathfrak{g}))$ for $\alpha \in P$, and that E_i^ℓ and F_i^ℓ commute with K_α for $i = 1, \ldots, n$ and $\alpha \in P$. Also, E_i commutes with F_j when $i \neq j$. That E_i^ℓ commutes with F_i follows from the case of $U_\epsilon(\mathfrak{sl}_2(k))$ (Exercise III.2.A); similarly, F_i^ℓ commutes with E_i. To see that, for $j = 1, \ldots, N$, $E_{\beta_j}^\ell$ commutes with $E_1, \ldots, E_n$ (and analogously that $F_{\beta_j}^\ell$ commutes with $F_1, \ldots, F_n$), requires somewhat more effort, which we relegate to an exercise (Exercise III.6.A).

2. It's clear from the definition of the B_W-action that Z_0 contains the monomials $M_{\ell r, \ell k, \ell\alpha}$. It is shown in [40, Proposition 3.3] that the algebra spanned by

these monomials is B_W-stable. Since this algebra contains the generators (2), it must equal Z_0, by definition of Z_0. Linear independence of the monomials $M_{\ell r, \ell k, \ell \alpha}$ follows from the PBW theorem, Theorem III.6.1(3).

3. First, consider the comultiplication Δ. We've got to show that $\Delta(Z_0) \subseteq Z_0 \otimes Z_0$, and since Δ is a homomorphism it's enough to treat the algebra generators of Z_0, namely $E_{\beta_i}^\ell$, $F_{\beta_i}^\ell$, $K_{\ell\alpha}$ for $1 \leq i \leq N$, $\alpha \in P$. The group-like elements are easily dealt with: $\Delta(K_{\ell\alpha}) = \Delta(K_\alpha^\ell) = \Delta(K_\alpha)^\ell = K_\alpha^\ell \otimes K_\alpha^\ell = K_{\ell\alpha} \otimes K_{\ell\alpha}$ for $\alpha \in P$. Next look at a simple root vector E_j, $1 \leq j \leq n$. Since $a := E_j \otimes 1$ and $b := K_{\alpha_j} \otimes E_j$ satisfy $ab = \epsilon^{-2d_i} ba$ in $\check{U}_\epsilon(\mathfrak{g}) \otimes \check{U}_\epsilon(\mathfrak{g})$, it follows from the q-binomial theorem (I.6.1) that

$$\Delta(E_j^\ell) \quad = \quad E_j^\ell \otimes 1 + K_{\alpha_j}^\ell \otimes E_j^\ell \in Z_0. \tag{3}$$

Similar remarks apply to F_j^ℓ, $j = 1, \ldots, n$.

But unfortunately it's not so easy to handle the generators corresponding to non-simple roots – for these one has to make use of the Poisson structure of Z_0, as we'll explain below in (III.6.5) – see in particular the Remark at the end of that paragraph.

For the antipode S, the story is similar – E_i^ℓ, F_i^ℓ, K_α^ℓ for $1 \leq i \leq n$, $\alpha \in P$ are routinely disposed of, but again the non-simple roots require special treatment.

4. This is immediate from 2 and the PBW theorem, Theorem III.6.1(3). $\quad\square$

III.6.3. The irreducible representations of $\check{U}_\epsilon(\mathfrak{g})$.
The results of the previous paragraph show that we are in the setting of (III.1.1) – thus hypotheses (**H**) of (III.1.1) hold, with $A = \check{U}_\epsilon(\mathfrak{g})$. As a consequence of Kaplansky's structure theorem for primitive PI rings (I.13.3), every irreducible $\check{U}_\epsilon(\mathfrak{g})$-module is finite dimensional over k and is annihilated by a maximal ideal of Z_0, Proposition III.1.1(4). Let's express this another way, which will be fundamental to our approach in later sections: *The study of the irreducible $U := \check{U}_\epsilon(\mathfrak{g})$-modules is equivalent to studying the irreducible modules for the bundle of finite dimensional k-algebras*

$$\mathcal{B}_U \quad := \quad \{U/\mathfrak{m}U : \mathfrak{m} \in \text{maxspec}\, Z_0\}.$$

Suppose for simplicity that k is algebraically closed. Then for all $\mathfrak{m} \in \text{maxspec}\, Z_0$, we have $Z_0/\mathfrak{m} \cong k$ by the Nullstellensatz, so Theorem III.6.2(4) shows that

$$\dim_k U/\mathfrak{m}U \quad = \quad \ell^{\dim \mathfrak{g}}$$

for all $\mathfrak{m} \in \text{maxspec}\, Z_0$.

For instance, when $\mathfrak{g} = \mathfrak{sl}_2(\mathbb{C})$, we've seen in (III.2.4) that there are 3 isomorphism classes of algebras in $\mathcal{B}_U$, each of dimension ℓ^3.

III.6.4. The restricted quantized enveloping algebra. Recall from (III.4.4) that, amongst the algebras in $\mathcal{B}_U$, one of particular significance is the *restricted quantized enveloping algebra of* $\mathfrak{g}$ *at the root ϵ of unity*, $\overline{U}$. This is the member of $\mathcal{B}_U$ supported by the augmentation ideal of Z_0,

$$\mathfrak{m}_0 \quad = \quad \langle E_{\beta_j}^\ell,\ F_{\beta_j}^\ell,\ K_\alpha^\ell - 1 : 1 \leq j \leq N,\ \alpha \in P \rangle. \tag{4}$$

Note that $\overline{U}$ is a Hopf algebra of dimension $\ell^{\dim \mathfrak{g}}$. An analogous Hopf algebra of the same dimension can be constructed from $U_\epsilon(\mathfrak{g})$, and it is isomorphic – as an algebra – to $\overline{U}$ (Exercise III.6.C). When $\mathfrak{g} = \mathfrak{sl}_2(\mathbb{C})$ the defining relations for $\overline{U_\epsilon(\mathfrak{g})}$ have been written down in (III.4.4), and it's a simple matter to use (4) to write down a set of defining relations for arbitrary $\mathfrak{g}$.

One should be aware that there is a very close relationship between $\overline{U}$ and the *restricted quantized enveloping algebra* $\widehat{U}_\epsilon(\mathfrak{g})$ defined by Lusztig, in [**143**] for the simply laced case, and in [**144**] in general, as a key tool in working towards a character formula for simply connected semisimple groups in positive characteristic. (This programme is discussed in more detail in (III.10.6).) When ϵ is a primitive pth root of unity for a prime p, $\widehat{U}_\epsilon(\mathfrak{g})$ is a characteristic 0 analogue of the first Frobenius kernel of the corresponding group G in characteristic p. In fact there is an epimorphism of Hopf algebras from $\widehat{U}_\epsilon(\mathfrak{g})$ onto $\overline{U}$, with kernel generated by a central idempotent. Indeed $\widehat{U}_\epsilon(\mathfrak{g})$ has dimension $2^n \ell^{\dim(\mathfrak{g})}$, (where as usual n is the rank of $\mathfrak{g}$), and essentially the only difference between $\widehat{U}_\epsilon(\mathfrak{g})$ and $\overline{U}$ is that, for $1 \leq i \leq n$, $K_i^{2\ell} = 1$ in the former, as compared with $K_i^\ell = 1$ in the latter.

III.6.5. Poisson structure on $Z(\check{U}_\epsilon(\mathfrak{g}))$. For the remainder of the chapter, assume that $k = \mathbb{C}$.

From the structure of $\check{U}_\epsilon(\mathfrak{g})$ in (III.6.1) it's clear that $Z(\check{U}_\epsilon(\mathfrak{g}))$ admits a structure of Poisson algebra by the recipe in (III.5.6). Here, h is the element $q - \epsilon$ of $\mathcal{A} = k[q^{\pm 1}]$. We collect here the key facts about this Poisson structure. First,

$$Z_0 \quad \text{is a Poisson subalgebra of } Z(\check{U}_\epsilon(\mathfrak{g})). \tag{5}$$

In fact, Z_0 is the smallest subalgebra of $Z(\check{U}_\epsilon(\mathfrak{g}))$ closed under the Poisson bracket and containing the algebra W generated by the elements K_α^ℓ, E_i^ℓ, F_i^ℓ, ($\alpha \in P$; $1 \leq i \leq n$) of (III.6.2)(2). Thus Proposition III.5.7 applies to Z_0, enabling us to complete the proof of Theorem III.6.2(3) – as noted already in (III.6.2), W is a sub-Hopf algebra of $\check{U}_\epsilon(\mathfrak{g})$, so by Proposition III.5.7, Z_0 is a sub-Hopf algebra of $\check{U}_\epsilon(\mathfrak{g})$, and the Hopf operations are compatible with the Poisson bracket. That is, in the language of (III.5.3),

$$\text{maxspec } Z_0 \quad \text{is a Poisson algebraic group } H. \tag{6}$$

Next, we want to identify the group H in question. Let G be the connected, simply connected, semisimple group over $\mathbb{C}$ whose Lie algebra is $\mathfrak{g}$. Let T be a maximal torus in G, and let $B^+ = N^+T$ be a Borel subgroup containing T. Let B^- be the (unique) Borel subgroup of G with $B^+ \cap B^- = T$. Define H to be the kernel of the projection $\mu \circ (\pi^- \times \pi^+) : B^- \times B^+ \longrightarrow T : (x, y) \mapsto \pi^-(x)\pi^+(y)$, where $\pi^\pm : B^\pm \longrightarrow T$ are the canonical epimorphisms. Then we can write

$$H = (N^- \times N^+) \rtimes T \equiv \{(x, y, t) : x \in N^-,\ y \in N^+,\ t \in T\}, \qquad (7)$$

via the map $(t^{-1}x, ty) \mapsto (x, y, t)$. The key to the Poisson structure on H is a group $\widetilde{G}$ of automorphisms of Z_0 constructed in [40, §3.5]. It's proved in [44, Proposition 16.1] that

$$\text{the symplectic leaves in maxspec } Z_0 \text{ coincide with the } \widetilde{G}\text{-orbits.} \qquad (8)$$

This leaves the problem of identifying the $\widetilde{G}$-orbits. Let

$$G^0 := N^- T N^+ = B^- B^+,$$

the *big cell* in G, a (dense) open subset of G. Let σ be the restriction to H of the morphism $B^- \times B^+ \longrightarrow G : (b^-, b^+) \mapsto (b^-)^{-1}b^+$, so that $\operatorname{im}\sigma = G^0$, and

$$\sigma : H \longrightarrow G^0 : (x, y, t) \mapsto x^{-1}t^2 y. \qquad (9)$$

Clearly, σ is an unramified covering of G^0 of degree 2^n. The vector fields on G^0 defined by the Chevalley generators of $\mathfrak{g}$ correspond, when lifted via σ to H, to derivations whose exponentials generate the group $\widetilde{G}$. (This explains the terminology *quantum coadjoint action* for the action of $\widetilde{G}$ on H.) From this it follows immediately [44, Theorem 16.2] that the $\widetilde{G}$-orbits on H are the sets $\sigma^{-1}(\mathcal{C} \cap G^0)$, where $\mathcal{C}$ is a conjugacy class of a non-central element in G. Under the identification (6), the subgroups N^-, N^+ and T of H correspond respectively to the maximal ideal spectra of

$$Z_0^- := k[F_{\beta_1}^\ell, \ldots, F_{\beta_N}^\ell], \quad Z_0^+ := k[E_{\beta_1}^\ell, \ldots, E_{\beta_N}^\ell], \quad Z_0^0 := k[K_{\varpi_1}^{\pm\ell}, \ldots, K_{\varpi_n}^{\pm\ell}].$$

For example, the elements of T are those maximal ideals of Z_0 containing $E_{\beta_j}^\ell - 1$ and $F_{\beta_j}^\ell - 1$, $(1 \leq j \leq N)$; and similarly for N^- and N^+. Thus in the light of Proposition III.5.8, we can analyse the algebras in $\mathcal{B}_{\tilde{U}_\epsilon(\mathfrak{g})}$ by considering the conjugacy classes of G and applying σ^{-1}.

Remark. When $k = \mathbb{C}$ we've shown above, just before (6), that the "Hopf centre" Z_0 of $U_\epsilon(\mathfrak{g})$ is a Hopf subalgebra. For a general field k of characteristic 0 containing an ℓth root ϵ of 1 we can now obtain the same conclusion by (a) reducing to the case where k has (at most) countable dimension over $\mathbb{Q}$, because $U_\epsilon(\mathfrak{g})$ is finitely related, so that we can work over such a field and then tensor up; and then (b) noting that Z_0 is the intersection of two Hopf algebras, namely $U_\epsilon(\mathfrak{g})$ and $Z_0 \otimes_k \mathbb{C}$, so that Z_0 itself is a Hopf algebra.

III.6.6. PI-degree and fully Azumaya locus. Let G_{reg} denote the set of *regular* elements of G; that is, G_{reg} is the union of the conjugacy classes of *maximal* dimension, namely $2N$. Note that, by basic properties of conjugacy classes of semisimple groups, [100]

$$G_{\text{reg}} \text{ is open in } G \text{ and } \text{codim}(G \setminus G_{\text{reg}}) = 3. \tag{10}$$

Since σ is a finite morphism,

$$H_{\text{reg}} \quad := \quad \sigma^{-1}(G^0 \cap G_{\text{reg}})$$

is the union of the symplectic leaves of H of maximal dimension, and this dimension is $2N$. Moreover, since σ is a finite morphism,

$$H_{\text{reg}} \text{ is open in } H \text{ and } \text{codim}(H \setminus H_{\text{reg}}) = 3. \tag{11}$$

It is shown in [44, Theorem 24.1] that every $\widetilde{G}$-orbit in H_{reg} contains a point $\mathfrak{m} \in \text{maxspec } Z_0 = H$ such that

$$\textit{every irreducible } \check{U}_\epsilon(\mathfrak{g})/\mathfrak{m}\check{U}_\epsilon(\mathfrak{g})\textit{-module has dimension } \ell^N. \tag{12}$$

But, by Proposition III.5.8, (12) remains true if $\mathfrak{m}$ is replaced by $g.\mathfrak{m}$ for any $g \in \widetilde{G}$. That is, (12) is true *for all* $\mathfrak{m} \in H_{\text{reg}}$.

We can draw several consequences from the above. First, from (11) and (12),

$$\cap\{M \in \text{maxspec } \check{U}_\epsilon(\mathfrak{g}) : \check{U}_\epsilon(\mathfrak{g})/M \cong M_{\ell^N}(\mathbb{C})\} = 0. \tag{13}$$

Therefore, by (13) and Lemma III.1.2,

Theorem 1. *With the above notation and hypotheses,*

$$\text{PI-deg}(\check{U}_\epsilon(\mathfrak{g})) = \ell^N. \quad \square \tag{14}$$

Recall from (III.4.10) that the *fully Azumaya locus* $\mathcal{F}_{\check{U}_\epsilon(\mathfrak{g})}$ (of Z_0 with respect to $\check{U}_\epsilon(\mathfrak{g})$) consists of precisely those maximal ideals $\mathfrak{m}$ of Z_0 such that every irreducible $\check{U}_\epsilon(\mathfrak{g})/\mathfrak{m}\check{U}_\epsilon(\mathfrak{g})$-module has dimension ℓ^N. Thus, from (12) and (13),

$$H_{\text{reg}} \subseteq \mathcal{F}_{\check{U}_\epsilon(\mathfrak{g})}. \tag{15}$$

In particular, from (11),

$$\text{codim}(\text{maxspec } Z_0 \setminus \mathcal{F}_{\check{U}_\epsilon(\mathfrak{g})}) \geq 3. \tag{16}$$

This will have an important consequence in Chapter III.8.

In fact the inclusion (15) is an equality, [22, Theorem 4.17]. But the proof of this requires a somewhat delicate analysis of the structure of baby Verma modules which we won't get involved in here. We simply record the

Theorem 2. *The fully Azumaya locus* $\mathcal{F}_{\check{U}_\epsilon(\mathfrak{g})}$ *equals* H_{reg}. $\quad \square$

NOTES

Most of the material in this chapter is taken from the papers of De Concini, Kac and Procesi, specifically [44], [40], and [42]. Theorem 2 of (III.6.6) is taken from [22].

EXERCISES

Exercise III.6.A. Let R be a ring, $x \in R$ and σ an automorphism of R.

(i) (Recall the definition of skew derivation from (I.1.11).) Show that the map

$$\mathrm{ad}_\sigma x : R \longrightarrow R$$
$$r \longmapsto xr - \sigma(r)x$$

defines a left σ-derivation of R (cf. (II.5.4)).

(ii) Suppose that R is a k-algebra with $\sigma(x) = q^{-1}x$ for some $q \in k^\times$. Show that, for $m \geq 1$ and $r \in R$,

$$
\begin{aligned}
(\mathrm{ad}_\sigma x)^m(r) &= \sum_{i=0}^m \binom{m}{i}_q (-1)^i (\rho_x \sigma)^i \lambda_x^{m-i}(r) \\
&= \sum_{i=0}^m \binom{m}{i}_q (-1)^i q^{-i(2m-i-1)/2} x^{m-i} \sigma^i(r) x^i.
\end{aligned}
$$

To do this, write $\mathrm{ad}_\sigma x$ as $\lambda_x + (-\rho_x\sigma)$, where λ_x and ρ_x denote, respectively, left and right multiplication by x. Check that $\lambda_x(-\rho_x\sigma) = q(-\rho_x\sigma)\lambda_x$, and apply the q-binomial formula (I.6.1) to $(\mathrm{ad}_\sigma x)^m$.

(iii) Continue with the notation of (ii), but assume now that q is an ℓth root of 1 satisfying the conditions (1) from the start of this chapter. Deduce that, for $r \in R$,

$$(\mathrm{ad}_\sigma x)^\ell(r) = x^\ell r - \sigma^\ell(r)x^\ell.$$

(iv) Now let $R = U_q(\mathfrak{g})$ or $R = \breve{U}_\epsilon(\mathfrak{g})$, and let σ_i be the automorphism of R defined by $\sigma_i(r) = K_i^{-1}rK_i$, for $i = 1,\dots,n$. (Recall that here, as usual, K_i denotes K_{α_i}.) Use (ii) with $x = E_i$ and $\sigma = \sigma_i$ (thus $q = q_i^2$ or ϵ^{2d_i}) to show that the quantized Serre relations of (I.6.3) can be written as

$$(\mathrm{ad}_{\sigma_i} E_i)^{1-a_{ij}}(E_j) = 0 \qquad (i \neq j),$$

and similarly for F_i and F_j. (Recall the formula (I.6.1)(1) relating the two types of q-binomial coefficients.)

(v) Deduce from (iii) and (iv) and the observations in (III.6.2) that, for $i = 1,\dots,n$, E_i^ℓ and F_i^ℓ are in the centre of $\breve{U}_\epsilon(\mathfrak{g})$. (Note from the conditions on ℓ that $1 - a_{ij} \leq \ell$ for all i,j.)

(vi) Use the action of the braid group to complete the proof of Theorem III.6.2(1) from (v).

Exercise III.6.B. Prove (III.6.2)(3).

Exercise III.6.C. The restricted quantized enveloping algebra $\overline{U}$ discussed in (III.6.4) can be defined for any $U_\epsilon(\mathfrak{g}; M)$ as follows:

$$\overline{U_\epsilon(\mathfrak{g}; M)} \quad := \quad U_\epsilon(\mathfrak{g}; M)/\langle E^\ell_{\beta_j}, \; F^\ell_{\beta_j}, \; K^\ell_\alpha - 1 : 1 \leq j \leq N, \; \alpha \in M\rangle.$$

Prove that these algebras (for fixed $\mathfrak{g}$) are isomorphic to each other as k-algebras (but not as Hopf algebras). In particular,

$$\overline{U_\epsilon(\mathfrak{g})} \quad \cong \quad \overline{\check{U}_\epsilon(\mathfrak{g})}.$$

Exercise III.6.D. Complete the details of the argument given in Remark III.6.5.

STRUCTURE AND REPRESENTATIONS OF $\mathcal{O}_\epsilon(G)$

In this chapter we show that $\mathcal{O}_\epsilon(G)$ also falls into the PI Hopf triple pattern. This time, the Hopf centre Z_0 is a copy of the classical coordinate ring $\mathcal{O}(G)$, Theorem III.7.2. The finite dimensional representation theory of $\mathcal{O}_\epsilon(G)$ is much better understood than that of $U_\epsilon(\mathfrak{g})$, in part because the Poisson geometry of $\mathcal{O}(G)$ is easier than the corresponding geometry for $U_\epsilon(\mathfrak{g})$, (which, as we saw in the last chapter, essentially hinges on the conjugacy classes of G), and in part because the irreducible representations of $\overline{\mathcal{O}_\epsilon(G)}$ are all of dimension 1, so that (III.4.11) can be applied. We explain the geometric aspects in (III.7.8), and discuss the irreducible $\mathcal{O}_\epsilon(G)$-modules in (III.7.7) and (III.7.9).

III.7.1. Notation and definitions. Let G be a connected, simply connected complex semisimple group with Lie algebra $\mathfrak{g}$. In [43] De Concini and Lyubashenko construct an integral form $\Gamma(\mathfrak{g})$ of the quantized enveloping algebra $U_q(\mathfrak{g})$. Now $\Gamma(\mathfrak{g})$ is a Hopf R-algebra, where $R := \mathbb{Q}[q, q^{-1}]$, and $\Gamma(\mathfrak{g})$ admits as modules integral lattices in the highest weight modules $V(\lambda)$ for $U_q(\mathfrak{g})$ discussed in (I.6). One can thus construct an R-subalgebra of the Hopf dual of $\Gamma(\mathfrak{g})$ in a way exactly analogous to that used in (I.7.5) to construct $\mathcal{O}_q(G)$, so obtaining a Hopf R-algebra $R_q(G)$. Now let $\ell \geq 3$ be an odd positive integer, prime to 3 if $\mathfrak{g}$ contains a G_2-component, and let ϵ be a complex primitive ℓ^{th} root of 1. Let $p_\ell(q) \in R$ be the ℓth cyclotomic polynomial, so that $R/p_\ell(q)R \cong \mathbb{Q}(\epsilon)$ and $p_\ell(q)R_q(G)$ is a Hopf ideal of $R_q(G)$. The factor algebra $R_q(G)/p_\ell(q)R_q(G)$ is denoted $\mathcal{O}_\epsilon(G)_{\mathbb{Q}(\epsilon)}$ and is called *the quantized coordinate algebra of G over $\mathbb{Q}(\epsilon)$ at the root of unity ϵ.*

In the same way as for $\mathcal{O}_\epsilon(G)_{\mathbb{Q}(\epsilon)}$ we can form the $\mathbb{Q}(\epsilon)$-Hopf algebra $\Gamma_\epsilon(\mathfrak{g}) := \Gamma(\mathfrak{g})/p_\ell(q)\Gamma(\mathfrak{g})$.

Proposition. [43] *There is a perfect Hopf pairing between $R_q(G)$ and $\Gamma(\mathfrak{g})$, which induces a perfect Hopf pairing between $\mathcal{O}_\epsilon(G)_{\mathbb{Q}(\epsilon)}$ and $\Gamma_\epsilon(\mathfrak{g})$. In particular, $\mathcal{O}_\epsilon(G)_{\mathbb{Q}(\epsilon)} \subseteq \Gamma_\epsilon(\mathfrak{g})^\circ$.* $\square$

If k is any field containing $\mathbb{Q}(\epsilon)$ we can obtain a k-form of $\mathcal{O}_\epsilon(G)$, namely $\mathcal{O}_\epsilon(G)_k := \mathcal{O}_\epsilon(G) \otimes_{\mathbb{Q}(\epsilon)} k$. When $k = \mathbb{C}$ we shall simply write $\mathcal{O}_\epsilon(G)$ for $\mathcal{O}_\epsilon(G)_\mathbb{C}$, and refer to this algebra as *the quantized coordinate algebra of G at the root of unity ϵ.* To keep things as simple as possible, this is the algebra we'll usually work with.

An analogous construction with specialisations to fields of positive characteristic is given in [7].

When we take $G = SL_n(\mathbb{C})$ in the above construction we get the Hopf $\mathbb{C}$-algebra $\mathcal{O}_\epsilon(SL_n(\mathbb{C}))$ with generators $\{X_{ij} : 1 \leq i, j \leq n\}$ and relations as in the generic case (I.2.4), but with q replaced with ϵ; the Hopf structure is unchanged from the generic case. (The method of [**7**, Appendix] can be used to confirm this.) In particular, when $G = SL_2(\mathbb{C})$ we get the $\mathbb{C}$-algebra with generators a, b, c and d and relations as in (III.3.1).

III.7.2. Finiteness over the centre. We recall a few more details of the construction of $\mathcal{O}_\epsilon(G)$ referred to in (III.7.1). Let $R = \mathbb{Q}[q^{\pm 1}]$, so $\mathcal{O}_\epsilon(G)$ is constructed by first defining an R-integral form $\Gamma(\mathfrak{g})$ of $U_q(\mathfrak{g})_k$, where $k = \mathbb{Q}(q)$. Namely, $\Gamma(\mathfrak{g})$ is the R-subalgebra of $U_q(\mathfrak{g})_k$ generated by

$$K_{\alpha_i}^{-1} \qquad\qquad\qquad\qquad (1 \leq i \leq n),$$

$$\binom{K_{\alpha_i}; 0}{t} := \prod_{s=1}^{t}\left(\frac{K_{\alpha_i} q_i^{-s+1} - 1}{q_i^s - 1}\right) \qquad (t \geq 1,\ 1 \leq i \leq n),$$

$$E_i^{(t)} := \frac{E_i^t}{[t]_{q_i}!} \qquad\qquad\qquad (t \geq 1,\ 1 \leq i \leq n),$$

$$F_i^{(t)} := \frac{F_i^t}{[t]_{q_i}!} \qquad\qquad\qquad (t \geq 1,\ 1 \leq i \leq n).$$

Recall from (I.6.3) the abbreviation $q_i = q^{d_i}$ and from (I.6.7) the notations $[t]_u = \frac{u^t - u^{-t}}{u - u^{-1}}$ and $[t]_u! = [t]_u[t-1]_u \cdots [1]_u$. Now, as stated already in (III.7.1), $\Gamma(\mathfrak{g})$ admits as modules integral R-lattices in the highest weight modules $V(\lambda)$ for $U_q(\mathfrak{g})$ discussed in (I.6.12). One can thus use these lattices to construct an R-subalgebra of the Hopf dual of $\Gamma(\mathfrak{g})$, so obtaining a Hopf R-algebra $R_q(G)$, a subalgebra of $U_q(\mathfrak{g})^\circ$. It follows from results of Lusztig [**144**] on $\Gamma(\mathfrak{g})$-modules that the pairing $R_q(G) \otimes_R \Gamma(\mathfrak{g}) \longrightarrow R$ is nondegenerate. Now $R_q(G)$ and $\Gamma(\mathfrak{g})$ are free R-modules having dual bases with respect to the pairing, so that, when we factor by the ideal generated by the ℓth cyclotomic polynomial $p_\ell(q)$, we get a nondegenerate pairing

$$R_q(G)/\langle p_\ell(q)\rangle \otimes_{\mathbb{Q}(\epsilon)} \Gamma(\mathfrak{g})/\langle p_\ell(q)\rangle \longrightarrow \mathbb{Q}(\epsilon);$$

that is, writing $\Gamma_\epsilon(\mathfrak{g})$ for $\Gamma(\mathfrak{g})/\langle p_\ell(q)\rangle$, there is a nondegenerate pairing

$$\mathcal{O}_\epsilon(G)_{\mathbb{Q}(\epsilon)} \otimes_{\mathbb{Q}(\epsilon)} \Gamma_\epsilon(\mathfrak{g}) \longrightarrow \mathbb{Q}(\epsilon). \tag{1}$$

From this set-up one deduces the

Theorem. (De Concini-Lyubashenko)
1. $\mathcal{O}_\epsilon(G)$ contains a central sub-Hopf algebra (isomorphic to) $\mathcal{O}(G)$.
2. $\mathcal{O}_\epsilon(G)$ is a finitely generated projective $\mathcal{O}(G)$-module of rank $\ell^{\dim G}$.

Sketch proof of 1. [**43**] **Step 1.** Introduce notation for elements of $U(\mathfrak{g})_{\mathbb{Q}(\epsilon)}$ (the $\mathbb{Q}(\epsilon)$-form of the ordinary enveloping algebra of $\mathfrak{g}$) analogous to the definitions above for $\Gamma_\epsilon(\mathfrak{g})$ – thus we set $e_i^{(m)} := e_i^m/m!$ and $f_i^{(m)} := f_i^m/m!$, while $\binom{h_i}{m} :=$ $h_i(h_i - 1)\cdots(h_i - m + 1)/m!$. Define a $\mathbb{Q}(\epsilon)$-algebra homomorphism, (using the Serre generators (I.5.5) for $\mathfrak{g}$),

$$\phi : \Gamma_\epsilon(\mathfrak{g}) \longrightarrow U(\mathfrak{g})_{\mathbb{Q}(\epsilon)}$$

with, for $i = 1, \ldots, n$,

$$\phi(E_i^{(t)}) = \begin{cases} e_i^{(t/\ell)} & \text{if } \ell \mid t \\ 0 & \text{otherwise,} \end{cases}$$

$$\phi(F_i^{(t)}) = \begin{cases} f_i^{(t/\ell)} & \text{if } \ell \mid t \\ 0 & \text{otherwise,} \end{cases}$$

$$\phi\left(\binom{K_{\alpha_i}; 0}{t}\right) = \begin{cases} \binom{h_i}{t/\ell} & \text{if } \ell \mid t \\ 0 & \text{otherwise,} \end{cases}$$

$$\phi(K_{\alpha_i}^{-1}) = 1.$$

One checks [**43**, Theorem 6.3] that ϕ is a well defined homomorphism mapping $\Gamma_\epsilon(\mathfrak{g})$ onto $U(\mathfrak{g})_{\mathbb{Q}(\epsilon)}$ (Exercise III.7.A).

Step 2. Dualising the epimorphism ϕ yields a Hopf algebra injection

$$\eta : \mathbb{Q}(\epsilon)(G) \longrightarrow \Gamma_\epsilon(\mathfrak{g})^\circ,$$

since $\mathbb{Q}(\epsilon)(G)$, the $\mathbb{Q}(\epsilon)$-form of $\mathcal{O}(G)$, embeds in $U(\mathfrak{g})_{\mathbb{Q}(\epsilon)}^\circ$ as in (I.9.23), Note (b). (This much does not need algebraic closure.)

Step 3. Recall that, by our construction using the nondegenerate pairing (1),

$$\mathcal{O}_\epsilon(G) \subseteq \Gamma_\epsilon(\mathfrak{g})^\circ.$$

Thus, to complete the proof, two facts need to be checked: First, that

$$\operatorname{im}\eta \quad \subseteq \quad \mathcal{O}_\epsilon(G); \tag{2}$$

and second, that

$$\operatorname{im}\eta \quad \subseteq \quad Z(\Gamma_\epsilon(\mathfrak{g})^\circ). \tag{3}$$

For (2), recall that the elements of $\mathcal{O}_\epsilon(G)$ are obtained by specialisation of q from elements of $R_q(G)$; see (1). So we need to know that the elements of $\operatorname{im}\eta$ can be so realised. Complete reducibility of $U(\mathfrak{g})_{\mathbb{Q}(\epsilon)}$-modules means that we just need to be sure that each finite dimensional irreducible $U(\mathfrak{g})_{\mathbb{Q}(\epsilon)}$-module occurs as the specialisation $q \longrightarrow \epsilon$ of a $\Gamma(\mathfrak{g})$-module. The recipe for this is the following: Let $V(\lambda)$ be the irreducible $U(\mathfrak{g})_{\mathbb{Q}(\epsilon)}$-module whose highest weight is the dominant integral weight λ. Then take the irreducible $U_q(\mathfrak{g})$-module $V(\ell\lambda)$, let $v \in V(\ell\lambda)$ have weight $\ell\lambda$, and take $M = \Gamma(\mathfrak{g})v$.

For (3), set $I = \ker \phi$. Suppose we show that, for all $u \in \Gamma_\epsilon(\mathfrak{g})$,

$$\Delta(u) - \Delta^{\mathrm{op}}(u) \in I \otimes I. \tag{4}$$

Then, for all $f \in \operatorname{im} \eta$, $g \in \Gamma_\epsilon(\mathfrak{g})^\circ$, $u \in \Gamma_\epsilon(\mathfrak{g})$,

$$(f * g)(u) \quad = \quad \sum_u f(u_1)g(u_2) \quad = \quad \sum_u f(u_2)g(u_1)$$

by (4), since $f(I) = 0$. Since $(g * f)(u) = \sum_u f(u_2)g(u_1)$, this will prove (3). Also, a simple calculation confirms that, because Δ and Δ^{op} are algebra homomorphisms, it's enough to check (4) for u equal to a generator E_i, $E_i^{(\ell)}$, F_i, $F_i^{(\ell)}$, $K_{\alpha_i}^{\pm 1}$, $\binom{K_{\alpha_i};0}{\ell}$, $(1 \leq i \leq n)$ of $\Gamma_\epsilon(\mathfrak{g})$. For these values of u, (4) is a straightforward calculation (Exercise III.7.B), and so (3) follows. $\square$

We'll prove part 2 of the theorem in (III.7.4).

III.7.3. Dual of the multiplication map on the big cell. Let B^- and B^+ be Borel subgroups of G with $B^- \cap B^+ = T$, T being a maximal torus of G. Fundamental to the study of both $\mathcal{O}_q(G)$ and $\mathcal{O}_\epsilon(G)$ is the quantization of the dual of the multiplication morphism

$$m : B^+ \times B^- \longrightarrow G. \tag{5}$$

Written in the dual language of coordinate rings, this is the homomorphism

$$(\pi_1 \otimes \pi_2) \circ \Delta : \mathcal{O}(G) \longrightarrow \mathcal{O}(G) \otimes \mathcal{O}(G) \longrightarrow \mathcal{O}(B^+) \otimes \mathcal{O}(B^-),$$

where π_i, $(i = 1, 2)$ are restriction maps. If we imitate this in the quantized setting we get an algebra homomorphism

$$\mu := (\pi_1 \otimes \pi_2) \circ \Delta : \mathcal{O}_q(G) \longrightarrow \mathcal{O}_q(G) \otimes \mathcal{O}_q(G) \longrightarrow \mathcal{O}_q(B^+) \otimes \mathcal{O}_q(B^-), \tag{6}$$

where we've written $\mathcal{O}_q(B^\pm)$ to denote the images of $\mathcal{O}_q(G)$ under the restrictions of functionals to the Hopf subalgebras $U_q^{\geq 0}(\mathfrak{g})$ and $U_q^{\leq 0}(\mathfrak{g})$ of $U_q(\mathfrak{g})$.

Recall the definition (I.9.22) of a perfect Hopf pairing. A crucial fact which now comes into play, and is *peculiar to the quantized setting*, is the following: There is a perfect Hopf pairing

$$U_q^{\geq 0}(\mathfrak{g})^{\mathrm{op}} \times U_q^{\leq 0}(\mathfrak{g}) \longrightarrow k = \mathbb{Q}(q) \tag{7}$$

(Theorem I.8.10 and Proposition I.8.12). This can be extended to a perfect Hopf pairing between $U_{\bar{q}}^{\geq 0}(\mathfrak{g})^{\mathrm{op}}$ and $\check{U}_{\bar{q}}^{\leq 0}(\mathfrak{g}) = U_{\bar{q}}^{\leq 0}(\mathfrak{g}; P)$, thanks to which we can identify $\check{U}_{\bar{q}}^{\leq 0}(\mathfrak{g})$ with a subalgebra of $U_{\bar{q}}^{\geq 0}(\mathfrak{g})^\circ$ (note that $U_{\bar{q}}^{\geq 0}(\mathfrak{g})^\circ$ and $(U_{\bar{q}}^{\geq 0}(\mathfrak{g})^{\mathrm{op}})^\circ$ are identical as algebras, although not as Hopf algebras). An analogous perfect pairing between $\check{U}_{\bar{q}}^{\geq 0}(\mathfrak{g})$ and $U_{\bar{q}}^{\leq 0}(\mathfrak{g})^{\mathrm{op}}$ allows us to identify $\check{U}_{\bar{q}}^{\geq 0}(\mathfrak{g})$ with a subalgebra of $U_{\bar{q}}^{\leq 0}(\mathfrak{g})^\circ$.

It turns out, in fact, that after the above identifications,

$$\mathrm{im}\,\pi_1 \quad \subseteq \quad \check{U}_{\bar{q}}^{\leq 0}(\mathfrak{g})$$

and

$$\mathrm{im}\,\pi_2 \quad \subseteq \quad \check{U}_{\bar{q}}^{\geq 0}(\mathfrak{g}).$$

So we can rewrite (6) as

$$\mu : \mathcal{O}_q(G) \longrightarrow \check{U}_{\bar{q}}^{\leq 0}(\mathfrak{g}) \otimes \check{U}_{\bar{q}}^{\geq 0}(\mathfrak{g}). \tag{8}$$

It's not hard to see that the image of μ is contained in the subalgebra D generated by

$$\{1 \otimes E_{\beta_j}, \ F_{\beta_j} \otimes 1, \ K_\lambda \otimes K_{-\lambda} : 1 \leq j \leq N, \ \lambda \in P\}. \tag{9}$$

Indeed, after the inversion of a suitable normal element d in $\mathcal{O}_q(G)$, we get

$$\mu\big(\mathcal{O}_q(G)[d^{-1}]\big) \quad = \quad D. \tag{10}$$

All of the above can also be carried out at the level of the integral forms $R_q(G)$, $\Gamma(\mathfrak{g})$ (and related forms $\Gamma^{\geq 0}(\mathfrak{g})$, $\Gamma^{\leq 0}(\mathfrak{g})$). In this way we get an injective homomorphism of R-forms,

$$\widetilde{\mu} : R_q(G) \longrightarrow \Gamma^{\leq 0}(\mathfrak{g}) \otimes_R \Gamma^{\geq 0}(\mathfrak{g}),$$

and, $\widetilde{\mu}$ being an R-algebra homomorphism, it factors through $\langle p_\ell(q)\rangle$ to yield an injective homomorphism

$$\widehat{\mu} : \mathcal{O}_\epsilon(G) \longrightarrow \check{U}_{\bar{\epsilon}}^{\leq 0}(\mathfrak{g}) \otimes \check{U}_{\bar{\epsilon}}^{\geq 0}(\mathfrak{g}). \tag{11}$$

It's a simple matter to calculate the image of $\mathcal{O}_\epsilon(G)$ and of $\mathcal{O}(G)$ under $\widehat{\mu}$, at least after localizing at the image d_ϵ of d. Thus, we can assume without loss that $d_\epsilon \in \mathcal{O}(G)$. More precisely, the equation

$$d_\epsilon \quad = \quad 0$$

defines the complement of the *big cell* of G, where the latter is by definition the image of the multiplication morphism (5). Then

$$\widehat{\mu}\big(\mathcal{O}_\epsilon(G)[d_\epsilon^{-1}]\big) \quad = \quad A_\epsilon, \tag{12}$$

where A_ϵ is the subalgebra of $\check{U}_\epsilon^{\leq 0}(\mathfrak{g}) \otimes \check{U}_\epsilon^{\geq 0}(\mathfrak{g})$ generated by the images of the elements (9); and

$$\hat{\mu}(\mathcal{O}(G)[d_\epsilon^{-1}]) = \mathbb{C}\langle 1 \otimes E_{\beta_j}^\ell, \ F_{\beta_j}^\ell \otimes 1, \ K_{\ell\lambda} \otimes K_{-\ell\lambda} : 1 \leq j \leq N, \ \lambda \in P\rangle. \quad (13)$$

De Concini and Lyubashenko show in [**43**, (1.10) and (7.1)] that for $i = 1, \dots, n$ there are elements $n_i \in N_G(T)$ with the following properties:

(1) $n_i T = s_i$ in $N_G(T)/T = W$;

(2) There is a function $\hat{n}_i \in \mathcal{O}_\epsilon(G)^*$ such that the left action $n_i \triangleleft$ on $\mathcal{O}_\epsilon(G)$ given by (I.9.16) satisfies

$$\hat{n}_i \triangleleft (zu) \quad = \quad (\hat{n}_i \triangleleft z)(\hat{n}_i \triangleleft u) \quad\quad (14)$$

for $z \in \mathcal{O}(G)$, $u \in \mathcal{O}_\epsilon(G)$;

(3) $\hat{n}_i \triangleleft$ restricted to $\mathcal{O}(G)$ is the right winding automorphism corresponding to evaluation of $\mathcal{O}(G)$ at n_i (I.9.25).

Since the right winding automorphisms of $\mathcal{O}(G)$ induce the right translations by elements of G on maxspec $\mathcal{O}(G) = G$, the map $\hat{n}_i \triangleleft : \mathcal{O}_\epsilon(G) \longrightarrow \mathcal{O}_\epsilon(G)$ can be viewed as a *quantization* of right multiplication on G by n_i^{-1}.

(4) Given $w \in W$ with reduced expression as a product of simple reflections $w = s_{i_1} s_{i_2} \dots s_{i_k}$, we can define

$$\hat{n}_w \quad = \quad \hat{n}_{s_{i_1}} \cdots \hat{n}_{s_{i_k}}.$$

Clearly this operator on $\mathcal{O}_\epsilon(G)$ satisfies the same equation (14) as $\hat{n}_i$, and is a quantization of right multiplication on G by n_w^{-1}, where $n_w = n_{i_1} \cdots n_{i_k}$.

(5)

$$G \quad = \quad \bigcup_{w \in W} n_w B^+ B^-. \quad\quad (15)$$

To see this, start from the normal form for elements of G stemming from the Bruhat decomposition: every element g of G can be expressed, for a unique $w \in W$, as a product

$$g \quad = \quad u n_w b$$

with $u, b \in B^-$ and $n_w^{-1} u n_w \in B^+$. Thus

$$g \quad = \quad n_w (n_w^{-1} u n_w) b \in n_w B^+ B^-$$

as required.

(6) By (15), if $\mathfrak{m} \in$ maxspec $\mathcal{O}(G)$ we can find $w \in W$ such that $d_w := \hat{n}_w \triangleleft d_\epsilon$ does not belong to $\mathfrak{m}$.

III.7.4. Proof of Theorem III.7.2(2). First, observe that $\mathcal{O}_\epsilon(G)[d_\epsilon^{-1}]$ is a finitely generated module over $\mathcal{O}(G)[d_\epsilon^{-1}]$ by (III.7.3)(12) and (13). Hence, for each $w \in W$, $\mathcal{O}_\epsilon(G)[d_w^{-1}]$ is a finitely generated $\mathcal{O}(G)[d_w^{-1}]$-module, by (III.7.3)(2), (3) and (4). It follows at once from (III.7.3)(6) and Exercise III.7.C that $\mathcal{O}_\epsilon(G)$ is a finitely generated $\mathcal{O}(G)$-module.

Finally, $\mathcal{O}(G) \subseteq Z(\mathcal{O}_\epsilon(G)) \subseteq \mathcal{O}_\epsilon(G)$ satisfies all the conditions (III.4.1) required of a PI Hopf triple. (Note that $\mathcal{O}_\epsilon(G)$ is a domain thanks to (III.7.3)(12).) Thus Corollary III.4.7 applies, and we conclude that $\mathcal{O}_\epsilon(G)$ is a projective $\mathcal{O}(G)$-module. (Alternatively, projectivity can be read off directly from the discussion in (III.7.3), since being a finitely generated projective module over a commutative noetherian ring is also a local property.) Since, looking at (III.7.3)(12) and (13), $\mathcal{O}_\epsilon(G)[d_\epsilon^{-1}]$ is visibly free of rank $\ell^{\dim G}$ over $\mathcal{O}(G)[d_\epsilon^{-1}]$, the statement regarding rank is also clear. $\square$

III.7.5. Global dimension. Here is another important but easy application of the isomorphism (III.7.3)(12):

Theorem. *The global dimension of $\mathcal{O}_\epsilon(G)$ is finite.*

Proof. Since d_ϵ defines the complement to the big cell B^+B^- in G, d_ϵ does not belong to the augmentation ideal $\mathfrak{m}_0$ of $\mathcal{O}(G)$. Being an iterated skew polynomial algebra, A_ϵ has finite global dimension (cf. [**158**, Theorem 7.5.3]). Thus, by (III.7.3)(12),

$$\text{gl.dim}\, \mathcal{O}_\epsilon(G)[d_\epsilon^{-1}] \quad < \quad \infty.$$

But the trivial $\mathcal{O}_\epsilon(G)$-module survives as a module over $\mathcal{O}_\epsilon(G)[d_\epsilon^{-1}]$, by our choice of d_ϵ. Hence it has a finite resolution by projective $\mathcal{O}_\epsilon(G)[d_\epsilon^{-1}]$-modules. Since $\mathcal{O}_\epsilon(G)[d_\epsilon^{-1}]$ is a flat $\mathcal{O}_\epsilon(G)$-module, and $\mathcal{O}_\epsilon(G)$ is noetherian, it follows that the trivial module has finite projective dimension over $\mathcal{O}_\epsilon(G)$. But in fact it is a basic property of Hopf algebras, which we'll prove in (III.8.2(2)), that the projective dimension of the trivial module bounds the global dimension. So the theorem follows. $\square$

III.7.6. The case $G = SL_n(\mathbb{C})$. As noted in (III.7.1), when we take $G = SL_n(\mathbb{C})$ in the construction (III.7.2), we get the Hopf $\mathbb{C}$-algebra $\mathcal{O}_\epsilon(SL_n(\mathbb{C}))$ with generators $\{X_{ij} : 1 \leq i, j \leq n\}$ and relations as in the generic case (I.2.4), but with q replaced with ϵ; the Hopf structure is unchanged from the generic case. Therefore it's not hard to calculate directly (Exercise III.7.D) that

$$Z_0 \quad := \quad \mathbb{C}[X_{ij}^\ell : 1 \leq i, j \leq n]/\langle D_\ell - 1 \rangle$$

(with $D_\ell := \det[X_{ij}^\ell]$) is a central subalgebra of $\mathcal{O}_\epsilon(G)$, and

$$Z_0 \quad \text{is a sub-Hopf algebra of } \mathcal{O}_\epsilon(G), \text{ with } Z_0 \cong \mathcal{O}(G).$$

Thus one has a direct and more elementary proof of Theorem III.7.2 in this case.

III.7.7. One dimensional representations of $\mathcal{O}_\epsilon(G)$. $\mathcal{O}_\epsilon(G)$ admits a torus T of representations of dimension 1, which we now describe. Recall first from (III.4.3), (III.4.4) the formalism for studying irreducible modules for a PI Hopf triple. Namely, there is a projection map

$$\pi' : \operatorname{maxspec} \mathcal{O}_\epsilon(G) \longrightarrow \operatorname{maxspec} \mathcal{O}(G)$$
$$M \longmapsto M \cap \mathcal{O}(G)$$

and a bundle of algebras $\mathcal{B}_{\mathcal{O}_\epsilon(G)}$, each of dimension $\ell^{\dim G}$ by Theorem III.7.2(2), such that each irreducible $\mathcal{O}_\epsilon(G)$-module is a module for exactly one of the algebras $\mathcal{O}_\epsilon(G)/\mathfrak{m}\mathcal{O}_\epsilon(G)$ in $\mathcal{B}_{\mathcal{O}_\epsilon(G)}$. (Here, $\mathfrak{m} \in \operatorname{maxspec} \mathcal{O}(G)$.) Since $\operatorname{maxspec} \mathcal{O}(G) \equiv G$, we can write

$$\operatorname{maxspec} \mathcal{O}(G) \quad = \quad \{\mathfrak{m}_g : g \in G\},$$

where $\mathfrak{m}_g$ is the maximal ideal of $\mathcal{O}(G)$ of functions vanishing at g, and define

$$\mathcal{O}_\epsilon(G)(g) \quad := \quad \mathcal{O}_\epsilon(G)/\mathfrak{m}_g\mathcal{O}_\epsilon(G) \quad \in \quad \mathcal{B}_{\mathcal{O}_\epsilon(G)}.$$

Recall also from (I.9.24) that the group of one-dimensional representations of $\mathcal{O}_\epsilon(G)$ is just the character group $X(\mathcal{O}_\epsilon(G))$, and can equally well be regarded as the group of group-like elements of $\mathcal{O}_\epsilon(G)^\circ$.

Proposition. (De Concini-Lyubashenko) *Keep the above notation, and let $n = \operatorname{rank}(G)$.*

1. If $g \in T$, then $\mathcal{O}_\epsilon(G)(g)$ admits a one-dimensional representation.

2. If $g \in T$, then every irreducible representation of $\mathcal{O}_\epsilon(G)(g)$ is one-dimensional.

3. $X(\mathcal{O}_\epsilon(G)) \cong (\mathbb{C}/2\pi i\ell\mathbb{Z})^n$, in such a way that the irreducible representations of $\mathcal{O}_\epsilon(G)(1_G)$ correspond to $(2\pi i\mathbb{Z}/2\pi i\ell\mathbb{Z})^n$.

4. Each algebra $\mathcal{O}_\epsilon(G)(t)$, $(t \in T)$, admits exactly ℓ^n distinct one-dimensional representations. $\square$

Remarks. 1. The proof in [**43**, (9.2)] proceeds by constructing some primitive elements H_i in $\Gamma_\epsilon(\mathfrak{g})$, obtained as scalar multiples of the images of the elements $K_i^\ell - 1$ of $\Gamma(\mathfrak{g})$, for $i = 1, \ldots, n$. The operators $\exp(\sum_i \lambda_i H_i)$, for $\lambda_i \in \mathbb{C}$, then yield characters of $\mathcal{O}_\epsilon(G)$, which afford the representations promised in 1.

2. If $G = SL_{n+1}(\mathbb{C})$ then it's not hard to prove the proposition directly from the presentation discussed in (III.7.6), (Exercise III.7.E).

3. Apart from the intrinsic interest for the representation theory of $\mathcal{O}_\epsilon(G)$, the importance of the proposition lies in the fact that, as in (I.9.25), the torus T acts as left and right winding automorphisms of $\mathcal{O}_\epsilon(G)$. In the case of $G = SL_{n+1}(\mathbb{C})$, these actions are easy to describe (Exercise III.7.F).

III.7.8. Poisson bracket and symplectic leaves. Recall from (III.7.2) that $\mathcal{O}_\epsilon(G)$ is constructed from $R_q(G)$ by factoring out $\langle p_\ell(q) \rangle$. This is exactly the type of construction discussed in (III.5.6), and so yields a Poisson bracket $\{\ ,\ \}$ on $Z(\mathcal{O}_\epsilon(G))$. Calculations in the dual setting of $\Gamma(\mathfrak{g})$ show that $\mathcal{O}(G)$ is a Poisson subalgebra of $Z(\mathcal{O}_\epsilon(G))$, [**43**, Proposition 8.4], and indeed is a Poisson group in the sense of (III.5.3). (The details for $G = SL_2(\mathbb{C})$ are discussed in Example III.5.5(2) and in Exercise III.5.D.) Notice that Proposition III.7.7 tells us that the (right or left) winding automorphism τ of $\mathcal{O}(G)$ determined by an element t of T extends to a (right or left) winding automorphism of $\mathcal{O}_\epsilon(G)$, which we shall here also denote by τ. One finds that for $f, g \in \mathcal{O}(G)$, $\{f, g\}(t) = 0$ for all $t \in T$. It follows that the tori T of right and left winding automorphisms of $\mathcal{O}_\epsilon(G)$ which we have just constructed act, in their restriction to $\mathcal{O}(G)$, as Poisson automorphisms of $\mathcal{O}(G)$. (This is [**43**, Proposition 8.7], but the straightforward calculation is sketched in Exercise III.7.G.)

Thus, we can define an equivalence relation $\sim$ on G by first setting, for $g, h \in G$,

$$g \sim\sim h \quad \Longleftrightarrow \quad \begin{cases} g, h & \text{in the same symplectic leaf,} \\ \text{or} \quad g = th & \text{for some } t \in T, \\ \text{or} \quad g = ht & \text{for some } t \in T, \end{cases}$$

and then taking $\sim$ to be the equivalence relation generated by $\sim\sim$. It follows from the above discussion and Corollary III.5.8 that, for $g, g' \in G$,

$$g \sim g' \quad \Longrightarrow \quad \mathcal{O}_\epsilon(G)(g) \cong \mathcal{O}_\epsilon(G)(g'). \tag{16}$$

It's therefore important to determine the $\sim$-equivalence classes in G. This has been done by Hodges and Levasseur [**94**, Appendix A]. The very pretty answer is as follows. Recall that there are earmarked Borel subgroups B^- and B^+ of G determined by the construction of $\mathcal{O}_\epsilon(G)$, with $B^+ \cap B^- = T$. There is a *(double) Bruhat decomposition* of G indexed by $W \times W$,

$$G = \bigsqcup_{(w_1, w_2) \in W \times W} X_{(w_1, w_2)},$$

where

$$X_{(w_1, w_2)} := B^+ \dot{w}_1 B^+ \cap B^- \dot{w}_2 B^- \tag{17}$$

and, for $w \in W$, $\dot{w} \in N_G(T)$ denotes a coset representative for $w \in N_G(T)/T$. Hodges and Levasseur also prove that

Each leaf in $X_{(w_1, w_2)}$ has dimension $\ell(w_1) + \ell(w_2) + \text{rank}_\mathfrak{h}(w_1 - w_2)$, $\qquad$ (18)

T acts transitively on the leaves in $X_{(w_1, w_2)}$ $\qquad$ (19)

and, (recalling that $n = \text{rank}(G)$),

$$\text{if } L \in X_{(w_1, w_2)}, \ T/\text{Stab}_T(L) \text{ is a torus of rank } n - \text{rank}_{\mathfrak{h}}(w_1 - w_2). \tag{20}$$

Note from (19) that

$$\text{The sets } X_{(w_1, w_2)} \text{ are the } \sim\text{-equivalence classes in } G. \tag{21}$$

We recall also at this point the discussion of the function $s : W \longrightarrow \mathbb{Z}$ of (II.4.13). Thus, for $w \in W$, $s(w) := \text{rank}_{\mathfrak{h}}(\text{id} - w)$. Clearly,

$$s(w_1 w_2^{-1}) \quad = \quad \text{rank}_{\mathfrak{h}}(w_1 - w_2)$$

for $w_1, w_2 \in W$. For later convenience, we set

$$m(w_1, w_2) \quad = \quad \ell(w_1) + \ell(w_2) + s(w_1 w_2^{-1}) \tag{22}$$

for $w_1, w_2 \in W$. It is necessarily the case from (18) (but can be proved directly from properties of Coxeter groups) that $m(w_1, w_2)$ is even.

Each $X_{(w_1, w_2)}$ is open in its closure. In particular,

$$\begin{aligned} X_{(w_0, w_0)} \quad &= \quad B^+ \dot{w}_0 B^+ \ \cap \ B^- \dot{w}_0 B^- \\ &= \quad B^+ \dot{w}_0 B^+ \dot{w}_0 \dot{w}_0 \ \cap \ \dot{w}_0 \dot{w}_0 B^- \dot{w}_0 B^- \\ &= \quad B^+ B^- \dot{w}_0 \ \cap \ \dot{w}_0 B^+ B^-, \end{aligned}$$

an intersection of two translates of the big cell, is open (and so dense) in G. Notice also that (18) includes the fact, noted above, that $T = X_{(1,1)}$ consists of trivial leaves.

There is a partial order $\leq$ on W, called the *Bruhat order*: for $w_1, w_2 \in W$, $w_1 \leq w_2$ if and only if a reduced expression for w_1 as a product of simple reflections can be got from one for w_2 by deleting some reflections from the expression [**116**, Corollary A.1.8]. This induces a Bruhat ordering on $W \times W$, whereby $(w_1, w_2) \leq (w'_1, w'_2)$ if and only if $w_1 \leq w'_1$ and $w_2 \leq w'_2$. Then

$$X_{(w_1, w_2)} \subseteq \overline{X_{(w'_1, w'_2)}} \quad \Longleftrightarrow \quad (w_1, w_2) \leq (w'_1, w'_2).$$

III.7.9. Irreducible $\mathcal{O}_\epsilon(G)$-modules. In this paragraph we impose a further restriction on ℓ – namely, as well as assuming that ℓ is odd, and prime to 3 when $\mathfrak{g}$ has a factor of type G_2, we'll assume that ℓ is *good*. That is, if $\sum_{i=1}^n a_i \alpha_i$ is the highest root of $\mathfrak{g}$, then ℓ is prime to all the a_i.

The first point to observe is that $\mathcal{O}_\epsilon(G)$ satisfies hypothesis (26) of (III.4.11) – by Proposition III.7.7(2) and (3) every irreducible $\overline{\mathcal{O}_\epsilon(G)}$-module has dimension

1, and $X(\overline{\mathcal{O}_\epsilon(G)}) \cong (\mathbb{Z}/\ell\mathbb{Z})^{\oplus n}$. Moreover, by (III.7.8)(16) and (21), there are at most $|W|^2$ isomorphism classes of algebras $\mathcal{O}_\epsilon(G)(g)$, $g \in G$. By Proposition and Corollary III.4.11, for each $g \in G$ all the irreducible $\mathcal{O}_\epsilon(G)(g)$-modules have the same dimension, and the number of them is a divisor of ℓ^n.

At one extreme are the algebras $\mathcal{O}_\epsilon(G)(t)$, for $t \in T$. These are the algebras for points in $X_{(1,1)}$ – they are all isomorphic to $\overline{\mathcal{O}_\epsilon(G)} = \mathcal{O}_\epsilon(G)(1)$. For the other strata the details have been worked out by De Concini and Procesi [45]. The results are a striking illustration of how the geometry of the strata as in (III.7.8)(18) and (19) is translated into the algebra of representations:

Theorem 1. *Let* $(w_1, w_2) \in W \times W$ *and let* $g \in X_{(w_1,w_2)}$.
   1. *Every irreducible* $\mathcal{O}_\epsilon(G)(g)$*-module has dimension* $\ell^{m(w_1,w_2)/2}$.
   2. *The number of irreducible* $\mathcal{O}_\epsilon(G)(g)$*-modules is* $\ell^{n-s(w_1 w_2^{-1})}$.

Proof. [45, Section 4] □

Notice that the maximum value of the dimension in 1 is ℓ^N, which is attained when (but *not* only when) $w_1 = w_2 = w_0$. Thus, by Lemma III.1.2(2) and Theorem I.13.5,

$$\text{PI-deg}\,\mathcal{O}_\epsilon(G) \quad = \quad \ell^N. \tag{23}$$

Consider g in the dense open set $X_{(w_0,w_0)}$. By the above, $\mathcal{O}_\epsilon(G)(g)$ has ℓ^n irreducible modules, each of dimension ℓ^N. Thus

$$\dim_{\mathbb{C}} \mathcal{O}_\epsilon(G)(g)/J(\mathcal{O}_\epsilon(G)(g)) = \ell^n \cdot \ell^{2N} = \ell^{\dim G} = \dim_{\mathbb{C}} \mathcal{O}_\epsilon(G)(g).$$

That is, for $g \in X_{(w_0,w_0)}$,

$$\mathcal{O}_\epsilon(G)(g) \quad \cong \quad M_{\ell^N}(\mathbb{C})^{\oplus \ell^n}. \tag{24}$$

It follows at once from (23) and Theorem III.7.9(1) that the fully Azumaya locus $\mathcal{F}_{\mathcal{O}_\epsilon(G)}$ is the union of those strata $X_{(w_1,w_2)}$ for which $m(w_1, w_2) = 2N$. In other words:

Theorem 2. *The fully Azumaya locus* $\mathcal{F}_{\mathcal{O}_\epsilon(G)}$ *consists of those maximal ideals* $\mathfrak{m}_g$ *of* $Z_0 = \mathcal{O}(G)$ *with* $g \in X_{(w_1,w_2)}$, *where*

$$\ell(w_1) + \ell(w_2) + s(w_1 w_2^{-1}) \quad = \quad 2N. \quad \square$$

For example, in the case of $SL_2(\mathbb{C})$,

$$\mathcal{F}_{\mathcal{O}_\epsilon(SL_2(k))} = X_{(w_0,w_0)} \cup X_{(w_0,1)} \cup X_{(1,w_0)} = G \setminus T.$$

This agrees with our calculations in Chapter III.3 (see (III.3.3)(3)) and in Exercise III.5.G. Each stratum $X_{(w_1,w_2)}$ has non-empty intersection with the big cell $B^+ B^-$

of G, so in view of (III.7.8)(16) and (21) when studying the algebras $\mathcal{O}_\epsilon(G)(g)$ we can restrict attention to $g \in B^+B^-$. Then (III.7.3)(11), (12) and (13) factor through $\mathfrak{m}_g\mathcal{O}_\epsilon(G)$ and we deduce that

$$\mathcal{O}_\epsilon(G)/\mathfrak{m}_g\mathcal{O}_\epsilon(G) \;\cong\; A_\epsilon/\widehat{\mu}(\mathfrak{m}_g)A_\epsilon.$$

Note that from the description of A_ϵ in (III.7.3) it's easy to get an explicit presentation of the right hand side of the above isomorphism.

III.7.10. Duality of the restricted algebras. A presentation of the sort we've just been discussing can be written down in particular for the restricted algebra $\overline{\mathcal{O}_\epsilon(G)} = \mathcal{O}_\epsilon(G)(1)$. It's natural to hope that this Hopf algebra of dimension $\ell^{\dim G}$ is the Hopf dual of the restricted quantized enveloping algebra $\overline{U_\epsilon(\mathfrak{g})}$ of (III.6.4), which does at least have the same dimension $\ell^{\dim \mathfrak{g}}$. That this hope is correct is probably well-known, but the result does not seem to be recorded in the literature. Thus we state here the

Theorem. *Continue with the notation and hypotheses of (III.7.1). Let $\mathfrak{g}$ denote the Lie algebra of G. Then*

$$\overline{\mathcal{O}_\epsilon(G)} \;\cong\; \overline{U_\epsilon(\mathfrak{g})}^*$$

as Hopf algebras.

Sketch of Proof. (Gordon) One first shows that the non-degenerate pairing

$$\mathcal{O}_\epsilon(G)_{\mathbb{Q}(\epsilon)} \otimes_{\mathbb{Q}(\epsilon)} \Gamma_\epsilon(\mathfrak{g}) \longrightarrow \mathbb{Q}(\epsilon)$$

of (III.7.2)(1) induces a Hopf pairing

$$\overline{\mathcal{O}_\epsilon(G)}_{\mathbb{Q}(\epsilon)} \otimes_{\mathbb{Q}(\epsilon)} \widehat{U}_\epsilon(\mathfrak{g}) \longrightarrow \mathbb{Q}(\epsilon),$$

where $\widehat{U}_\epsilon(\mathfrak{g})$ is the Lusztig form of the restricted quantized enveloping algebra discussed in (III.6.4), and which occurs as the subalgebra of $\Gamma_\epsilon(\mathfrak{g})$ generated by $\{E_i, F_i, K_i : i = 1, \dots, n\}$. This therefore yields a Hopf algebra homomorphism from $\overline{\mathcal{O}_\epsilon(G)}_{\mathbb{Q}(\epsilon)}$ to $\widehat{U}_\epsilon(\mathfrak{g})^*$, and in fact all the functionals on $\widehat{U}_\epsilon(\mathfrak{g})$ obtained via this map have $\langle K^\ell - 1 : i = 1, \dots, n \rangle$ in their kernels, so that we get a Hopf algebra map from $\overline{\mathcal{O}_\epsilon(G)}_{\mathbb{Q}(\epsilon)}$ to $\overline{U_\epsilon(\mathfrak{g})}^*$. The problem is to show that this is bijective, and by comparing dimensions it's enough to prove that the map is onto. To do this, one observes that every matrix coefficient in $\overline{U_\epsilon(\mathfrak{g})}^*$ arises as a matrix coefficent of a projective $\overline{U_\epsilon(\mathfrak{g})}$-module. Then one applies some representation theory to see that every projective module is the image of a tensor product of Weyl modules. Matrix coefficients of the latter are easily seen to be in the image of $\overline{\mathcal{O}_\epsilon(G)}_{\mathbb{Q}(\epsilon)}$ thanks to the construction of $\mathcal{O}_\epsilon(G)$ as a dual of $\Gamma_\epsilon(\mathfrak{g})$. This gives the result. $\square$

III.7.11. Freeness over Z_0. By contrast with the situation for $U_\epsilon(\mathfrak{g})$ (Theorem III.6.2(4)), and also that for $U(\mathfrak{g})$ in positive characteristic (Theorem I.13.2), it's by no means obvious that $\mathcal{O}_\epsilon(G)$ is a *free* module over $\mathcal{O}(G)$. Nevertheless, this is in fact the case:

Theorem. (Brown-Gordon-Stafford) *Continue with the notation and hypotheses of (III.7.1). Then $\mathcal{O}_\epsilon(G)$ is a free $\mathcal{O}(G)$-module of rank $\ell^{\dim G}$.*

Proof. [**24**] This hinges on two facts from algebraic K-theory, for which we need to recall the K-*group* $K_0(R)$ of finitely generated projective modules over the ring R. (This is by definition the free abelian group on isomorphism classes $[P]$ of such R-modules, factored by the subgroup generated by all elements of the form $[P \oplus Q] - [P] - [Q]$, where P and Q are finitely generated projective R-modules.)

First, by a result of [**152**] (see also [**159**]), for G simply connected semisimple as in our case,

$$K_0(\mathcal{O}(G)) \quad = \quad \langle\, [\mathcal{O}(G)]\, \rangle \quad \cong \quad \mathbb{Z}. \tag{25}$$

Recall that, for a commutative ring R, $K_0(R) = \langle [R] \rangle$ if and only if every finitely generated projective R-module P is *stably free*, meaning that $P \oplus R^{\oplus n} \cong R^{\oplus m}$ for suitable non-negative integers n and m. Now $\mathcal{O}_\epsilon(G)$ is a finitely generated projective $\mathcal{O}(G)$-module by Theorem III.7.2(2), so that, by (25),

$$\mathcal{O}_\epsilon(G) \oplus \mathcal{O}(G)^{\oplus n} \quad \cong \quad \mathcal{O}(G)^{\oplus m} \tag{26}$$

as $\mathcal{O}(G)$-modules, for integers m and n with $m - n = \ell^{\dim G}$.

To complete the proof we appeal to the second fact: Since the Krull dimension of $\mathcal{O}(G)$, which is the dimension of the algebraic variety G, namely $\dim G$, is less than $\ell^{\dim G}$, the Cancellation Theorem for K_0 of commutative noetherian rings [**10**, Corollary IV.3.5] permits us to cancel $\mathcal{O}(G)^{\oplus n}$ from each side of (26), so proving the theorem. $\quad\square$

Remarks. 1. The above proof doesn't provide a basis for $\mathcal{O}_\epsilon(G)$ as $\mathcal{O}(G)$-module. For the case $G = SL_2(\mathbb{C})$, such a basis has been given in [**39**].

2. We'll discuss this question of freeness for PI Hopf triples further in Chapter III.10.

NOTES

Except where otherwise stated in the text, this chapter is based on [**43**] and [**45**]. Theorem III.7.5 is taken from [**21**], where it was also shown, with essentially the same proof, that $\mathcal{O}_q(G)$ has finite global dimension. There has been further progress in understanding the representation theory of the algebras $\mathcal{O}_\epsilon(G)$, in [**23**], which will be discussed briefly in (III.10.5).

EXERCISES

Exercise III.7.A. Check that ϕ is an algebra epimorphism in Step 1 of the proof of Theorem III.7.2.

Exercise III.7.B. Prove (III.7.2)(4).

Exercise III.7.C. Let R be a commutative domain and M an R-module. Suppose that there is a finite set $\{d_1, \dots, d_t\}$ of non-zero elements of R such that $\sum_i d_i R = R$ and $M[d_i^{-1}]$ is a finitely generated $R[d_i^{-1}]$-module for each $i = 1, \dots, t$. Show that M is a finitely generated R-module.

Exercise III.7.D. Prove for $G = SL_n(\mathbb{C})$ the properties of Z_0, as suggested in (III.7.6).

Exercise III.7.E. Give a direct proof of Proposition III.7.7 for $G = SL_{n+1}(\mathbb{C})$, as follows.

    (i) Show that if V is a 1-dimensional $\mathcal{O}_\epsilon(G)$-module, then $X_{ij} V = 0$ for all $i \neq j$.

    (ii) Deduce parts 2 and 3 of the proposition for $G = SL_{n+1}(\mathbb{C})$.

Exercise III.7.F. For $\mathcal{O}_\epsilon(SL_{n+1}(\mathbb{C}))$, describe the action of the tori of left and right winding automorphisms (Remark III.7.7(3)). Compare with the generic case discussed in (II.1.18).

Exercise III.7.G. Let H be a Poisson group and let $h \in H$ be such that $\{f, g\}(h) = 0$ for all $f, g \in \mathcal{O}(H)$. Show that the (right or left) winding automorphism τ of $\mathcal{O}(H)$ determined by h is a Poisson automorphism. [Hint: We have to check that $\tau(\{f, g\}) = \{\tau(f), \tau(g)\}$. Do this (say for τ *right* winding) by writing τ as $\mu \circ (1 \otimes h) \circ \Delta$ and making use of the compatibility of the Poisson bracket with the coproduct.]

HOMOLOGICAL PROPERTIES
AND THE AZUMAYA LOCUS

III.8.1. The main aim of this chapter is to prove a variant of Theorem III.1.8, which gives sufficient conditions, for an algebra A which is a finite module over its centre Z, for the Azumaya locus to coincide with the locus of non-singular points of maxspec Z. The variant we have in mind exploits the fact that we are concerned exclusively with Hopf algebras in these notes; so we can use some homological properties of Hopf algebras to avoid direct discussion of Auslander-regularity and so give a relatively elementary proof of the theorem. The homological properties are of independent interest, and it is with them that we'll begin, after first stating the precise theorem we'll prove. Recall that $\mathcal{A}_A$ denotes the *Azumaya locus* of the algebra A, (III.1.7), and that $\mathcal{S}_Z$ denotes the *singular locus* of the affine commutative algebra Z (III.1.8), so that

$$\mathcal{S}_Z := \{\mathfrak{m} \in \text{maxspec}\, Z : \text{gl.dim}(Z_{\mathfrak{m}}) = \infty\}.$$

Theorem. *Let A be a prime affine algebra over a field k. Suppose that:*
    (A) *A is a finitely generated module over its centre Z;*
    (B) *There is a fixed non-negative integer n such that* p.dim $V = n$ *for all irreducible A-modules V;*
    (C) *$A_{\mathfrak{p}}$ is Azumaya over $Z_{\mathfrak{p}}$ for all height one primes $\mathfrak{p}$ of Z.*
    Then
$$\mathcal{A}_A = (\text{maxspec}\, Z) \setminus \mathcal{S}_Z.$$

III.8.2. Global dimension of Hopf algebras. In this paragraph, let H be an arbitrary Hopf algebra. Recall (Exercise III.8.A) that if A and B are H-modules then so is $\text{Hom}_k(A, B)$, with

$$(h.f)(a) \quad = \quad \sum_{(h)} h_1.[f(S(h_2).a)]$$

for $h \in H$, $a \in A$, $f \in \text{Hom}_k(A, B)$.

Lemma. *Let A, B and C be H-modules. Then the natural adjoint isomorphism $\phi : \mathrm{Hom}_k(A,\ \mathrm{Hom}_k(B,C)) \longrightarrow \mathrm{Hom}_k(A \otimes B,\ C)$ restricts to an isomorphism*

$$\mathrm{Hom}_H(A,\ \mathrm{Hom}_k(B,C)) \ \cong \ \mathrm{Hom}_H(A \otimes B,\ C).$$

Proof. Write $\phi(f) = \overline{f}$, so that $\overline{f}(a \otimes b) = f(a)(b)$. One has to check that f is H-linear if and only if $\overline{f}$ is – see Exercise III.8.B(i). □

Corollary. *Let A, B and C be H-modules. Then, for all $i \geq 0$,*

$$\mathrm{Ext}_H^i(A,\ \mathrm{Hom}_k(B,C)) \ \cong \ \mathrm{Ext}_H^i(A \otimes B,\ C).$$

Proof. Fix an H-projective resolution $\cdots \longrightarrow P_i \longrightarrow \cdots \longrightarrow P_0 \longrightarrow A \longrightarrow 0$ of A. Then the modules $P_i \otimes B$ are H-projective by the lemma (with $A = P_i$, see Exercise III.8.B(ii)). Thus

$$\cdots \longrightarrow P_i \otimes B \longrightarrow \cdots \longrightarrow P_0 \otimes B \longrightarrow A \otimes B \longrightarrow 0 \tag{1}$$

is an H-projective resolution of $A \otimes B$. The lemma now gives a commutative diagram of complexes

$$
\begin{array}{ccccc}
\mathrm{Hom}_H(P_0 \otimes B,\ C) & \longrightarrow \cdots \longrightarrow & \mathrm{Hom}_H(P_i \otimes B,\ C) & \longrightarrow & \cdots \\
\cong \downarrow & & \cong \downarrow & & \\
\mathrm{Hom}_H(P_0,\ \mathrm{Hom}_k(B,C)) & \longrightarrow \cdots \longrightarrow & \mathrm{Hom}_H(P_i,\ \mathrm{Hom}_k(B,C)) & \longrightarrow & \cdots
\end{array}
$$

The homology of the top row is $\mathrm{Ext}_H^*(A \otimes B,\ C)$, while the homology of the bottom row is $\mathrm{Ext}_H^*(A,\ \mathrm{Hom}(B,C))$. □

A result of Bhatwadekar [14] and Goodearl [67] (cf. [158, Corollary 7.1.14]) asserts that if R is a (two-sided) noetherian ring of finite global dimension then the global dimension of R is $\sup\{\mathrm{p.dim}_R(V) : {}_RV \text{ simple}\}$. For a (not necessarily noetherian) Hopf algebra H we can do much better. Thus it follows immediately from the corollary (or indeed just from (1) with $A = k$) that, for all H-modules B,

$$\mathrm{p.dim}_H(B) \leq \mathrm{p.dim}_H({}_Hk).$$

Thus

$$\mathrm{gl.dim}(H) \ = \ \mathrm{p.dim}_H({}_Hk), \tag{2}$$

a fact noted by Pareigis [179] under an extra hypothesis, and later by Lorenz and Lorenz [141]. We shall need a more precise statement than the inequality above, namely

Proposition. *Let H be a Hopf algebra. Suppose that* $\mathrm{gl.dim}(H) = n < \infty$. *If V is any non-zero finite dimensional H-module, then* $\mathrm{p.dim}_H(V) = n$.

Proof. Suppose that $_HV$ is non-zero and $\mathrm{p.dim}_H(V) < n$. Then the corollary applied with $A = V$ and $B = V^*$ shows that

$$\mathrm{p.dim}_H(V \otimes V^*) < n. \tag{3}$$

Fix a basis $\{v_i\}$ of V and corresponding dual basis $\{w_i\}$ of V^*, and define the coevaluation map

$$\delta : k \longrightarrow V \otimes V^*$$

$$\alpha \longmapsto \sum_i \alpha v_i \otimes w_i.$$

Then δ is a homomorphism of H-modules (Exercise III.8.C), so there is an exact sequence of finite dimensional H-modules

$$0 \longrightarrow k \longrightarrow V \otimes V^* \longrightarrow B \longrightarrow 0, \tag{4}$$

and clearly $B \neq 0$ by (3). Now (3) and (4) together with standard properties of the homological dimensions of short exact sequence of modules show that $\mathrm{p.dim}_H(B) = n + 1$. This is a contradiction, and so no such V can exist. $\square$

III.8.3. Centrally Macaulay rings. A ring R which is a finitely generated module over its noetherian centre and satisfies the conclusion of Proposition III.8.2 is called *homologically homogeneous* in [**25**], where such rings are shown to have many desirable properties, generalising to a noncommutative setting much of the theory of commutative noetherian regular rings. Homologically homogeneous rings are examples of the *Auslander-regular Macaulay rings* of (I.15.3) and (I.15.4) – in fact, the two notions coincide for rings which are module finite over affine centres [**218**].

Recall that a commutative noetherian local ring R is Cohen-Macaulay if and only if it has an ideal I which is generated by an R-sequence, such that R/I is artinian, [**120**, page 95]. If R is *regular* – that is, has finite global dimension – then R is Cohen-Macaulay, and in fact we can take $I = J(R)$ to be the co-artinian ideal generated by an R-sequence in this case, [**120**, Theorem 170]. Here we shall prove the "correct" generalisation of this well-known fact to the class of rings module finite over their noetherian centres – namely, we'll show that homologically homogeneous rings are Cohen-Macaulay modules over their centres. This is the key to completing the proof of Theorem III.8.1.

Let R be a ring which is a finitely generated module over its centre Z, and let I be an ideal of R. A *Z-sequence in I on R* is a sequence $(x_1, \dots, x_t)$ of elements of $Z \cap I$ such that

1. x_i is not a zero divisor modulo $\sum_{j=1}^{i-1} x_j R$, for $i = 1, \dots, t$.

If, in addition,

2. t is the greatest integer with this property,

then $(x_1, \dots, x_t)$ is called a *maximal Z-sequence in I on R*. Of course, if R is noetherian, such maximal sequences exist. It is not hard to confirm that any two such maximal sequences have the same length [**120**, Theorem 121]. The common length t is called the *grade* $g(I)$ of I. If R is as above and M is a maximal ideal of R, then

$$g(M) \quad \leq \quad \text{height}(M), \tag{5}$$

(Exercise III.8.D). When this inequality is an equality for every maximal ideal M, we say that R is *centrally Macaulay*.

We need two basic lemmas about Z-sequences.

Lemma 1. *Let M be a maximal ideal of the noetherian ring R which is a finitely generated module over its local centre Z. Then the Z-sequence $(x_1, \dots, x_t)$ in M on R is maximal if and only if $R/\sum_{j=1}^{t} x_j R$ has non-zero socle.*

Proof. ($\Longleftarrow$) Suppose that $R/\sum_{j=1}^{t} x_j R$ has non-zero socle, where $(x_1, \dots, x_t)$ is a Z-sequence in M on R. Set $\mathfrak{m} = M \cap Z$, so $\mathfrak{m}$ is the maximal ideal of Z by [**158**, Corollary 10.2.10(iv)]. The socle of $R/\sum_{j=1}^{t} x_j R$ is annihilated by $\mathfrak{m}$ (using the same reference again). In particular, $\mathfrak{m}$ consists of zero divisors *modulo* $\sum_{j=1}^{t} x_j R$. So t is maximal.

($\Longrightarrow$) Suppose that $(x_1, \dots, x_t)$ is a maximal Z-sequence in M on R. So $\mathfrak{m}$ consists of zero divisors *modulo* $\sum_{j=1}^{t} x_j R$. By [**120**, Theorem 80], there exists $y \in R \backslash \sum_{j=1}^{t} x_j R$ with $\mathfrak{m}y \subseteq \sum_{j=1}^{t} x_j R$. Since $R/\mathfrak{m}R$ is artinian, the factor module $\left(Ry + \sum_{j=1}^{t} x_j R\right)/\sum_{j=1}^{t} x_j R$ is contained in the socle of $R/\sum_{j=1}^{t} x_j R$. $\quad \Box$

Lemma 2. *Let R, Z and I be as in the definition above, with $(x_1, \dots, x_t)$ a Z-sequence in I on R. Then $\text{p.dim}_R(R/\sum_{j=1}^{t} x_j R) = t$.*

Proof. This is an easy induction on t – see Exercise III.8.E. $\quad \Box$

We are now ready to prove the

Theorem. *Let R be a noetherian ring which is a finitely generated module over its local centre Z. Suppose that $\text{gl.dim}(R) = n < \infty$ and that $\text{p.dim}_R(V) = n$ for every irreducible R-module V. Then R is centrally Macaulay with classical Krull dimension n.*

For the proof, we need to know that the finitely generated R-modules of projective dimension n are *precisely* those with non-zero socle. This is not hard to deduce from the following

Lemma 3. 1. *Let T be a ring with* $\mathrm{l.gl.dim}(T) = \mathrm{p.dim}_T(V) = n < \infty$ *for every irreducible left T-module V. Let X be any left T-module with non-zero socle. Then* $\mathrm{p.dim}_T(X) = n$.

2. *Let R be as in the theorem. Let P be a prime ideal of R and suppose that P is not maximal. Then* $\mathrm{p.dim}_R(R/P) \leq n - 1$.

Proof. Exercise III.8.F.   □

Corollary (to Lemma 3). *Let R be a noetherian ring which is a finitely generated module over its centre Z, and let M be a maximal ideal of R. Then*

$$\mathrm{height}(M) \quad \leq \quad \mathrm{gl.dim}(R).$$

Proof. We may clearly assume (by localizing at $M \cap Z$) that Z is local, and that $\mathrm{gl.dim}(R) = n$ is finite. The result can now easily be deduced from the lemma by induction on n, by localizing at $Z \setminus (P \cap Z)$ for any non-maximal prime P of R. □

Proof of the theorem. Let R be as in the theorem, let $(x_1, \dots, x_t)$ be a maximal Z-sequence in the maximal ideal M of R, and write $I = \sum_{j=1}^{t} x_j R$. By Lemma 2,

$$\mathrm{p.dim}_R(R/I) = t.$$

But Lemma 1 and Lemma 3(1) imply that

$$\mathrm{p.dim}_R(R/I) = n,$$

so that

$$t = n. \tag{6}$$

By (5) and the Corollary to Lemma 3(2),

$$t \quad \leq \quad \mathrm{height}(M) \quad \leq \quad n. \tag{7}$$

Combining (6) and (7) yields the result.   □

III.8.4. Proof of Theorem III.8.1. Let A be a prime affine algebra over a field k and suppose that A satisfies hypotheses (A) and (B) of Theorem III.8.1. By the Artin-Tate Lemma (I.13.4), Z is k-affine and so both Z and A are noetherian rings. By a result of Rainwater [**185**], $\mathrm{gl.dim}(A) = n < \infty$, (although when A is a Hopf algebra this also follows from (III.8.2)(2)). Recall that we denote the Azumaya locus of A by $\mathcal{A}_A$ and the singular locus of Z by $\mathcal{S}_Z$.

Lemma 1. *With the above hypotheses and notation,*

$$\mathcal{A}_A \subseteq (\text{maxspec } Z) \setminus \mathcal{S}_Z.$$

Proof. This is proved in Lemma III.1.8. □

As pointed out in (III.1.8), the inclusion of the lemma is strict in general – thus Theorem III.8.1 is false without hypothesis (C). We bring hypothesis (C) into play with

Lemma 2. *Let A be a prime k-algebra with centre Z, satisfying hypotheses* (A) *and* (C) *of Theorem III.8.1. Suppose that A is a projective Z-module. Then A is Azumaya over Z.*

Proof. By Example III.1.3(3) we need to show that $A_{\mathfrak{m}}$ is Azumaya over $Z_{\mathfrak{m}}$ for all maximal ideals $\mathfrak{m}$ of Z. So there is no harm in assuming that Z is local, as we now do. Thus A is a free Z-module, of rank r, say. Then $A \otimes_Z A^{\mathrm{op}}$ and $E := \mathrm{End}_Z(A)$ are both free Z-modules of rank r^2.

Let $f : A \otimes_Z A^{\mathrm{op}} \longrightarrow E$ be the natural map. For any height one prime $\mathfrak{p}$ of Z there is a commutative diagram

$$
\begin{array}{ccc}
A \otimes_Z A^{\mathrm{op}} & \xrightarrow{\ f\ } & E \\
{\scriptstyle \lambda_{\mathfrak{p}}}\big\downarrow & & \big\downarrow \\
A_{\mathfrak{p}} \otimes_Z A_{\mathfrak{p}}^{\mathrm{op}} & \xrightarrow{\ f_{\mathfrak{p}}\ } & \mathrm{End}_{Z_{\mathfrak{p}}}(A_{\mathfrak{p}})
\end{array}
$$

Here, the natural localization map $\lambda_{\mathfrak{p}}$ is injective because $A \otimes_Z A^{\mathrm{op}}$ is Z-torsion free, and $f_{\mathfrak{p}}$ is an isomorphism by hypothesis. Hence f must be injective, so that $\mathrm{im}\, f$ is Z-free of rank r^2. We can therefore define a Z-monomorphism $g : E \longrightarrow E$ with $\mathrm{im}\, g = \mathrm{im}\, f$, and the above diagram shows that $g_{\mathfrak{p}}$ is a bijection on $E_{\mathfrak{p}}$ for every height one prime $\mathfrak{p}$ of Z. This means that $\det g$ is invertible in $Z_{\mathfrak{p}}$ for every such $\mathfrak{p}$, so that $\det g$ is not contained in any height one prime $\mathfrak{p}$ of Z. It follows from the Principal Ideal Theorem [**120**, Theorem 142] that $\det g$ is invertible; so g is invertible. In other words, $\mathrm{im}\, f = \mathrm{im}\, g = E$. Therefore f is surjective and A is Azumaya. □

Proof of the theorem. Suppose that A, Z and k satisfy the hypotheses of Theorem III.8.1. In view of Lemma 1 we only have to prove that

$$(\text{maxspec } Z) \setminus \mathcal{S}_Z \subseteq \mathcal{A}_A. \tag{8}$$

So choose $\mathfrak{m} \in \text{maxspec } Z$ with $Z_{\mathfrak{m}}$ of finite global dimension. Note that hypothesis (B) of the theorem is preserved when we localize at $\mathfrak{m}$; for if V is any irreducible A-module annihilated by $\mathfrak{m}$ then all Ext_A-groups with V in an argument are $(Z \setminus \mathfrak{m})$-torsion free. By Theorem III.8.3 applied to $A_{\mathfrak{m}}$, this ring is centrally Macaulay.

That is, $A_{\mathfrak{m}}$ is a Cohen-Macaulay $Z_{\mathfrak{m}}$-module. But, by the Auslander-Buchsbaum depth theorem [**120**, Theorem 173], a Cohen-Macaulay module over a regular local ring is free. By Lemma 2, $A_{\mathfrak{m}}$ is Azumaya over $Z_{\mathfrak{m}}$, and (8) is proved. $\square$

Finally, combining Theorem III.8.1 and Proposition III.8.2, we immediately get the result which we'll apply to our classes of Hopf algebras:

Corollary. *Let H be a prime Hopf k-algebra which has finite global dimension and which is a finitely generated module over its affine centre Z. Suppose that $H_{\mathfrak{p}}$ is Azumaya over $Z_{\mathfrak{p}}$ for every height one prime $\mathfrak{p}$ of Z. Then*

$$(\text{maxspec } Z) \setminus \mathcal{S}_Z \;\; = \;\; \mathcal{A}_A. \;\; \square$$

III.8.5. Applications to PI Hopf triples.
1. Quantized function algebras $\mathcal{O}_{\epsilon}(G)$. We operate here with the notation and hypotheses of Chapter III.7. Write Z for $Z(\mathcal{O}_{\epsilon}(G))$. Let $\mathcal{A}'$ be the non-Azumaya locus of $\mathcal{O}_{\epsilon}(G)$. That is, by Theorem III.1.6, since PI-deg $\mathcal{O}_{\epsilon}(G) = \ell^N$ (see (III.7.9)(23)),

$$\mathcal{A}' \;\; = \;\; \{\mathfrak{m} \in \text{maxspec } Z : \mathcal{O}_{\epsilon}(G)/\mathfrak{m}\mathcal{O}_{\epsilon}(G) \text{ has an irreducible}$$
$$\text{module of dimension } < \ell^N \}.$$

We know that $\mathcal{A}'$ is a proper closed subset of maxspec Z, by Theorem III.1.7. To apply Corollary III.8.4 we need information about the codimension of $\mathcal{A}'$. This we can get from (III.7.9), Theorem 1. Let $\mu' : \text{maxspec } Z \longrightarrow \text{maxspec } Z_0 : M \mapsto M \cap Z_0$, as in (III.4.3). Then by (III.7.9), Theorem 1,

$$\mathcal{A}' \;\; = \;\; \mu'^{-1}\Big(\bigcup X_{(w_1, w_2)} : (w_1, w_2) \in W \times W, \; m(w_1, w_2) \leq 2N - 2\Big). \quad (9)$$

By (III.7.8)(18) and (19), for $(w_1, w_2) \in W \times W$ as in (9) above,

$$\dim X_{(w_1, w_2)} \;\; = \;\; n + m(w_1, w_2) \;\; \leq \;\; n + 2N - 2 \;\; = \;\; \dim G - 2. \quad (10)$$

Since $|W \times W| < \infty$ and μ' is a finite morphism, it follows from (9) and (10) that

$$\dim \mathcal{A}' \;\; \leq \;\; \dim Z - 2. \quad (11)$$

By (11) and Theorem III.7.5, $\mathcal{O}_{\epsilon}(G)$ satisfies all the hypotheses of Corollary III.8.4. Thus we can conclude:

Theorem 1. [**21**, Theorem 4.5] *The set of singular points of $Z(\mathcal{O}_{\epsilon}(G))$ is precisely the set of non-Azumaya points.* $\square$

2. Quantized enveloping algebras $\check{U}_\epsilon(\mathfrak{g})$. For these algebras our work is already done. The condition on the codimension of the non-Azumaya locus is satisfied by (III.6.6)(16). That

$$\text{gl.dim } \check{U}_\epsilon(\mathfrak{g}) < \infty$$

follows from Theorem III.6.1(5) and the filtered-graded result Theorem I.12.14. Since $\check{U}_\epsilon(\mathfrak{g})$ is a domain by Lemma I.12.12, Corollary III.8.4 applies once more in this case to give the following confirmation of a conjecture of De Concini and Kac [**40**, Conjecture 5.2(c)].

Theorem 2. [**21**, Theorem 4.3] *The set of singular points of $Z(\check{U}_\epsilon(\mathfrak{g}))$ is precisely the set of non-Azumaya points.* □

3. Enveloping algebras in characteristic p. A result paralleling Theorems 1 and 2 holds also for enveloping algebras of semisimple Lie algebras $\mathfrak{g} = \text{Lie}(G)$ over fields of positive characteristic. This verifies a conjecture of Premet [**184**]. Here, in a very similar way to the case of $\check{U}_\epsilon(\mathfrak{g})$, the "symplectic leaves" correspond to the "coadjoint orbits" of G on $\mathfrak{g}^*$. The Azumaya points correspond to the *regular* orbits in $\mathfrak{g}^*$ – that is, those with centraliser in G of minimal dimension (namely, the rank of G). Since the regular elements of $\mathfrak{g}^*$ form an open subset of $\mathfrak{g}^*$ with complement of codimension 3, the result follows from Corollary III.8.4 as above. Some minor restrictions on the prime p are required – see [**21**] for details.

NOTES

Theorem III.8.1 is a variant of a theorem proved in [**21**]. An earlier version, for ℕ-graded algebras, can be found in [**131**]. The homological properties of Hopf algebras outlined in (III.8.2) and the applications given in (III.8.5) are also taken from [**21**]. The elementary approach to homological aspects adopted here makes use of ideas from [**25**], which generalised work on local orders in [**206**]. Considerably more powerful and sophisticated techniques are available today and should certainly be brought into play in efforts to raise our understanding of the homological aspects of Hopf algebras. For example, [**198**] is a key reference in the development of techniques which don't require finiteness over the centre as a hypothesis, and more recently, papers such as [**213**] have shifted the language of choice once more, to that of derived categories and dualizing complexes.

EXERCISES

Exercise III.8.A. Let H be a Hopf k-algebra and let M, N be H-modules.
    (i) Show that $\text{Hom}_k(M, N)$ is a (left) H-module via Δ and

$$((h \otimes t)f)(m) \quad := \quad hf\big(S(t)m\big)$$

for $h, t \in H$, $f \in \text{Hom}_k(M, N)$ and $m \in M$.
    (ii) Show that $N \otimes M^* \cong \text{Hom}_k(M, N)$ as left H-modules.

Exercise III.8.B. Let A, B, C be H-modules, H a Hopf algebra.

(i) Show that $\operatorname{Hom}_H(A, \operatorname{Hom}_k(B, C)) \cong \operatorname{Hom}_H(A \otimes B, C)$.

(ii) Deduce from (i) that if P is a projective H-module and B is any H-module then $P \otimes B$ is H-projective.

(iii) Deduce from (i) that $\operatorname{Ext}^i_H(A \otimes B, C) \cong \operatorname{Ext}^i_H(A, C \otimes B^*)$ for all $i \geq 0$.

Exercise III.8.C. Prove that the coevaluation map δ of the proof of Proposition III.8.2 is an H-homomorphism. Complete the details of the proof.

Exercise III.8.D. Prove (III.8.3)(5): The grade $g(M) \leq \operatorname{height}(M)$ for a maximal ideal M of a noetherian ring R which is a finitely generated module over its centre.

Exercise III.8.E. Prove Lemma III.8.3(2).

Exercise III.8.F. Prove Lemma III.8.3(3).

CHAPTER III.9

MÜLLER'S THEOREM AND BLOCKS

In this chapter we investigate the blocks of maximal ideals of an algebra A satisfying condition (**H**) of (III.1.1), and then apply our conclusions to the particular classes of PI Hopf triples introduced in preceding chapters. The starting point for our discussion is the circle of ideas reviewed in Appendix I.16. In the first 3 paragraphs we specialise the concepts introduced there to the case of noetherian rings which are finite modules over their centres; in (III.9.2) and (III.9.3) we state and then prove the theorem of Müller referred to in the chapter title. Finally in paragraphs (III.9.4) – (III.9.6) we examine the consequences of this theorem for our classes of PI Hopf triple of particular interest.

III.9.1. Rings finite over their centres. Let A be a noetherian k-algebra. As Example I.16.3(3) illustrates, the block decomposition defined in (I.16.2) is not useful when there are no non-split extensions of finite dimensional A-modules. Reversing this maxim, we should expect the block decomposition to be of most value when – in an appropriate sense – there are no non-trivial extensions of finite dimensional A-modules by finitely generated A-modules which are *not* finite dimensional. This is the case when A satisfies the *second layer condition*. We won't discuss this concept here, referring the reader to [**83**, Chapter 11] for details. Instead we consider a special case which will be sufficient for our applications. Recall that a module W over a ring R is an *essential extension* of an R-submodule V if $V \cap X \neq 0$ for every non-zero submodule X of W. Recall also that an ideal I of a noetherian ring R has the *strong left Artin-Rees property* if, for all finitely generated left R-modules B and submodules A, there exists a nonnegative integer j with

$$I^i B \cap A \quad = \quad I^{i-j}(I^j B \cap A)$$

for all $i \geq j$. The *weak* left Artin-Rees property requires only that the inclusion $I^n B \cap A \subseteq IA$ holds for some n.

Lemma. *Let R be a noetherian ring.*

    1. *If I is an ideal of R which is generated by elements in the centre of R, then I has the strong left Artin-Rees property.*

    2. *For an ideal I of R, the following statements are equivalent:*

(i) *I has the weak left Artin-Rees property.*

(ii) *Let W be a finitely generated left R-module containing an essential sub-module V with $IV = 0$. Then there exists $n \geq 1$ such that $I^n W = 0$.*

Proof. 1. This is proved in [**181**, Theorem 11.2.2]. The argument is sketched in Exercise III.9.A.

2. (i) $\Rightarrow$ (ii): This is clear.

(ii) $\Rightarrow$ (i): Suppose that (ii) holds and let $A \subseteq B$ be left R-modules with B finitely generated. Choose a submodule C of B maximal with respect to the property $A \cap C = IA$. Then A/IA is mapped isomorphically by the canonical homomorphism onto an essential submodule of B/C. So (ii) implies that $I^n(B/C) = 0$ for some $n > 0$, or in other words, $I^n B \subseteq C$. Hence $A \cap I^n B \subseteq IA$, and (i) is proved. $\square$

The following consequence of the lemma makes apparent the value of determining the blocks of maximal ideals for an algebra satisfying (**H**).

Proposition. *Let A be a k-algebra satisfying hypotheses (**H**) of (III.1.1), and let V be an irreducible A-module. Then the injective hull $E_A(V)$ of V is locally finite dimensional.*

Proof. Recall that, by Proposition III.1.1(4), $\dim_k(V) < \infty$; indeed, by the same result, if $\mathfrak{m} = \mathrm{ann}_{Z_0}(V)$, then $\dim_k(A/\mathfrak{m}A) < \infty$. Let W be a finitely generated submodule of $E_A(V)$. Then the lemma shows that some power of $\mathfrak{m}A$ annihilates W, so proving the result. $\square$

III.9.2. Müller's theorem. Suppose that A is a k-algebra satisfying (**H**). As just noted in the previous proof, Proposition III.1.1(4) ensures that all the maximal ideals of A have finite codimension over k. Let $M, N \in \mathrm{maxspec}\, A$ with $M \rightsquigarrow N$, so that $B := (M \cap N)/MN \neq 0$. Let Z be the centre of A and set $\mathfrak{m} = M \cap Z$ and $\mathfrak{n} = N \cap Z$. Thus $\mathfrak{m}B = 0 = B\mathfrak{n}$, and since $\mathfrak{m}$ and $\mathfrak{n}$ are maximal ideals of Z this is only possible if $\mathfrak{m} = \mathfrak{n}$. Thus maximal ideals of A can belong to the same block only if they have the same intersection with the centre of A. The point of Müller's theorem is that the converse of this observation is also true:

Theorem. *Let A be a k-algebra satisfying hypotheses (**H**), and let Z be the centre of A. Then maximal ideals M and N of A belong to the same block if and only if $M \cap Z = N \cap Z$.* $\square$

To prove this result we need two lemmas. The first is straightforward from the definitions; the second is the key to the involvement of the Artin-Rees property here. As is the case with almost all the results in the current section, the first of these lemmas is valid under weaker hypotheses than those stated here.

Lemma 1. *Let A be a k-algebra satisfying hypotheses (**H**) of (III.1.1). Let I be an ideal of A generated by central elements, set $\overline{A} = A/I$, and let $\pi : A \longrightarrow \overline{A}$ be the canonical epimorphism. Let $M \in \mathrm{maxspec}\, A$ with $I \subseteq M$. Let $\mathcal{B}_M \subseteq \mathrm{maxspec}\, A$*

and $\mathcal{B}_{\overline{M}} \subseteq \mathrm{maxspec}\,\overline{A}$ be the blocks of M in A and of $\overline{M} = M/I$ in $\overline{A}$ respectively. Then all the maximal ideals in $\mathcal{B}_M$ contain I, and

$$
\begin{aligned}
\mathcal{B}_{\overline{M}} &= \pi(\mathcal{B}_M) \\
\mathcal{B}_M &= \pi^{-1}(\mathcal{B}_{\overline{M}}).
\end{aligned}
$$

Proof. As above, we use overbars to denote images under π. It suffices to prove the following for any distinct $L, N \in \mathrm{maxspec}\,A$:

(1) If $I \subseteq L \cap N$ and $\overline{L} \rightsquigarrow \overline{N}$, then $L \rightsquigarrow N$.

(2) If $L \rightsquigarrow N$ and either $I \subseteq L$ or $I \subseteq N$, then $I \subseteq L \cap N$ and $\overline{L} \rightsquigarrow \overline{N}$.

Statement (1) is clear.

To prove (2), suppose that $L \rightsquigarrow N$ and either $I \subseteq L$ or $I \subseteq N$. Thus $B := (L \cap N)/LN \neq 0$, and since $IB = 0$ if and only if $BI = 0$, we must have $I \subseteq L \cap N$. Since $B' := (LN + I)/LN$ is centrally generated,

$$
L \subseteq \mathrm{l.ann}(B') = \mathrm{r.ann}(B') \supseteq N.
$$

Then since L and N are distinct maximal ideals, B' must be zero, so that $\overline{L} \rightsquigarrow \overline{N}$, as required. $\square$

Lemma 2. *Let R be a ring which is a finitely generated module over its noetherian centre Z, and let I be an ideal of Z. Then there is a positive integer n such that the centre of $R/I^n R$ is contained in $(Z + IR)/I^n R$.*

Proof. Let $R = \sum_{i=1}^{t} Z r_i$, and let B be the Z-module $R^{\oplus t}$. Define a Z-module homomorphism

$$
\begin{aligned}
f : R &\longrightarrow B \\
r &\longmapsto (rr_1 - r_1 r, \dots, rr_t - r_t r),
\end{aligned}
$$

so that $\ker f = Z$. Since I has the left Artin-Rees property by Lemma III.9.1(1), there exists $n \geq 1$ such that

$$
f(R) \cap I^n B \subseteq I f(R) = f(IR). \tag{1}
$$

Now let $u + I^n R \in Z(R/I^n R)$. Then $f(u) \in I^n B$, so that (1) forces $f(u) = f(w)$ for some $w \in IR$. Thus

$$
u - w \in \ker f = Z,
$$

and

$$
u + I^n R \in (Z + IR)/I^n R,
$$

as claimed. $\square$

Proof of Müller's theorem. Let A be a k-algebra which is a finite module over its centre Z. We've already noted above that if maximal ideals M and N of A are in the same block, then $M \cap Z = N \cap Z$. Our task, then, is to prove the converse. So we let $\mathfrak{m}$ be a maximal ideal of Z and aim to show that all maximal ideals of A containing $\mathfrak{m}$ lie in the same block. By Lemma 1 and (I.16.1) this is equivalent to the following statement:

$$\text{for some } n \geq 1, \ A/\mathfrak{m}^n A \text{ is an indecomposable ring.} \tag{2}$$

We claim that (2) is true when n is the integer afforded by Lemma 2. For, with this choice of n, let e be a central idempotent of $A/\mathfrak{m}^n A$. By Lemma 2,

$$e = z + \gamma + \mathfrak{m}^n A$$

for some $z \in Z$ and $\gamma \in \mathfrak{m}A$. Since $e = e^2$,

$$z(1-z) = z - z^2 \in \mathfrak{m}A \cap Z = \mathfrak{m}.$$

But $Z/\mathfrak{m}$ is a field, so either $z \in \mathfrak{m}$ or $1 - z \in \mathfrak{m}$. This means that either e or $1 - e$ is a nilpotent idempotent, and hence is 0, proving (2). □

III.9.3. Corollaries and examples.

Example 1. Let k be an algebraically closed field and

$$A = \begin{pmatrix} k[X] & \langle X \rangle \\ k[X] & k[X] \end{pmatrix} \subseteq M_2(k[X])$$

as in Example III.1.2(2). Then the maximal ideals

$$M = \begin{pmatrix} \langle X \rangle & \langle X \rangle \\ k[X] & k[X] \end{pmatrix} \quad \text{and} \quad N = \begin{pmatrix} k[X] & \langle X \rangle \\ k[X] & \langle X \rangle \end{pmatrix}$$

form a block, whereas every other maximal ideal has form $\langle X - \lambda \rangle A$ for some non-zero scalar λ, and so lies in a singleton block.

For the applications to representation theory we restate Müller's theorem as follows:

Theorem. *Let A be a k-algebra satisfying hypotheses (**H**), with centre Z. Let Z_0 be a subalgebra of Z over which A is a finitely generated module. Let $\mathfrak{m} \in \text{maxspec } Z_0$. Then two maximal ideals $M/\mathfrak{m}H$ and $N/\mathfrak{m}H$ of $H/\mathfrak{m}H$ are in the same block of $H/\mathfrak{m}H$ if and only if $M \cap Z = N \cap Z$.*

Proof. This is immediate from Lemma 1 of (III.9.2), and Theorem III.9.2 itself. □

Given the description in (I.16.1) of the blocks of maximal ideals of an Artin algebra afforded by the intersection of the ideals with the centre of the ring, we can reformulate the above theorem in a way which makes no reference to blocks:

Corollary. *Let A, Z and Z_0 be as in the theorem and let $\mathfrak{m} \in \operatorname{maxspec} Z_0$. Then every primitive central idempotent of $A/\mathfrak{m}A$ lies in the subalgebra $(Z + \mathfrak{m}A)/\mathfrak{m}A$.*

Proof. Set $\overline{A} = A/\mathfrak{m}A$ and $\overline{Z} = (Z+\mathfrak{m}A)/\mathfrak{m}A$, and write $1_{\overline{Z}} = f_1 + \cdots + f_r$ where the f_i are orthogonal primitive idempotents of $\overline{Z}$. It suffices to show that the f_i remain primitive as central idempotents of $\overline{A}$.

Since $\overline{Z}f_i$ is an indecomposable commutative Artinian ring, it must be local. Its unique maximal ideal is of the form $\mathfrak{m}_i \overline{Z} f_i$ where $\mathfrak{m}_i$ is a maximal ideal of Z containing $Z \cap \mathfrak{m}A$. Note that $\mathfrak{m}_i \overline{Z} = \mathfrak{m}_i \overline{Z} f_i + \overline{Z}(1 - f_i)$. If $\overline{M} = M/\mathfrak{m}A$ and $\overline{N} = N/\mathfrak{m}A$ are any maximal ideals of $\overline{A}$ that contain $1 - f_i$, then $\overline{M} \cap \overline{Z} = \mathfrak{m}_i \overline{Z} = \overline{N} \cap \overline{Z}$ and consequently $M \cap Z = \mathfrak{m}_i = N \cap Z$. By the theorem, $\overline{M}$ and $\overline{N}$ are in the same block of $\overline{A}$. This shows that all maximal ideals of $\overline{A}f_i$ are in the same block of that ring. It follows from (I.16.1) that $\overline{A}f_i$ is an indecomposable ring, that is, f_i is a primitive central idempotent of $\overline{A}$, as desired. $\square$

Example 2. Notwithstanding the corollary, it is *not* true in general that with A, Z, Z_0 and $\mathfrak{m}$ as stated,

$$Z(A/\mathfrak{m}A) \quad = \quad (Z + \mathfrak{m}A)/\mathfrak{m}A.$$

For example, let $A = U(\mathfrak{sl}_2(k))$ with $\operatorname{char}(k) = 2$, and take $Z_0 = Z = Z(A)$. As noted in Example III.1.7(3),

$$Z = k[e^2, f^2, h].$$

If $\mathfrak{m}$ is the maximal ideal $\langle e^2, f^2, h \rangle$ of Z, then $(Z + \mathfrak{m}A)/\mathfrak{m}A = k$, whereas $Z(A/\mathfrak{m}A) = A/\mathfrak{m}A$ is 4-dimensional over k.

III.9.4. Applications to PI Hopf triples. As in (III.4.1) and (III.4.4), let

$$Z_0 \subseteq Z = Z(H) \subseteq H \tag{3}$$

be a PI Hopf triple over the algebraically closed field k, with associated bundle of finite dimensional k-algebras

$$\mathcal{B}_H \quad = \quad \{H/\mathfrak{m}H : \mathfrak{m} \in \operatorname{maxspec} Z_0\}.$$

There are two questions which present themselves at this point:

Q1. What are the blocks of $\operatorname{maxspec} H$?

Q2. For each algebra $H/\mathfrak{m}H \in \mathcal{B}_H$, what are the blocks of $H/\mathfrak{m}H$?

It's important to appreciate that these questions are essentially identical! This is a consequence of Lemma 1 of (III.9.2). Thus, if $M \in \operatorname{maxspec} H$, then $\mathfrak{m} := M \cap Z_0 \in \operatorname{maxspec} Z_0$, and the block of $\operatorname{maxspec} H$ containing M consists of the inverse images in H of the maximal ideals $N/\mathfrak{m}H$ of $H/\mathfrak{m}H$ such that $N/\mathfrak{m}H$ is in the block of $M/\mathfrak{m}H$.

Moreover, once we've answered (Q1) and (Q2), we know most of what there is to know about the extensions of irreducible H-modules. For, if V is an irreducible H-module, then its injective hull $E_H(V)$ is locally finite dimensional over k by Proposition III.9.1, so $E_H(V)$ is the union of finite dimensional modules whose composition factors all have annihilators in the block of $\mathrm{ann}_H(V)$.

Theorem III.9.3 tells us that, to answer (Q1) and (Q2), our aim should be:

$$\textit{Understand the algebras} \quad Z/\mathfrak{m}Z, \quad \mathfrak{m} \in \mathrm{maxspec}\, Z_0. \qquad (4)$$

There are two subtleties we should note about this aim: First, it's not *a priori* clear that

$$\mathfrak{m}Z \quad = \quad \mathfrak{m}H \cap Z \qquad (5)$$

for an arbitrary PI Hopf triple. (By Proposition III.1.1(5) for the pair $Z \subseteq H$ it's easy to see that $(\mathfrak{m}H \cap Z)/\mathfrak{m}Z$ is a nilpotent ideal of $Z/\mathfrak{m}Z$.) Equivalently, it's not clear that $Z/\mathfrak{m}Z$ embeds in $Z(H/\mathfrak{m}H)$. However (5) is true, by a trivial calculation, when Z is a Z-module summand of H, so it holds when k has characteristic 0, since in that case Z is the image of H under the reduced trace map, a homomorphism of Z-modules; and it's also true for $H = U(\mathfrak{g})$, for $\mathfrak{g}$ semisimple over k of positive characteristic $p > 0$, by an unpublished argument of Premet.

Second, as we've noted in (III.9.3), it's *not* true in general that $(Z+\mathfrak{m}H)/\mathfrak{m}H$ equals $Z(H/\mathfrak{m}H)$. What is necessarily true, by Corollary III.9.3, is that the subalgebra $(Z + \mathfrak{m}H)/\mathfrak{m}H$ contains all the central idempotents of $H/\mathfrak{m}H$.

Why, despite these warnings, do we prefer to state our aim as in (4)? Because this aim is achievable – as we'll now explain, for the Hopf algebras of interest to us enough is known about Z to enable the factors $Z/\mathfrak{m}Z, (\mathfrak{m} \in \mathrm{maxspec}\, Z_0)$, to be calculated.

III.9.5. The centre. There is a common pattern to the structure of the centre $Z(H)$ when H is $\mathcal{O}_\epsilon(G)$, $\breve{U}_\epsilon(\mathfrak{g})$, or $U(\mathfrak{g})$ for $\mathfrak{g}$ semisimple in characteristic p. In each case there are two "obvious" subalgebras of the centre – first, there is the Hopf centre Z_0 which we have already defined and studied, and which, we recall, has the same Krull dimension as the whole algebra. Second, there is the central subalgebra got by specialisation of the centre of the corresponding generic algebra, that is (respectively) from $Z(\mathcal{O}_q(G))$, $Z(\breve{U}_q(\mathfrak{g}))$, and from the centre of the corresponding enveloping algebra in characteristic 0.

Now in each case there is a result which (roughly speaking) says that the whole centre is generated by these two subalgebras, and generated in a (geometrically or algebraically) attractive way: geometrically, maxspec Z is the fibre product of the maxspecs of the two smaller pieces; algebraically, the centre is the tensor product over the intersection. More precisely, we have

Theorem. 1. (Enriques, [**43**, Appendix])

$$Z(\mathcal{O}_\epsilon(G)) \cong Z_q \otimes_{Z_q \cap \mathcal{O}(G)} \mathcal{O}(G),$$

where Z_q is a specific affine algebra of Krull dimension $2n$.

2. For a reductive Lie algebra $\mathfrak{g}$ in characteristic $p > 0$ satisfying certain mild necessary conditions,

$$Z(U(\mathfrak{g})) \cong Z_1 \otimes_{Z_0 \cap Z_1} Z_0,$$

where Z_1 is isomorphic to $S(\mathfrak{h})^W$ via the Harish-Chandra map. The algebra Z_1 is a polynomial algebra on $\mathrm{rank}(G)$ variables.

3. (De Concini-Kac-Procesi, [**42**, Theorem 6.4])

$$Z(\check{U}_\epsilon(\mathfrak{g})) \cong Z_1 \otimes_{Z_0 \cap Z_1} Z_0,$$

where Z_1 is the specialisation of $Z(\check{U}_q(\mathfrak{g}))$. There is a Harish-Chandra-style description of Z_1. $\square$

Remarks. 1. The result for characteristic p enveloping algebras originates with Veldkamp [**207**], and has subsequently been improved by a sequence of authors. A detailed treatment of the most general result is in [**22**, 3.2].

2. $Z(U_\epsilon(\mathfrak{g}; M))$ for $M \neq P$ is in general *not* generated by its subalgebras Z_0 and Z_1, [**42**, Remark 6.4(a)].

III.9.6. The number of blocks. Armed with Theorem III.9.5 we can hope to calculate the algebras $Z/\mathfrak{m}Z$ of (III.9.4)(4), for Z equal to the centre of members of any of the three classes of algebra H listed above, and $\mathfrak{m} \in \mathrm{maxspec}\, Z_0$. Then Müller's theorem tells us that

$$|\{\text{Blocks of } H/\mathfrak{m}H\}| = |\mathrm{maxspec}(Z/\mathfrak{m}Z)|. \tag{6}$$

This calculation has been carried out in each case. Here are the outcomes regarding the blocks of the quantized algebras. First, to state the result for function algebras, recall from (I.5.3) that the *fundamental weights* of the complex semisimple Lie algebra $\mathfrak{g}$ of rank n are denoted by $\varpi_1, \ldots, \varpi_n$; and that, for w in the Weyl group W, $s(w)$ equals the minimum length of an expression for w as a product of arbitrary reflections in W, (II.4.13). (In general, $s(w)$ is less than the length $\ell(w)$ as defined in (I.5.1).) Finally, recall that w_0 denotes the unique *longest element* of W (with respect to $\ell(-)$), as in (I.5.1).

Theorem 1. *Let G be a complex semisimple group of rank n, ϵ a primitive ℓ^{th} root of unity, with $\ell \geq 3$ odd, and prime to 3 if G has a factor of type G_2. Let $(w_1, w_2) \in W \times W$ and let $g \in X_{(w_1, w_2)}$. Let d be the cardinality of the set*

$$\{1 \leq i \leq n : w_0 w_1, \, w_0 w_2 \in \mathrm{Stab}_W(\varpi_i)\}.$$

Then $\mathcal{O}_\epsilon(G)(g)$ has exactly ℓ^d blocks, each containing exactly $\ell^{n - s(w_1 w_2^{-1}) - d}$ irreducible modules. $\square$

Examples. 1. If $w_1 = w_2 = w_0$, then $w_0 w_1 = w_0 w_2 = 1_W$, so $d = n$. We recover the fact, which we already know from (III.7.9)(24), that in this open dense set the blocks are singletons.

  2. If $w_1 = w_2 = 1_W$, then $w_0 w_1 = w_0 w_2 = w_0$. Since w_0 doesn't fix *any* fundamental weight, $d = 0$ in this case. Thus, for $g \in T$ (for example $g = 1_G$), $\mathcal{O}_\epsilon(G)(g)$ has only one block containing all ℓ^n irreducible modules.

  3. Let $n = 1$, so $G = SL_2(\mathbb{C})$ and $W = \langle w_0 \rangle = C_2$. There are 4 cases, (w_0, w_0), $(w_0, 1)$, $(1, w_0)$ and $(1, 1)$. The first and last of these are covered by Examples 1 and 2. In both the other cases $d = 0$, since $\{w_0 w_1, w_0 w_2\} = \{1, w_0\}$. Thus there is 1 block in each of these cases, and this block contains 1 irreducible. This confirms the analysis of (III.3.2) and (III.3.3).

To state the result for $\check{U}_\epsilon(\mathfrak{g})$ requires some preparation. Recall from (III.6.5)(6) and (7) that maxspec $Z_0 = H \cong (N^- \times N^+) \rtimes T$, and the symplectic leaves in H are the inverse images under the map σ of (III.6.5)(9) of the non-central conjugacy classes of G. Every non-central conjugacy class in G meets B^- [**100**, (0.3)], so we can consider without loss of generality $h = nt \in N^- T = B^- \subseteq H$. Then T admits an action of $W = N_G(T)/T$ by conjugation, as does the set

$$R_t := \{f \in T : f^\ell \in W.t^2\}.$$

Let $\mathfrak{m}_h$ denote the maximal ideal of Z_0 corresponding to h, and let

$$d \quad = \quad d(h) \quad := \quad |W\text{-orbits in } R_t|.$$

Theorem 2. *Keep the notation just introduced. The number of blocks of the algebra* $\check{U}_\epsilon(\mathfrak{g})/\mathfrak{m}_h \check{U}_\epsilon(\mathfrak{g})$ *is* d. $\square$

Example 4. Let $\mathfrak{g} = \mathfrak{sl}_2(\mathbb{C})$ and suppose $h \in N^-$, so that $t = 1 \in T \cong \mathbb{C}^\times$. Thus R_t consists of the ℓ elements

$$\begin{pmatrix} \epsilon^i & 0 \\ 0 & \epsilon^{-i} \end{pmatrix} \qquad (0 \leq i < \ell)$$

of $T \subseteq SL_2(\mathbb{C})$. Of course $W \cong C_2$ acts by interchanging the non-zero entries, so there is one singleton W-conjugacy class, namely $\{1_T\}$, and $(\ell - 1)/2$ pairs $\left\{ \begin{pmatrix} \epsilon^{\pm i} & 0 \\ 0 & \epsilon^{\mp i} \end{pmatrix} \right\}$, $1 \leq i \leq (\ell - 1)/2$. Hence

$$|W\text{-orbits in } R_t| \quad = \quad (\ell + 1)/2.$$

This agrees with the calculations in Chapter III.2.

Notes

1. Theorems 1 and 2 of (III.9.6) are extracts from more powerful results [22], [23] which give complete descriptions of the algebras $Z/\mathfrak{m}Z$ for $\mathfrak{m} \in \operatorname{maxspec} Z_0$, not just the number of their primary components. Using these results and Theorem III.4.10 one can give, for $\mathfrak{m} \in \mathcal{F}_H$, the fully Azumaya locus, a complete description of the algebras $H/\mathfrak{m}H$, for $H = \mathcal{O}_\epsilon(G)$, $\check{U}_\epsilon(\mathfrak{g})$, or indeed for $H = U(\mathfrak{g})$ in positive characteristic.

2. [22] also contains the result exactly analogous to Theorems 1 and 2 for the case of $U(\mathfrak{g})$, $\mathfrak{g}$ a semisimple Lie algebra in characteristic $p > 0$. In this case the description of $Z/\mathfrak{m}Z$ was obtained in [160]. The resulting description of the blocks confirms a conjecture of Humphreys [101].

3. A very full discussion of the variants of the Artin-Rees property and their applications can be found in [181].

4. The original proof of Müller's theorem was in [165] (though he also published a less widely available version two years earlier in the Kasch-Pareigis Seminar). The present proof is influenced by that in [83].

5. References for the other results in this chapter are given at the appropriate points in the text above.

Exercises

Exercise III.9.A. Let I be an ideal of the noetherian ring R. (i) Define the *Rees ring* of I to be the subring $R^*(I)$ of the polynomial ring $R[X]$ consisting of polynomials in which the coefficient of X^m belongs to I^m, for all $m \geq 0$. Deduce from the Hilbert basis theorem that if I is generated by central elements then $R^*(I)$ is noetherian.

(ii) Show that if $R^*(I)$ is left noetherian then I has the strong left Artin-Rees property. [Hint: Let B be a finitely generated left R-module with submodule A. Form the $R^*(I)$-module

$$\widehat{B} := B + XIB + X^2I^2B + \cdots,$$

and note that $\widehat{B}$ is a finitely generated $R^*(I)$-module. Now consider the submodule

$$\widehat{A} := A + X(IB \cap A) + X^2(I^2B \cap A) + \cdots$$

of $\widehat{B}$.]

Exercise III.9.B. Check the claims made regarding $U(\mathfrak{sl}_2(k))$ over a field k of characteristic 2 in Example 2 of (III.9.3).

Exercise III.9.C. Confirm directly that $\overline{U_\epsilon(\mathfrak{sl}_2(k))}$ has exactly $(\ell + 1)/2$ blocks, as predicted in Example 4 of (III.9.6), and list the simple modules in each block, by using the calculations from Chapter III.2, specifically Case (A) of Theorem III.2.2, Case I of (III.2.3) and Exercise III.2.D.

PROBLEMS AND PERSPECTIVES

In (III.10.1)–(III.10.3) we discuss problems concerning the PI Hopf triples of (III.4.1)(1). So throughout these three paragraphs let

$$Z_0 \subseteq Z \subseteq H \qquad (1)$$

be a PI Hopf triple over the algebraically closed field k. We then briefly consider more general PI Hopf algebras in (III.10.4), before moving on to more specialised questions about quantized function algebras and quantized enveloping algebras in the remainder of the chapter. In the cases where there has been progress on a question since it was stated in the lectures at Barcelona we have chosen, for the sake of historical accuracy, first to state the question as it was given in the lectures, and then to discuss subsequent developments.

III.10.1. Number of isomorphism classes in $\mathcal{B}_H$. Recall that $\mathcal{B}_H$ is the bundle of finite dimensional algebras $\{H/\mathfrak{m}H : \mathfrak{m} \in \text{maxspec } Z_0\}$.

Q.1: Is the number of isomorphism classes in $\mathcal{B}_H$ finite?

We don't even have a complete answer to (Q.1) for the three key classes $U(\mathfrak{g})$ (in characteristic $p > 0$), $\check{U}_\epsilon(\mathfrak{g})$ and $\mathcal{O}_\epsilon(G)$. First, $\mathcal{O}_\epsilon(G)$, where the answer is clearly "Yes" by Corollary III.5.8 combined with (III.7.8):

$$|\mathcal{B}_H| \quad \leq \quad |W|^2.$$

For $U(\mathfrak{g})$ when $\mathfrak{g}$ is reductive over a field k of positive characteristic p the answer is again "Yes", but this seems to be non-trivial: First, recall that $Z_0 \cong \mathcal{O}(\mathfrak{g}^*)$. A theorem of Kac and Weisfeiler (see [**108**, 7.4]) says that $U(\mathfrak{g})/\mathfrak{m}_\chi U(\mathfrak{g})$, for $\chi \in \mathfrak{g}^*$, is Morita equivalent to $U(\mathfrak{g}')/\mathfrak{m}_\eta U(\mathfrak{g}')$ for $\mathfrak{g}'$ reductive with $\dim_k \mathfrak{g}' \leq \dim_k \mathfrak{g}$ and η a *nilpotent* element of $\mathfrak{g}'^*$. Next, the isomorphism class of the factor algebra is invariant under the coadjoint action of the group G'; and there are only finitely many G'-conjugacy classes of nilpotent elements of $\mathfrak{g}'^*$ [**100**]. (Think of the Jordan canonical form when $\mathfrak{g}' = \mathfrak{sl}_n(k)$, though the result for general reductive $\mathfrak{g}'$ is deeper than this.) Thus there are only finitely many Morita equivalence classes. Since the algebras involved have k-dimension bounded by $p^{\dim \mathfrak{g}}$, there can only be finitely many isomorphism classes, (Exercise III.10.A).

It's not clear whether a similar argument can be pushed through in the case of $\check{U}_\epsilon(\mathfrak{g})$. There *is* an analogue for the Kac-Weisfeiler theorem, due in this case to De Concini and Kac [41] (and see [30]) but the result is not so strong (outside type A_n) as in the modular Lie algebras setting, and in any case there is no analogue at present for "reductive" Lie algebras in the quantum world – perhaps there should be? Anyway, the answer is sufficiently unclear to state:

Q.2: What is the answer to (Q.1) for $H = \check{U}_\epsilon(\mathfrak{g})$?

Nor do we know what happens for modular enveloping algebras of non-reductive Lie algebras. For example, the answer to the following question is probably much easier than the corresponding one for reductive algebras:

Q.3: What is the answer to (Q.1) for $U(\mathfrak{g})$, $\mathfrak{g}$ solvable?

Note that if we relax our demands on H and Z_0 and require them only to be (respectively) a bialgebra and a central sub-bialgebra, then (Q.1) is easily seen to have a negative answer, (Exercise III.10.E).

This is perhaps also a good point to mention that there are obviously unsatisfactory aspects of the argument using Hamiltonian flows along symplectic leaves which is used to answer (Q.1) for $\mathcal{O}_\epsilon(G)$. For a start, its analytic nature imposes constraints on the base fields we can work over. We therefore ask:

Q.4: Can a purely algebraic approach to (III.5.8) be found?

As a (possibly easier) variant of (Q.4) we ask whether a version of Proposition III.5.8 can be proved for algebras over fields of positive characteristic.

III.10.2. Freeness over the Hopf centre. Recall that when $H = U(\mathfrak{g})$, $\check{U}_\epsilon(\mathfrak{g})$, or $\mathcal{O}_\epsilon(G)$, H is a free Z_0-module, by Theorems I.13.2, III.6.2 and III.7.11. So we ask:

Q.5: Is H always a free Z_0-module?

Let $B \subseteq A$ be affine commutative Hopf k-algebras. It's not true in general that A is a free B-module, by a result of Oberst and Schneider [163, 3.5.2]. Indeed $\mathcal{O}(SL_{2n}(k))$ is not free over $\mathcal{O}(PSL_{2n}(k))$, [209]. So in the unlikely event that (Q.5) has a positive answer, this will depend on choosing Z_0 maximal (say, amongst normal commutative sub-Hopf algebras of H).

Even for those types of Hopf triple for which the answer to (Q.5) is known to be "yes", it's by no means the end of the story – for we'd like to have an explicit Z_0-basis for H, and – even – to express H as a *crossed product* over Z_0, usually written $H = Z_0 * \overline{H}$. In this case, the extension $Z_0 \subseteq H$ is called *cleft* by Hopf algebraists. See [163, Chapter 8] for a discussion of when an extension of Hopf algebras is cleft; for the specific case of PI Hopf triples, see also [192] (although the reader should be aware that this paper includes (Lemma 8) a (false) proof of the (false) result that a PI Hopf triple is always a crossed product $Z_0 * \overline{H}$). In view of Theorem III.7.11, we ask in particular

Q.6: Is $\mathcal{O}_\epsilon(G)$ a cleft extension of $\mathcal{O}(G)$?

III.10.3. Frobenius extensions. Recall that we showed in Corollary III.4.7 that (1) is a Frobenius extension of H by Z_0. The *Nakayama automorphism* μ of a (finite dimensional) Frobenius algebra A with associative form (,) is defined by

$$(x, y) \quad = \quad (y, \mu(x))$$

for all $x, y \in A$. This yields a well-defined algebra automorphism of A; if a different nondegenerate associative form is used in the definition of μ, the automorphism changes by an inner automorphism, (Exercise III.10.B(i)). If $\mu = \mathrm{id}_A$ then A is called a *symmetric algebra*. The Nakayama automorphism of A carries information about its representation theory – for example, if P is the projective indecomposable A-module with top V, then the socle of P is ${}^\mu V$. This concept can be extended to all *free central* Frobenius extensions $B \subseteq A$, (by which of course we mean that this is a Frobenius extension with A a free B-module and B central in A), to give a B-algebra automorphism μ of A, Exercise III.10.B(ii). Of course μ then factors through each of the ideals $\mathfrak{m}A$, where $\mathfrak{m} \in \mathrm{maxspec}\, B$, to give the Nakayama automorphisms of all the Frobenius algebras $A/\mathfrak{m}A$. In particular all this can be carried through for our PI Hopf triple (1). So we ask:

Q.7: What is the Nakayama automorphism in the case of $U(\mathfrak{g})$, of $\check{U}_\epsilon(\mathfrak{g})$ and of $\mathcal{O}_\epsilon(G)$?

The answer for $U(\mathfrak{g})$ (at least at the level of $\overline{U(\mathfrak{g})}$) has been calculated – see [55]. In recent unpublished work [88] I. Gordon has calculated μ in the cases of $\check{U}_\epsilon(\mathfrak{g})$ and of $\mathcal{O}_\epsilon(G)$, so that (Q.7) as it's stated above has now been answered. Namely, the Nakayama automorphism of $\check{U}_\epsilon(\mathfrak{g})$ is the identity, (so that the bundle of finite dimensional algebras consists in this case of symmetric algebras); whereas for $\mathcal{O}_\epsilon(G)$ the Nakayama automorphism is a (non-trivial) winding automorphism. But there is probably an answer to (Q.7) at the more "abstract" level of PI Hopf triples, and it's in that spirit that we propose (Q.7).

III.10.4. Hopf algebras satisfying a polynomial identity. Let H be an affine prime k-Hopf algebra satisfying a polynomial identity. It's *not* true that H is always a PI Hopf triple, Exercise III.10.C. But it seemed possible that the following question had a positive answer:

Q.8: Let H be an affine k-Hopf algebra satisfying a polynomial identity. Does H contain a normal commutative sub-Hopf algebra Z_0 with H a finitely generated Z_0-module? In particular, is H noetherian?

The first part of (Q.8) has now been answered in the negative by E. Letzter [135]. However Letzter's example is not prime (in fact, not semiprime), so it seems worthwhile to ask

Q.8′: What happens in (Q.8) if H is required to be prime?

As of December 2001, this question appears to have been answered in the negative, thanks to an example constructed by Gelaki and Letzter (see http://arXiv.org/abs/math.QA/0112038).

(Q.8) was asked in [18]. In that paper, the idea of a PI Hopf triple was introduced, and (in passing) a proof was sketched that *every PI Hopf triple H has finite injective dimension* [18, 2.3]. This sketch is reproduced in Exercise III.10.D. Just as finite global dimension implied Auslander-regularity in (III.8.3), one can then deduce that H is *Auslander-Gorenstein*. That H has finite injective dimension of course generalises the well-known theorem of Larson and Sweedler that a finite dimensional Hopf algebra is Frobenius [163, 2.1.3]. This circle of ideas naturally prompts the following question, also asked in [18]:

Q.9: Let H be an affine k-Hopf algebra satisfying a polynomial identity. Does H have finite injective dimension?

Considerable progress on this question has been made in the past two years – for example, (Q.9) has a positive answer if H is finitely generated as a module over its centre [213]. But the general case remains open. In effect, (Q.9) has been reduced to deciding whether every H as there admits a dualising complex.

III.10.5. Representation theory of $\mathcal{O}_\epsilon(G)$. The results in Chapter III.7 together with Theorem 1 of (III.9.6) give only a taste of what's known here. First, note that the stratification of the finite dimensional representation theory into T-orbits of symplectic leaves was developed by De Concini and Procesi [45] into a description of $\mathcal{O}_\epsilon(G)$ as a collection of Azumaya algebra factors, one such factor for each $(w_1, w_2) \in W \times W$. While this description allowed the determination of the irreducible modules, it didn't reveal the structure of the algebras $\mathcal{O}_\epsilon(G)(g)$, in particular their block structure and representation type. These aspects were studied by Gordon in his PhD thesis, and subsequently by Brown and Gordon – the strongest results available to date are in the paper [23], whose main conclusions are summarised in the following

Theorem. *Let G be a connected, simply connected complex semisimple group with Lie algebra $\mathfrak{g}$ and Weyl group W. Let G have rank n and let N be its number of positive roots. Let $\ell \geq 3$ be an odd positive integer, prime to 3 if $\mathfrak{g}$ contains a G_2-component, and let ϵ be a primitive ℓ^{th} root of 1. Let $(w_1, w_2) \in W \times W$, and let $g \in X_{(w_1, w_2)} \subseteq G$. Recall the definition of $s(-)$ in (II.4.13).*

1. $\mathcal{O}_\epsilon(G)(g)$ has finite representation type if

$$\ell(w_1) + \ell(w_2) \quad > \quad 2N - 2,$$

and is wild otherwise. In particular, (by (III.7.9), Theorem 2), finite type implies Azumaya.

2. If $\mathfrak{m}_g \in \mathcal{F}_{\mathcal{O}_\epsilon(G)}$, (so that $\ell(w_1) + \ell(w_2) + s(w_1w_2^{-1}) = 2N$ by (III.7.9), Theorem 2), then, letting $s = s(w_1w_2^{-1})$,

$$\mathcal{O}_\epsilon(G)(g) \quad \cong \quad \bigoplus_{i=1}^{\ell^{n-s}} M_{\ell^N}\left(\frac{k[X_1,\ldots,X_s]}{\langle X_1^\ell,\ldots,X_s^\ell\rangle}\right).$$

3. Let d be the integer defined in (III.9.6), Theorem 1. Each of the blocks of $\mathcal{O}_\epsilon(G)(g)$ is a "multiply-edged" Cayley graph of a certain elementary abelian ℓ-group of order ℓ^{n-s-d}. $\square$

Remarks. 1. Note that 2 generalises the isomorphism (III.7.9)(24) which describes the semisimple artinian algebras in the fully Azumaya locus.

2. The final part of the theorem may indicate how the results of [19] should be strengthened. (We use without giving definitions here some terminology – links between prime ideals, the graph of links, and cliques of prime ideals – generalising the concepts concerning blocks of maximal ideals introduced in Appendix I.16. Full details of these ideas can be found in [83], for example.) In [19], information was found on the cliques and link graph of $\mathcal{O}_q(G)$, for semisimple G and generic q, but the precise "size" and "shape" of the cliques of $\mathcal{O}_q(G)$ remain unknown. We suspect that part 3 of the above theorem is the "ℓ-shadow" of the as yet unknown generic result. That is, in $\mathcal{O}_q(G)$, the graph of links of each clique of primitive ideals in $X_{(w_1,w_2)}$ should be the Cayley graph of a free abelian group of rank $n - s - d$. But we don't understand why this should be so. Therefore:

Q.10: Understand better the connections between the structure and representation theory of $\mathcal{O}_q(G)$ and $\mathcal{O}_\epsilon(G)$.

3. Obviously, one would like to have results like part 2 of the theorem "beyond the Azumaya locus". A test question in this direction is:

Q.11: What is the minimal data on $(w_1, w_2) \in W \times W$ needed to determine the isomorphism type of $\mathcal{O}_\epsilon(G)$?

III.10.6. Lusztig's conjecture. The eponymous conjecture was formulated in 1979 [142]. It gives an analogue, for the irreducible modules of a reductive algebraic group G over an algebraically closed field k of positive characteristic p, of Weyl's character formula in characteristic 0. Let $\mathfrak{g}$ be the Lie algebra of G. There is a very close connection between the irreducible G-modules and the irreducible *restricted* $\mathfrak{g}$-modules in the block $\mathcal{B}_k$ of the trivial module – that is, (certain of) the irreducible $\overline{U(\mathfrak{g})}$-modules, in our notation. Lusztig's conjecture was confirmed for $p \gg 0$ in [6] using the following strategy. Let h be the Coxeter number of $\mathfrak{g}$ and let ϵ be a primitive p^{th} root of 1.

Step 1: [6] There is a $\mathbb{Z}$-algebra B which is a finite module over $\mathbb{Z}$, such that

(i) $\mathcal{B}_k$ is Morita equivalent to $B \otimes_{\mathbb{Z}} k$ if $p > h$.

(ii) $B \otimes_{\mathbb{Z}} \mathbb{Q}(\epsilon)$ is Morita equivalent to the block of $\overline{U_\epsilon(\mathfrak{g})}$ containing k, if $p > h$.

Thus, from (i) and (ii), Lusztig's conjecture will be moved to an analogous question about the representation theory of quantum groups (in characteristic 0) at p^{th} roots of unity, *provided* $B \otimes_{\mathbb{Z}} k$ and $B \otimes_{\mathbb{Z}} \mathbb{Q}(\epsilon)$ have the same representation theory. And (by elementary arguments involving idempotents), this is indeed so, *provided* none of the denominators (in $\mathbb{Z}$) of the primitive idempotents of $B \otimes_{\mathbb{Z}} \mathbb{Q}(\epsilon)$ are divisible by p. It's here that the condition that $p \gg 0$ comes into play.

Step 2: [124] The desired result for quantum groups at p^{th} roots of 1 follows from the truth of an analogous conjecture for affine Lie algebras with negative levels.

Step 3: [121] The conjectured result for affine Lie algebras with negative levels is true.

In the light of the above we should state the well-known

Q.12: Find an approach to Step 1 which yields an explicit bound on p.

III.10.7. Lusztig's Hope. This was put forward in [145]. It concerns the bundle of algebras $\mathcal{B}_{U(\mathfrak{g})}$, for $\mathfrak{g}$ as in (III.10.6). Recall that these algebras are parametrised by $\mathfrak{g}^*$. For $\chi \in \mathfrak{g}^*$, (which by the Kac-Weisfeiler theorem of (III.10.1) we can assume to be nilpotent), Lusztig gives a conjectured value for the composition multiplicities n_{ij} of the irreducible modules for $U_\chi := U(\mathfrak{g})/\mathfrak{m}_\chi U(\mathfrak{g})$ in the projective indecomposable U_χ-modules. The conjectured description of the n_{ij} is geometric – it is proposed that the n_{ij} should be given by a form relating the K-group of T-equivariant coherent sheaves (for a torus T) on the space $\mathcal{B}_\chi$ of all Borel subalgebras of $\mathfrak{g} = \mathfrak{g}^*$ which contain χ with the corresponding K-group of a certain related (smooth) space, Λ_χ. When χ is *regular* – that is, having centraliser of minimal dimension – so that $\mathfrak{m}_\chi$ is in the fully Azumaya locus, $\mathcal{B}_\chi$ is a point and everything is trivial. When $\chi = 0$, $\mathcal{B}_\chi = G/B$, the flag variety; in this case $U_\chi = \overline{U(\mathfrak{g})}$ and the Hope reduces to Lusztig's Conjecture of (III.10.6).

Current work (by e.g. Jantzen, Gordon, Rumynin) is concentrated on the "next best" cases after the regular one, namely where the centraliser of χ has dimension $n + 2$; these are known as the *subregular* (nilpotent) orbits. Jantzen has shown by calculation that the Hope is true for these orbits in types A, B, D and E, but beyond this only isolated cases are known – for instance in type C_3 it's not even known how many U_χ-irreducibles there are for a subregular χ! Gordon and Rumynin have been working to find a deeper structural interpretation of Lusztig's Hope in terms of equivalences of derived categories, and have succeeded in doing so for χ subregular in type A. In general, there is some expectation that the Hope can be proved by a strategy akin to that used for Lusztig's Conjecture – namely, formulate an analogous conjecture for quantum groups at p^{th} roots of 1, show that

the Hope is equivalent to this new conjecture, and prove the latter by passage to
Kac-Moody Lie algebras. But this is all for the future.

EXERCISES

Exercise III.10.A. Let A be a finite dimensional algebra over the algebraically
closed field k. Show that there are only finitely many isomorphism classes of algebras Morita equivalent to A and of dimension less than n, for any given positive
integer n.

Exercise III.10.B. (i) Prove the properties of the Nakayama automorphism of
a Frobenius algebra stated in (III.10.3).

(ii) Work out the details of the definition and properties of the Nakayama automorphism for a free central Frobenius extension as outlined in (III.10.3). *Warning*: The conditions imposed there are probably too restrictive.

(iii) Answer (Q.7)!

Exercise III.10.C. Let G be the infinite dihedral group,

$$G \quad = \quad \langle a, b : b^2 = 1, \ bab = a^{-1} \rangle,$$

and let k be an (algebraically closed) field.

(i) Calculate $Z(kG)$ and so deduce that kG is a prime Hopf algebra satisfying
a polynomial identity.

(ii) Show that the biggest Hopf subalgebra of $Z(kG)$ is k, so that kG is not
a PI Hopf triple according to our definition.

Exercise III.10.D. Prove that if $Z_0 \subseteq Z \subseteq H$ is a PI Hopf triple, then H has
finite injective dimension (equal to the Krull dimension of H or equivalently of
Z_0), by means of the following steps. (We give an argument which works without
the hypothesis that H is prime; note that most of Step (i) is superfluous if Z_0 is
a domain.)

(i) Prove that a commutative affine Hopf k-algebra A is Gorenstein (i.e.,
has finite injective dimension). In characteristic 0, this is clear since the algebra is
reduced and hence smooth [**209**, Theorems 11.4, 11.6]. (Note that "reduced implies
smooth" is an easy application of winding automorphisms, provided we know that
the singular locus is a proper closed subset.) In characteristic $p > 0$, show that one
can find a power of the Frobenius homomorphism $a \mapsto a^p$ whose image C, say, is
reduced, and so smooth. Then $C \subseteq A$ is a Frobenius extension, by (III.4.7). Then
apply step (ii).

(ii) Show that if $B \subseteq D$ is any Frobenius extension then the (left) injective
dimension of D is bounded above by the left injective dimension of B.

(iii) Apply (ii) to the extension $Z_0 \subseteq H$. Prove equality of the injective and
Krull dimensions of H. If Z_0 is smooth, does H have finite global dimension?

Exercise III.10.E. (Gordon) This exercise concerns (Q.1). Let k be a field and let B be the k-algebra

$$B := k[X, Y, S : XY = SYX, \ [S, X] = [S, Y] = 0].$$

(i) Show that B is a bialgebra with $\Delta(X) = X \otimes X$, $\Delta(Y) = Y \otimes Y$, $\Delta(S) = S \otimes S$. (In fact B is a sub-bialgebra of the group algebra of the first Heisenberg group.)

   (ii) Show that $C := B/\langle X^2, Y^2 \rangle$ is a quotient bialgebra of B, and that C is a finite module over the sub-bialgebra of C generated by the image of S.

   (iii) (Here we abuse notation by writing S for the image of S in C.) Show that, for $\lambda, \mu \in k^\times$, $C_\lambda := C/\langle S - \lambda \rangle$ is isomorphic to C_μ if and only if $\lambda = \mu$ or $\lambda = \mu^{-1}$.

BIBLIOGRAPHY

1. E. Abe, *Hopf Algebras*, Cambridge Univ. Press, Cambridge, 1980.
2. G. Abrams and J. Haefner, *Primeness conditions for group graded rings*, in Ring Theory, Proc. Biennial Ohio State – Denison Conf. 1992, World Scientific, Singapore, 1993, pp. 1-19.
3. J. Alev and F. Dumas, *Sur le corps de fractions de certaines algèbres quantiques*, J. Algebra **170** (1994), 229-265.
4. J. Alev, A. I. Ooms, and M. Van den Bergh, *A class of counterexamples to the Gel'fand-Kirillov conjecture*, Trans. Amer. Math. Soc. **348** (1996), 1709-1716.
5. A. S. Amitsur, *Algebras over infinite fields*, Proc. Amer. Math. Soc. **7** (1956), 35-48.
6. H. H. Andersen, J. C. Jantzen, and W. Soergel, *Representations of quantum groups at a p-th root of unity and of semisimple groups in characteristic p: independence of p*, Astérisque **220** (1994), 1-321.
7. H. H. Andersen, P. Polo and K. Wen, *Representations of quantum algebras*, Invent. Math. **104** (1991), 1-59; *Addendum*, Invent. Math. **120** (1995), 409-410.
8. M. Artin, W. Schelter, and J. Tate, *Quantum deformations of GL_n*, Comm. Pure Appl. Math. **44** (1991), 879-895.
9. M. Auslander and O. Goldman, *Maximal orders*, Trans. Amer. Math. Soc. **97** (1960), 1-24.
10. H. Bass, *Algebraic K-Theory*, Benjamin, New York, 1968.
11. A. Bell and R. Farnsteiner, *On the theory of Frobenius extensions and its application to Lie superalgebras*, Trans. Amer. Math. Soc. **335** (1993), 407-424.
12. A. D. Bell and G. Sigurdsson, *Catenarity and Gel'fand-Kirillov dimension in Ore extensions*, J. Algebra **127** (1989), 409-425.
13. G. M. Bergman, *The diamond lemma for ring theory*, Advances in Math. **29** (1978), 178-218.
14. S. M. Bhatwadekar, *On the global dimension of some filtered algebras*, J. London Math. Soc. (2) **13** (1976), 239-248.
15. J.-E. Björk, *The Auslander condition on Noetherian rings*, in Sém. d'Algèbre P. Dubreil et M.-P. Malliavin 1987-88 (M.-P. Malliavin, ed.), Lecture Notes in Math. 1404, Springer-Verlag, Berlin, 1989, pp. 137–173.
16. W. Borho, P. Gabriel, and R. Rentschler, *Primideale in Einhüllenden auflösbarer Lie-Algebren*, Lecture Notes in Math. 357, Springer-Verlag, Berlin, 1973.
17. W. Borho and J. C. Jantzen, *Über primitive Ideale in der Einhüllenden einer halbeinfachen Lie-Algebra*, Invent. Math. **39** (1977), 1-53.
18. K. A. Brown, *Representation theory of Noetherian Hopf algebras satisfying a polynomial identity*, in Trends in the Representation Theory of Finite Dimensional Algebras (Seattle 1997) (E. L. Green and B. Huisgen-Zimmermann, Eds.), Contemp. Math. **229** (1998), 49-79.
19. K. A. Brown and K. R. Goodearl, *Prime spectra of quantum semisimple groups*, Trans. Amer. Math. Soc. **348** (1996), 2465-2502.
20. _____ , *A Hilbert basis theorem for quantum groups*, Bull. London Math. Soc. **29** (1997), 150-158.
21. _____ , *Homological aspects of noetherian PI Hopf algebras and irreducible modules of maximal dimension*, J. Algebra **198** (1997), 240-265.

22. K. A. Brown and I. Gordon, *The ramification of centres: Lie algebras in positive charac-teristic and quantized enveloping algebras*, Math. Zeit. **238** (2001), 733-779.

23. _____, *The ramifications of the centres: quantised function algebras at roots of unity*, Proc. London Math. Soc. (to appear).

24. K. A. Brown, I. Gordon, and J. T. Stafford, $\mathcal{O}_\epsilon(G)$ *is a free module over* $\mathcal{O}[G]$, Preprint (2000); available at http://arXiv.org/abs/math.QA/0007179.

25. K. A. Brown and C. R. Hajarnavis, *Homologically homogeneous rings*, Trans. Amer. Math. Soc. **281** (1984), 197-208.

26. J. L. Bueso, J. Gómez-Torrecillas, and F. J. Lobillo, *Re-filtering and exactness of the Gelfand-Kirillov dimension* (to appear).

27. P. Caldero, *Étude des q-commutations dans l'algèbre* $U_q(\mathbf{n}^+)$, J. Algebra **178** (1995), 444-457.

28. _____, *On the Gelfand-Kirillov conjecture for quantum algebras*, Proc. AMS **128** (2000), 943-951.

29. A. Cannas da Silva and A. Weinstein, *Geometric Models for Noncommutative Algebras*, Berkeley Math. Lecture Notes 10, Amer. Math. Soc., Providence, 1999.

30. N. Cantarini, *The quantized enveloping algebra* $\mathcal{U}_q(\mathrm{sl}(n))$ *at the roots of unity*, Communic. Math. Phys. **211** (2000), 207-230.

31. G. Cauchon, *Quotients premiers de* $O_q(\mathfrak{m}_n(k))$, J. Algebra **180** (1996), 530–545.

32. _____, *Effacement des dérivations. Spectres premiers et primitifs des algèbres quantiques* (to appear).

33. _____, *Spectre premier de* $\mathcal{O}_q(M_n(k))$. *Image canonique et séparation normale* (to appear).

34. V. Chari and A. Pressley, *A Guide to Quantum Groups*, Cambridge Univ. Press, Cambridge, 1994.

35. N. Chriss and V. Ginzburg, *Representation Theory and Complex Geometry*, Birkhäuser, Boston, 1997.

36. G. Cliff, *The division ring of quotients of the coordinate ring of the quantum general linear group*, J. London Math. Soc. (2) **51** (1995), 503-513.

37. L. Conlon, *Differentiable Manifolds: A First Course*, Birkhäuser, Boston, 1993.

38. L. J. Corwin, I. M. Gel'fand, and R. Goodman, *Quadratic algebras and skew-fields*, in Representation Theory and Analysis on Homogeneous Spaces (New Brunswick, NJ 1993), Contemp. Math. **177** (1994), 217-225.

39. L. Dabrowski, C. Reina, and A. Zampa, $A[SL_q(2)]$ *at roots of unity is a free module over* $A[SL(2)]$, math.QA/0004092 (2000).

40. C. De Concini and V. G. Kac, *Representations of quantum groups at roots of 1*, in Operator Algebras, Unitary Representations, Enveloping Algebras, and Invariant Theory (Paris 1989), Birkhäuser, Boston, 1990, pp. 471-506.

41. _____, *Representations of quantum groups at roots of 1: reduction to the exceptional case*, in Infinite Analysis, Parts A,B (Kyoto, 1991), World Scientific, River Edge, N.J., 1992, pp. 141-149.

42. C. De Concini, V. G. Kac, and C. Procesi, *Quantum coadjoint action*, J. Amer. Math. Soc. **5** (1992), 151-189.

43. C. De Concini and V. Lyubashenko, *Quantum function algebra at roots of 1*, Advances in Math. **108** (1994), 205-262.

44. C. De Concini and C. Procesi, *Quantum groups*, in D-Modules, Representation Theory, and Quantum Groups (Venezia, June 1992) (G. Zampieri, and A. D'Agnolo, eds.), Lecture Notes in Math. 1565, Springer-Verlag, Berlin, 1993, pp. 31-140.

45. _____, *Quantum Schubert cells and representations at roots of 1*, in Algebraic Groups and Lie Groups (G.I. Lehrer, ed.), Austral. Math. Soc. Lecture Series 9, Cambridge Univ. Press, Cambridge, 1997, pp. 127-160.

46. E. E. Demidov, *Modules over a quantum Weyl algebra*, Moscow Univ. Math. Bull. **48** (1993), 49-51.

47. _____, *Some aspects of the theory of quantum groups*, Russian Math. Surveys **48:6** (1993), 41-79.

48. J. Dixmier, *Représentations irréductibles des algèbres de Lie résolubles*, J. Math. Pures Appl. (9) **45** (1966), 1-66.

49. _____, *Idéaux primitifs dans les algèbres enveloppantes*, J. Algebra **48** (1977), 96–112.

50. V. G. Drinfel'd, *Hopf algebras and the quantum Yang-Baxter equation*, Soviet Math. Doklady **32** (1985), 254-258.

51. _____, *Quantum groups*, in Proc. Internat. Congress of Mathematicians (Berkeley 1986), I, Amer. Math. Soc., Providence, 1987, pp. 798-820.

52. _____, *Quantum groups*, J. Soviet Math. **41:2** (1988), 898-915.

53. E. K. Ekström, *The Auslander condition on graded and filtered Noetherian rings*, in Sém. d'Algèbre P. Dubreil et M.-P. Malliavin 1987-88 (M.-P. Malliavin, ed.), Lecture Notes in Math. 1404, Springer-Verlag, Berlin, 1989, pp. 220-245.

54. L. D. Faddeev, N. Yu. Reshetikhin, and L. A. Takhtajan, *Quantization of Lie groups and Lie algebras*, in Algebraic Analysis, Vol. I (M. Kashiwara and T. Kawai, eds.), Academic Press, Boston, 1988, pp. 129-139.

55. R. Farnsteiner and H. Strade, *Shapiro's lemma and its consequences in the cohomology theory of modular Lie algebras*, Math. Zeitschr. **206** (1991), 153-168.

56. F. Fauquant-Millet, *Quantification de la localisation de Dixmier de $U(sl_{n+1}(C))$*, J. Algebra **218** (1999), 93-116.

57. O. Gabber, *Equidimensionalité de la variété caractéristique*, Exposé de O. Gabber rédigé par T. Levasseur, Université de Paris VI (1982).

58. P. Gabriel, *Représentations des algèbres de Lie résolubles (d'après J. Dixmier)*, in Séminaire Bourbaki 1968/69, Lecture Notes in Math. 179, Springer-Verlag, Berlin, 1971, pp. 1-22.

59. I. M. Gel'fand and A. A. Kirillov, *Fields associated with enveloping algebras of Lie algebras*, Soviet Math. Doklady **7** (1966), 407-409.

60. _____, *Sur les corps Liés aux algèbres enveloppantes des algèbres de Lie*, Publ. Math. I.H.E.S. **31** (1966), 5-19.

61. _____, *The structure of the quotient field of the enveloping algebra of a semisimple Lie algebra*, Soviet Math. Doklady **9** (1968), 669-671.

62. _____, *The structure of the Lie field connected with a split semisimple Lie algebra*, Func. Anal. Applic. **3** (1969), 6-21.

63. M. Gerstenhaber, A. Giaquinto, and S. D. Schack, *Quantum symmetry*, in Quantum Groups (Leningrad 1990), Lecture Notes in Math. 1510, Springer-Verlag, Berlin, 1992, pp. 9-46.

64. A. Giaquinto, *Quantization of tensor representations and deformations of matrix bialgebras*, J. Pure Appl. Algebra **79** (1992), 169-190.

65. J. Gómez Torrecillas, *Gelfand-Kirillov dimension of multi-filtered algebras*, Proc. Edinburgh Math. Soc. (2) **42** (1999), 155-168.

66. J. Gómez Torrecillas and F. J. Lobillo, *Global homological dimension of multifiltered rings and quantized enveloping algebras*, J. Algebra **225** (2000), 522-533.

67. K. R. Goodearl, *Global dimension of differential operator rings. II*, Trans. Amer. Math. Soc. **209** (1975), 65-85.

68. _____, *Prime ideals in skew polynomial rings and quantized Weyl algebras*, J. Algebra **150** (1992), 324-377.

69. _____, *Uniform ranks of prime factors of skew polynomial rings*, in Ring Theory, Proc. Biennial Ohio State–Denison Conf., 1992 (S. K. Jain and S. T. Rizvi, eds.), World Scientific, Singapore, 1993, pp. 182-199.

70. _____, *Prime spectra of quantized coordinate rings*, in Interactions between Ring Theory and Representations of Algebras (Murcia 1998) (F. Van Oystaeyen and M. Saorín, eds.), Dekker, New York, 2000, pp. 205-237.

71. _____, *Quantized primitive ideal spaces as quotients of affine algebraic varieties*, in Quantum Groups and Lie Theory (A. Pressley, ed.), London Math. Soc. Lecture Note Series 290, Cambridge Univ. Press, Cambridge, 2002.

72. K. R. Goodearl and T. H. Lenagan, *Catenarity in quantum algebras*, J. Pure Appl. Alg. **111** (1996), 123-142.

73. _____, *Prime ideals in certain quantum determinantal rings*, in Interactions between Ring Theory and Representations of Algebras (Murcia 1998) (F. Van Oystaeyen and M. Saorín, eds.), Dekker, New York, 2000, pp. 239-251.

74. _____, *Quantum determinantal ideals*, Duke Math. J. **103** (2000), 165-190.

75. _____, *Winding-invariant prime ideals in quantum 3×3 matrices*, Preprint (2001); available at http://arXiv.org/abs/math.QA/0112051.

76. K. R. Goodearl, T. H. Lenagan, and L. Rigal, *The first fundamental theorem of coinvariant theory for the quantum general linear group*, Publ. Research Inst. Math. Sci. (Kyoto) **36** (2000), 269-296.

77. K. R. Goodearl and E. S. Letzter, *Prime ideals in skew and q-skew polynomial rings*, Memoirs Amer. Math. Soc. **521** (1994).

78. _____, *Prime factor algebras of the coordinate ring of quantum matrices*, Proc. Amer. Math. Soc. **121** (1994), 1017-1025.

79. _____, *Prime and primitive spectra of multiparameter quantum affine spaces*, in Trends in Ring Theory (Miskolc, 1996) (V. Dlab and L. Marki, eds.), Canad. Math. Soc. Conf. Proc. Series **22** (1998), 39-58.

80. _____, *The Dixmier-Moeglin equivalence in quantum coordinate rings and quantized Weyl algebras*, Trans. Amer. Math. Soc. **352** (2000), 1381-1403.

81. _____, *Quantum n-space as a quotient of classical n-space*, Trans. Amer. Math. Soc. **352** (2000), 5855-5876.

82. K. R. Goodearl and J. T. Stafford, *The graded version of Goldie's Theorem*, in Algebra and its Applications (Athens, Ohio, 1999) (D. V. Huynh, S. K. Jain, and S. R. López-Permouth, Eds.), Contemp. Math. **259** (2000), 237-240.

83. K. R. Goodearl and R. B. Warfield, Jr., *An Introduction to Noncommutative Rings*, Cambridge Univ. Press, Cambridge, 1989.

84. I. Gordon, *Quantised function algebras at roots of unity and path algebras*, J. Algebra **220** (1999), 381-395.

85. _____, *Cohomology of quantised function algebras at roots of unity*, Proc. London Math. Soc. (3) **80** (2000), 337-359.

86. _____, *Complexity of representations of quantised function algebras and representation type*, J. Algebra **233** (2000), 437-482.

87. _____, *Representations of semisimple Lie algebras in positive characteristic and quantised enveloping algebras at roots of unity*, in Quantum Groups and Lie Theory (A. Pressley, ed.), London Math. Soc. Lecture Note Series 290, Cambridge Univ. Press, Cambridge, 2002.

88. _____, *Nakayama automorphisms for quantum groups at roots of unity*, Univ. of Glasgow Preprint No. 2001/32; available at http://www.maths.gla.ac.uk/~ig.

89. R. Hartshorne, *Algebraic Geometry*, Grad. Texts in Math. 52, Springer-Verlag, New York, 1977.

90. T. Hayashi, *Sugawara operators and Kac-Kazhdan conjecture*, Invent. Math. **94** (1988), 13-52.

91. _____, *Quantum deformation of classical groups*, Publ. Research Inst. Math. Sci. (Kyoto) **28** (1992), 57-81.

92. Y. Hinohara, *Projective modules over semilocal rings*, Tôhoku Math. J. **14** (1962), 205-211.

93. G. Hochschild, *Algebraic groups and Hopf algebras*, Illinois J. Math. **14** (1970)), 52-65.

94. T. J. Hodges and T. Levasseur, *Primitive ideals of* $\mathbf{C}_q[SL(3)]$, Comm. Math. Phys. **156** (1993), 581-605.

95. _____, *Primitive ideals of* $\mathbf{C}_q[SL(n)]$, J. Algebra **168** (1994), 455-468.

96. T. J. Hodges, T. Levasseur, and M. Toro, *Algebraic structure of multi-parameter quantum groups*, Advances in Math. **126** (1997), 52-92.

97. C. Huh and C. O. Kim, *Gelfand-Kirillov dimension of skew polynomial rings of automorphism type*, Communic. in Algebra **24** (1996), 2317-2323.

98. J. E. Humphreys, *Introduction to Lie Algebras and Representation Theory*, Grad. Texts in Math. 9, Springer-Verlag, New York, 1972.

99. _____, *Linear Algebraic Groups*, Grad. Texts in Math. 21, Springer-Verlag, Berlin, 1975.

100. _____, *Conjugacy Classes in Semisimple Algebraic Groups*, Math. Surveys and Monographs 43, Amer. Math. Soc., Providence, 1995.

101. _____, *Modular representations of simple Lie algebras*, Bull. Amer. Math. Soc. **35** (1998), 105-122.

102. C. Ingalls, *Quantum toric varieties* (to appear).

103. K. Iohara and F. Malikov, *Rings of skew polynomials and Gel'fand-Kirillov conjecture for quantum groups*, Comm. Math. Phys. **164** (1994), 217-237.

104. R. S. Irving and L. W. Small, *On the characterization of primitive ideals in enveloping algebras*, Math. Z. **173** (1980), 217-221.

105. N. Jacobson, *Lie Algebras*, Wiley-Interscience, New York, 1962; Reprinted edition, Dover, New York, 1979.

106. H. P. Jakobsen and H. Zhang, *The center of the quantized matrix algebra*, J. Algebra **196** (1997), 458-474.

107. J. C. Jantzen, *Lectures on Quantum Groups*, Grad. Studies in Math. 6, Amer. Math. Soc., Providence, 1996.

108. _____, *Representations of Lie algebras in prime characteristic*, (Notes by I. Gordon), in Representation Theories and Algebraic Geometry (Montreal 1997), NATO Adv. Sci. Inst. Ser. C Math. Phys. Sci. 514, Kluwer, Dordrecht, 1998, pp. 185-235.

109. E. Jespers, *Simple graded rings*, Communic. in Algebra **21** (1993), 2437-2444.

110. M. Jimbo, *A q-difference analog of* $U(g)$ *and the Yang-Baxter equation*, Lett. Math. Phys. **10** (1985), 63-69.

111. S. Jøndrup, *Representations of skew polynomial algebras*, Proc. Amer. Math. Soc. **128** (2000), 1301-1305.

112. V. F. R. Jones, *Subfactors and Knots*, CBMS Regional Conf. Series 80, Amer. Math. Soc., Providence, 1991.

113. D. A. Jordan, *A simple localization of the quantized Weyl algebra*, J. Algebra **174** (1995), 267-281.

114. A. Joseph, *Proof of the Gelfand-Kirillov conjecture for solvable Lie algebras*, Proc. Amer. Math. Soc. **45** (1974), 1-10.

115. _____, *On the prime and primitive spectra of the algebra of functions on a quantum group*, J. Algebra **169** (1994), 441–511.

116. _____, *Quantum Groups and Their Primitive Ideals*, Ergebnisse der Math. (3) 29, Springer-Verlag, Berlin, 1995.

117. _____, *Sur une conjecture de Feigin*, C. R. Acad. Sci. Paris, Sér. I Math. **320** (1995), 1441-1444.

118. A. Joseph and G. Letzter, *Local finiteness of the adjoint action for quantized enveloping algebras*, J. Algebra **153** (1992), 289-318.

119. _____, *Separation of variables for quantized enveloping algebras*, Amer. J. Math. **116** (1994), 127-177.

120. I. Kaplansky, *Commutative Rings*, Allyn and Bacon, Boston, 1970.

121. M. Kashiwara and T. Tanisaki, *Kazhdan-Lusztig conjecture for affine Lie algebras with negative level I, II*, Duke Math. J. **77** (1995), 21-62; **84** (1996), 771-813.

122. C. Kassel, *Quantum Groups*, Grad. Texts in Math. 155, Springer-Verlag, New York, 1995.

123. C. Kassel, M. Rosso, and V. Turaev, *Quantum Groups and Knot Invariants*, Panoramas et Synthèses 5, Société Mathématique de France, Paris, 1997.

124. D. Kazhdan and G. Lusztig, *Tensor structures arising from affine Lie algebras. I, II, III, IV*, J. Amer. Math. Soc. **6** (1993), 905-1011; **7** (1994), 335-453.

125. A. Klimyk and K. Schmüdgen, *Quantum Groups and their Representations*, Springer-Verlag, Berlin, 1997.

126. L. I. Korogodski and Ya. S. Soibelman, *Algebras of Functions on Quantum Groups: Part I*, Math. Surveys and Monographs 56, Amer. Math. Soc., Providence, 1998.

127. G. Krause and T. H. Lenagan, *Growth of Algebras and Gelfand-Kirillov Dimension*, Pitman, Boston, 1985; Revised Ed., Grad. Studies in Math. 22, Amer. Math. Soc., Providence, 2000.

128. H. F. Kreimer and M. Takeuchi, *Hopf algebras and Galois extensions of an algebra*, Indiana Univ. Math. J. **30** (1981), 675-692.

129. P. P. Kulish and N. Yu. Reshetikhin, *Quantum linear problem for the sine-Gordon equation and higher representations*, (Russian), Zap. Nauchn. Sem. Leningrad. Otdel. Mat. Inst. Steklov **101** (1981), 101-110, 207.

130. L. A. Lambe and D. E. Radford, *Introduction to the Quantum Yang-Baxter Equation and Quantum Groups: An Algebraic Approach*, Kluwer, Dordrecht, 1997.

131. L. Le Bruyn, *Central singularities of quantum spaces*, J. Algebra **177** (1995), 142-153.

132. T. H. Lenagan, *Enveloping algebras of solvable Lie superalgebras are catenary*, Contemp. Math. **130** (1992), 231–236.

133. A. Leroy, J. Matczuk, and J. Okninski, *On the Gel'fand-Kirillov dimension of normal localizations and twisted polynomial rings*, Perspectives in Ring Theory (Antwerp 1987) (F. Van Oystaeyen and L. Le Bruyn, eds.), Kluwer, Dordrecht, 1988, pp. 205-214.

134. E. S. Letzter, *Remarks on the twisted adjoint representation of $R_q[G]$*, Communic. in Algebra **27** (1999), 1889-1893.

135. _____, *An affine PI Hopf algebra not finite over a normal commutative Hopf subalgebra*, Preprint (2001); available at http://www.math.temple.edu/~letzter.

136. E. S. Letzter and M. Lorenz, *Polycyclic-by-finite group algebras are catenary*, Math. Research Letters **6** (1999), 183-194.

137. T. Levasseur, *Some properties of noncommutative regular graded rings*, Glasgow Math. J. **34** (1992), 277-300.

138. T. Levasseur and J. T. Stafford, *The quantum coordinate ring of the special linear group*, J. Pure Applied Algebra **86** (1993), 181-186.

139. S. Z. Levendorskii and Ya. S. Soibelman, *Some applications of the quantum Weyl groups*, J. Geom. Phys. **7** (1990), 241-254.

140. _____, *Algebras of functions on compact quantum groups, Schubert cells and quantum tori*, Comm. Math. Phys. **139** (1991), 141-170.

141. M. E. Lorenz and M. Lorenz, *On crossed products of Hopf algebras*, Proc. Amer. Math. Soc. **123** (1995), 33-38.

142. G. Lusztig, *Some problems in the representation theory of finite Chevalley groups*, in The Santa Cruz Conference on Finite Groups (Santa Cruz, 1979), Proc. Symp. Pure Math. **37** (1980), 313-317.

143. _____, *Finite dimensional Hopf algebras arising from quantized universal enveloping algebras*, J. Amer. Math. Soc. **3** (1990), 257-296.

144. _____, *Quantum groups at roots of 1*, Geom. Dedicata **35** (1990), 89-114.

145. _____, *Periodic W-graphs*, Representation Theory **1** (1997), 207-279 (electronic).

146. S. Majid, *Foundations of Quantum Group Theory*, Cambridge Univ. Press, Cambridge, 1995.

147. G. Maltsiniotis, *Calcul différentiel quantique*, Groupe de travail, Université Paris VII (1992).

148. Yu. I. Manin, *Some remarks on Koszul algebras and quantum groups*, Ann. Inst. Fourier (Grenoble) **37** (1987), 191-205.

149. _____, *Quantum groups and non-commutative geometry*, Publ. Centre de Recherches Math., Univ. de Montreal, 1988.

150. _____, *Multiparametric quantum deformation of the general linear supergroup*, Comm. Math. Phys. **123** (1989), 163-175.

151. _____, *Topics in Noncommutative Geometry*, Princeton Univ. Press, Princeton, 1991.

152. R. Marlin, *Anneaux de Grothendieck des variétés de drapeaux*, Bull. Soc. Math. France **104** (1976), 337-348.

153. J. Matczuk, *The Gel'fand-Kirillov dimension of Poincaré-Birkhoff-Witt extensions*, in Perspectives in Ring Theory (Antwerp 1987) (F. Van Oystaeyen and L. Le Bruyn, eds.), Kluwer, Dordrecht, 1988, pp. 221-226.

154. H. Matsumura, *Commutative Algebra*, 2nd. Ed., Benjamin/Cummings, Reading, MA, 1980.

155. J. C. McConnell, *Representations of solvable Lie algebras and the Gelfand-Kirillov conjecture*, Proc. London Math. Soc. (3) **29** (1974), 453-484.

156. _____, *Quantum groups, filtered rings and Gel'fand-Kirillov dimension*, in Noncommutative Ring Theory (Athens, Ohio, 1989), Lecture Notes in Math. 1448, Springer-Verlag, Berlin, 1990, pp. 139-147.

157. J. C. McConnell and J. J. Pettit, *Crossed products and multiplicative analogues of Weyl algebras*, J. London Math. Soc. (2) **38** (1988), 47–55.

158. J. C. McConnell and J. C. Robson, *Noncommutative Noetherian Rings*, Wiley-Interscience, Chichester-New York, 1987; Reprinted with corrections, Grad. Studies in Math. 30, Amer. Math. Soc., Providence, 2001.

159. A. S. Merkur'ev, *Comparison of the equivariant and ordinary K-theory of algebraic varieties*, St. Petersburg Math. J. **9** (1998), 815-850.

160. I. Mirkovic and D. Rumynin, *Centers of reduced enveloping algebras*, Math. Zeitschr. **231** (1999), 123-132.

161. C. Moeglin, *Idéaux primitifs des algèbres enveloppantes*, J. Math. Pures Appl. **59** (1980), 265–336.

162. C. Moeglin and R. Rentschler, *Orbites d'un groupe algébrique dans l'espace des idéaux rationnels d'une algèbre enveloppante*, Bull. Soc. Math. France **109** (1981), 403–426.

163. S. Montgomery, *Hopf Algebras and Their Actions on Rings*, CBMS Regional Conf. Series in Math. 82, Amer. Math. Soc., Providence, 1993.

164. V. G. Mosin and A. N. Panov, *Division rings of quotients and central elements of multiparameter quantizations*, Sbornik: Mathematics **187:6** (1996), 835-855.

165. B. J. Müller, *Localization in non-commutative Noetherian rings*, Canad. J. Math. **28** (1976), 600-610.

166. D. Mumford, *Algebraic Geometry I, Complex Projective Varieties*, Grundlehren der math. Wissenschaften 221, Springer-Verlag, Berlin, 1976.

167. I. M. Musson, *Ring theoretic properties of the coordinate rings of quantum symplectic and Euclidean space*, in Ring Theory, Proc. Biennial Ohio State–Denison Conf., 1992 (S. K. Jain and S. T. Rizvi, eds.), World Scientific, Singapore, 1993, pp. 248-258.

168. ———, *On the Goldie quotient ring of the enveloping algebra of a classical simple Lie superalgebra*, J. Algebra **235** (2001), 203-213.

169. M. Nagata, *On the chain problem of prime ideals*, Nagoya Math. J. **10** (1956), 51-64.

170. C. Năstăsescu and F. Van Oystaeyen, *Graded and Filtered Rings and Modules*, Lecture Notes in Math. 758, Springer-Verlag, Berlin, 1979.

171. ———, *Graded Ring Theory*, North-Holland, Amsterdam, 1982.

172. M. H. A. Newman, *On theories with a combinatorial definition of "equivalence"*, Annals of Math. (2) **43** (1942), 223-243.

173. D. G. Northcott, *Affine Sets and Affine Groups*, London Math. Soc. Lecture Note Series 39, Cambridge Univ. Press, Cambridge, 1980.

174. S.-Q. Oh, *Primitive ideals of the coordinate ring of quantum symplectic space*, J. Algebra **174** (1995), 531-552.

175. ———, *Catenarity in a class of iterated skew polynomial rings*, Comm. Alg. **25** (1997), 37-49.

176. C. Ohn, *Quantum SL(3,C)'s with classical representation theory*, J. Alg. **213** (1999), 721-756.

177. A. N. Panov, *Skew fields of twisted rational functions and the skew field of rational functions on $GL_q(n, K)$*, St. Petersburg Math. J. **7** (1996), 129-143.

178. ———, *Fields of fractions of quantum solvable algebras*, J. Algebra **236** (2001), 110-121.

179. B. Pareigis, *On the cohomology of modules over Hopf algebras*, J. Algebra **22** (1972), 161-182.

180. B. Parshall and J.-P. Wang, *Quantum linear groups*, Memoirs Amer. Math. Soc. **439** (1991).

181. D. S. Passman, *The Algebraic Structure of Group Rings*, Wiley-Interscience, New York, 1977; Reprinted edition, Krieger, Malabar, FL, 1985.

182. ———, *Infinite Crossed Products*, Academic Press, New York, 1989.

183. R. S. Pierce, *Associative Algebras*, Grad. Texts in Math. 88, Springer-Verlag, New York, 1982.

184. A. Premet, *Irreducible modules of Lie algebras of reductive groups and the Kac-Weisfeiler conjecture*, Invent. Math. **121** (1995), 79-117.

185. J. Rainwater, *Global dimension of fully bounded Noetherian rings*, Communic. in Algebra **15** (1987), 2143-2156.

186. N. Yu. Reshetikhin, L. A. Takhtadjan, and L. D. Faddeev, *Quantization of Lie groups and Lie algebras*, Leningrad Math. J. **1** (1990), 193-225.

187. C. M. Ringel, *Hall algebras and quantum groups*, Invent. Math. **101** (1990), 583-591.

188. ———, *PBW-bases of quantum groups*, J. reine angew. Math. **470** (1996), 51-88.

189. M. Rosso, *Analogues de la forme de Killing et du théorème de Harish-Chandra pour les groupes quantiques*, Ann. Sci. École Norm. Sup. **23** (1990), 445-467.

190. J. J. Rotman, *An Introduction to Homological Algebra*, Academic Press, New York, 1979.

191. L. Rowen, *Ring Theory*, Vols. I,II, Academic Press, New York, 1988.

192. D. Rumynin, *Hopf-Galois extensions with central invariants and their geometric properties*, Algebras and Representation Theory 1 (1998), 353-381.

193. W. Schelter, *Non-commutative affine P.I. rings are catenary*, J. Algebra **51** (1978), 12-18.

194. J.-P. Serre, *Complex Semisimple Lie Algebras*, Springer-Verlag, Berlin, 1987.

195. G. Sigurdsson, *Differential operator rings whose prime factors have bounded Goldie dimension*, Archiv Math. **42** (1984), 348-353.

196. Ya. S. Soibelman, *Irreducible representations of the function algebra on the quantum group SU(n), and Schubert cells*, Soviet Math. Dokl. **40** (1990), 34-38.

197. _____, *The algebra of functions on a compact quantum group, and its representations*, Leningrad Math. J. **2** (1991), 161-178; *Correction*, (Russian), Algebra i Analiz **2** (1990), 256.

198. J. T. Stafford and J. J. Zhang, *Homological properties of (graded) noetherian PI rings*, J. Algebra **168** (1994), 988-1026.

199. B. Stenström, *Rings of Quotients*, Grundlehren der math. Wissenschaften 217, Springer-Verlag, Berlin, 1975.

200. A. Sudbury, *Consistent multiparameter quantisation of GL(n)*, J. Phys. A **23** (1990), L697-L704.

201. M. Sweedler, *Hopf Algebras*, Benjamin, New York, 1969.

202. M. Takeuchi, *Quantum orthogonal and symplectic groups and their embedding into quantum GL*, Proc. Japan Acad., Ser. A Math. Sci. **65** (1989), 55-58.

203. T. Tanisaki, *Killing forms, Harish-Chandra isomorphisms, and universal R-matrices for quantum algebras*, in Infinite Analysis, Part A,B (Kyoto 1991), World Scientific, Singapore, 1992, pp. 941-961.

204. P. Tauvel, *Sur les quotients premiers de l'algèbre enveloppante d'un algèbre de Lie résoluble*, Bull. Soc. Math. France **106** (1978), 177–205.

205. L. L. Vaksman and Ya. S. Soibelman, *Algebra of functions on the quantum group SU(2)*, Func. Anal. Applic. **22** (1988), 170-181.

206. W. V. Vasconcelos, *On quasi-local regular algebras*, in Convegno di Algebra Commutativa (Roma, 1971) Symp. Math. Vol. XI, Academic Press, London, 1973, pp. 11-22.

207. F. D. Veldkamp, *The center of the universal enveloping algebra of a Lie algebra in characteristic p*, Ann. Sci. École Norm. Sup. **5** (1972), 217-240.

208. N. Vonessen, *Actions of algebraic groups on the spectrum of rational ideals, II*, J. Algebra **208** (1998), 216-261.

209. W. C. Waterhouse, *Introduction to Affine Group Schemes*, Grad. Texts in Math. 66, Springer-Verlag, Berlin, 1979.

210. A. Weinstein, *The local structure of Poisson manifolds*, J. Diff. Geom. **18** (1983), 523-557.

211. S. L. Woronowicz, *Twisted SU(2) group. An example of a noncommutative differential calculus*, Publ. Res. Inst. Math. Sci. Kyoto **23** (1987), 117-181.

212. _____, *Compact matrix pseudogroups*, Communic. Math. Phys. **111** (1987), 613-665.

213. Q.-S. Wu and J. J. Zhang, *Homological identities for noncommutative rings*, J. Algebra (to appear).

214. S. Yammine, *Les théorèmes de Cohen-Seidenberg en algèbre non commutative*, in Séminaire d'Algèbre P. Dubreil 1977-78, Lecture Notes in Math. 740, Springer-Verlag, Berlin, 1979, pp. 120-169.

215. A. Yekutieli and J. J. Zhang, *Rings with Auslander dualizing complexes*, J. Algebra **213** (1999), 1–51.

216. A. Zaks, *Injective dimension of semi-primary rings*, J. Algebra **13** (1969), 73-86.

217. J. J. Zhang, *Connected graded Gorenstein algebras with enough normal elements*, J. Algebra **189** (1997), 390-405.

218. Y. Zhong, *Injective homogeneity and the Auslander-Gorenstein property*, Glasgow Math. J. **37** (1995), 191-204.

INDEX

Advanced Courses in Mathematics CRM Barcelona

Birkhäuser Verlag launches the new textbook series
Advanced Courses in Mathematics CRM Barcelona.
Since 1995 the Centre de Recerca Matemàtica (CRM) in
Barcelona has conducted a number of annual Summer
Schools at the post-doctoral or advanced graduate level.
Sponsored mainly by the European Community, these
Advanced Courses have usually been held at the CRM
in Bellaterra. The books in this new series consist
essentially of the expanded and embellished
material presented by the authors in
their lectures.

For orders originating from all over
the world except USA and Canada:
Birkhäuser Verlag AG
c/o Springer GmbH & Co
Haberstrasse 7
D-69126 Heidelberg
Tel.: ++49 / 6221 / 345 205
Fax: ++49 / 6221 / 345 229
e-mail: birkhauser@springer.de

For orders originating
in the USA and Canada:
Birkhäuser
333 Meadowland Parkway
USA-Secaucus, NJ 07094-2491
Fax: +1 201 348 4033
e-mail: orders@birkhauser.com

http://www.birkhauser.ch

Birkhäuser

Dwyer, W.G., University of Notre Dame, USA /
Henn, H.-W., University Louis Pasteur, Strasbourg, France

Homotopy Theoretic Methods in Group Cohomology

2001. 108 pages. Softcover
ISBN 3-7643-6605-2

This book looks at group cohomology with tools that come from
homotopy theory. These tools give both decomposition theorems
(which rely on homotopy colimits to obtain a description of the
cohomology of a group in terms of the cohomology of suitable subgroups) and global structure theorems (which exploit the action of
the ring of topological cohomology operations). The approach is
expository and thus suitable for graduate students and others who
would like an introduction to the subject that organizes and adds to
the relevant literature and leads to the frontier of current research.
The book should also be interesting to anyone who wishes to learn
some of the machinery of homotopy theory (simplicial sets, homotopy colimits, Lannes' T-functor, the theory of unstable modules over
the Steenrod algebra) by seeing how it is used in a practical setting.

CENTRE DE RECERCA MATEMÀTICA

Casacuberta, C. / Miró-Roig, R.M. / Verdera, J.,
Universitat Autònoma de Barcelona / **Xambó-
Descamps, S.**, Universitat Politecnica de Catalunya
(Eds.)

European Congress of Mathematics

Barcelona, July 10-14, 2000

Volume I
2001. 628 pages. Hardcover
ISBN 3-7643-6417-3
PM – Progress in Mathematics, Vol. 201

Volume II
2001. Approx. 650 pages. Hardcover
ISBN 3-7643-6418-1
PM – Progress in Mathematics, Vol. 202

Volume I of the proceedings of the third
European Congress of Mathematics pre-
sents the speeches delivered at the
Congress, the list of lectures, and short
summaries of the achievements of the prize
winners as well as papers by plenary and
parallel speakers.
Volume II collects articles by prize winners
and speakers of the mini-symposia.

Lectures in Mathematics ETH Zürich

Hélein, F., Constant Mean Curvature Surfaces, Harmonic Maps and Integrable Systems (2001)
ISBN 3-7643-6576-5

Kreiss, H.-O / Ulmer Busenhart, H., Time-dependent Partial Differential Equations and Their Numerical Solution (2001)
ISBN 3-7643-6125-5

Polterovich, L., The Geometry of the Group of Symplectic Diffeomorphisms (2001)
ISBN 3-7643-6432-7

Turaev, V., Introduction to Combinatorial Torsions (2001)
ISBN 3-7643-6403-3

Tian, G., Canonical Metrics in Kähler Geometry (2000)
ISBN 3-7643-6194-8

Le Gall, J.-F., Spatial Branching Processes, Random Snakes and Partial Differential Equations (1999)
ISBN 3-7643-6126-3

Jost, J., Nonpositive Curvature. Geometric and Analytic Aspects (1997)
ISBN 3-7643-5736-3

Newman, Ch.M., Topics in Disordered Systems (1997)
ISBN 3-7643-5777-0

Yor, M., Some Aspects of Brownian Motion. Part II: Some Recent Martingale Problems (1997)
ISBN 3-7643-5717-7

Carlson, J.F., Modules and Group Algebras (1996)
ISBN 3-7643-5389-9

Freidlin, M., Markov Processes and Differential Equations: Asymptotic Problems (1996)
ISBN 3-7643-5392-9

Simon, L., Theorems on Regularity and Singularity of Energy Minimizing Maps. based on lecture notes by Norbert Hungerbühler (1996)
ISBN 3-7643-5397-X

Holzapfel, R.P., The Ball and Some Hilbert Problems (1995)
ISBN 3-7643-2835-5

Baumslag, G., Topics in Combinatorial Group Theory (1993)
ISBN 3-7643-2921-1

Giaquinta, M., Introduction to Regularity Theory for Nonlinear Elliptic Systems (1993)
ISBN 3-7643-2879-7

Nevanlinna, O., Convergence of Iterations for Linear Equations (1993)
ISBN 3-7643-2865-7

LeVeque, R.J., Numerical Methods for Conservation Laws (1992)
Second Revised Edition. 5th printing 199
ISBN 3-7643-2723-5

Narasimhan, R., Compact Riemann Surfaces (1992, 2nd printing 1996)
ISBN 3-7643-2742-1

Tromba, A.J., Teichmüller Theory in Riemannan Geometry. Second Edition. Based on lecture notes by Jochen Denzler (1992)
ISBN 3-7643-2735-9

Bättig, D. / Knörrer, H., Singularitäten (1991)
ISBN 3-7643-2616-6

Boor, C. de, Splinefunktionen (1990)
ISBN 3-7643-2514-3

Monk, J.D., Cardinal Functions on Boolan Algebras (1990)

Department of Mathematics / Research Institute of Mathematics

Each year the Eidgenössische Technische Hochschule (ETH) at Zürich invites a selected group of mathematicians to give postgraduate seminars in various areas of pure and applied mathematics. These seminars are directed to an audience of many levels and backgrounds. Now some of the most successful lectures are being published for a wider audience through the Lectures in Mathematics, ETH Zürich series. Lively and informal in style, moderate in size and price, these books will appeal to professionals and students alike, bringing a quick understanding of some important areas of current research.

http://www.birkhauser.ch

For orders originating from all over the world except USA and Canada:

Birkhäuser Verlag AG
c/o Springer GmbH & Co
Haberstrasse 7, D-69126 Heidelberg
Fax: ++49 / 6221 / 345 229
e-mail: birkhauser@springer.de

For orders originating in the USA and Canada:

Birkhäuser
333 Meadowland Parkway
USA-Secaucus, NJ 07094-2491
Fax: +1 201 348 4033
e-mail: orders@birkhauser.com

Birkhäuser